Synthesis of Functional Nanomaterials for Electrochemical Energy Storage

Huan Pang · Xiaoyu Cao ·
Limin Zhu · Mingbo Zheng

Synthesis of Functional Nanomaterials for Electrochemical Energy Storage

Springer

Huan Pang
School of Chemistry and Chemical Engineering
Yangzhou University
Yangzhou, Jiangsu, China

Xiaoyu Cao
School of Chemistry and Chemical Engineering
Henan University of Technology
Zhengzhou, Henan, China

Limin Zhu
School of Chemistry and Chemical Engineering
Henan University of Technology
Zhengzhou, Henan, China

Mingbo Zheng
School of Chemistry and Chemical Engineering
Yangzhou University
Yangzhou, Jiangsu, China

ISBN 978-981-13-7371-8 ISBN 978-981-13-7372-5 (eBook)
https://doi.org/10.1007/978-981-13-7372-5

This Springer imprint is published by the registered company Springer Nature Singapore Pte Ltd.
The registered company address is: 152 Beach Road, #21-01/04 Gateway East, Singapore 189721, Singapore

Contents

Chapter 1
Introduction

Human imagination and dreams often give rise to novel science and technology. As a frontier of twenty-first century, nanotechnology was born out of such dreams (Manoharan 2008). Nanotechnology is one of the critical technologies covering various fields including physics, chemistry, materials science, engineering, biology, and even medicine (Jain and Jain 2017; Catherine and Olivier 2017). This technology utilizes the unique electronic, optical, or mechanical properties of nanomaterials which cannot be achieved in their bulk counterparts (Choi et al. 2016; Zhang 2015; Tanaka and Chujo 2015). Nanomaterials are the core components of nanotechnology which are defined as the understanding and control of matter at dimensions between 1 and 100 nm in principle (Murphy 2002).

Although human exposure to nanoparticles has taken place throughout human history, the concept of "nanometer" was first proposed by Nobel Laureate Richard Feynman in his visionary lecture "There is plenty of room at the bottom inspiring the concepts for the rapidly exploding research topic of nanotechnology", in which he introduced the concept of manipulating matter at the atomic level. Since the term "nanotechnology" had not appeared on the horizon, Feynman said: "What I want to talk about is the problem of manipulating and controlling things on a small scale… What I have demonstrated is that there is room—that you can decrease the size of things in a practical way… I will not discuss how we are going to do it, but only what is possible in principle… We are not doing it simply because we haven't yet gotten around to it" (Feynman 2018). Decades later, scientists have realized that the manipulation atoms, molecules, and clusters on the surface are feasible, while new fundamental physics governs the properties of nano-objects. There are significant differences in the definition of nanomaterials between agencies (Boverhof et al. 2015). According to ISO/TS 80,004, nanomaterials are defined as a "material with any external dimension in the nanoscale or having internal structure or surface structure in the nanoscale", while nanoscale is defined as the "length range approximately from 1 to 100 nm" (Hatto 2011). The European Commission adopted the following definition of nanomaterials: "natural, incidental or manufactured material containing particles, in an unbound state or as an aggregate or as an agglomerate and for 50% or

H. Pang et al., *Synthesis of Functional Nanomaterials for Electrochemical Energy Storage*,
https://doi.org/10.1007/978-981-13-7372-5_1

more of the particles in the number size distribution, one or more external dimensions is in the size range 1–100 nm. In specific cases and where warranted by concerns for the environment, health, safety, or competitiveness the number size distribution threshold of 50% may be replaced by a threshold between 1 and 50%" (Commission 2011).

The beginning of the twenty-first century has seen an increased interest in the emerging fields of modern nanoscience and nanotechnology (Kagan et al. 2016). Nanomaterials provide basic building blocks for the fabrication of various devices with desired functions and have become the foundation of remarkable industrial applications with exponential growth (Min et al. 2015; Sun et al. 2015b; Stark et al. 2015). Owing to their inherent shape effects and quantum size, they have many important applications ranging from electronics, catalysis, information processing, optoelectronics, environmental science, biomedical science, energy storage, and many other fields (Abe et al. 2016; Dutta and Datta 2014; Si et al. 2016; Candelaria et al. 2012; Wu et al. 2009; Hochbaum and Yang 2009; Sichert et al. 2015).

With the recent development of nanotechnology, a new scientific field of materials physics and chemistry emphasizing the rational synthesis of nanomaterials has emerged (Lin et al. 2012). Functional nanomaterials are especially an attractive topic because they enable the creation of materials with new or improved properties by mixing multiple constituents and exploiting synergistic effects such as electronic, optical, magnetic, catalytic properties or bioactivity, selective permeation, and adsorption (Gawande et al. 2016; Ouyang et al. 2015; Sampaio et al. 2015; Perreault et al. 2015; Tao et al. 2014; Gai et al. 2018). With a special property or several remarkable functions, functional nanomaterials are a type of high added-value materials possessing potential applications in various fields including catalysis, computing, photonics, energy, biology, and medicine (Rengan et al. 2015; Zhang and Lieber 2015; Carrow and Gaharwar 2015; Wei et al. 2017; Sobon 2015).

For example, functional nanomaterials have had an impact on medical devices such as drug delivery, systems diagnostic biosensors, and imaging probes (Biju 2014; Kumar et al. 2015; Li et al. 2015). The emerging field of nanobiotechnology holds the potential of revolutionizing biomedical and biology studies by employing new nanomaterial-based tools for investigative, diagnostic, and therapeutic techniques (Biju 2014; Wang et al. 2017; Nazir et al. 2014). Scientists have made great efforts in developing various kinds of nanomaterials and nanofabrication techniques in recent years. Their unique optical, magnetic, and mechanical properties of functional nanomaterials offer new opportunities for investigating complicated biological processes, which are hard to study by traditional strategies, suggesting exciting avenues in biological and biomedical fields (Barkalina et al. 2014). Nanomaterials provide almost unlimited combinations of various compositions, sizes, dimensions, and shapes of materials, which can be tailored to couple different biomolecules in order to develop nanoprobes with desired properties (Shao et al. 2015; Zhang et al. 2015b). Nanomaterials also have dramatical improvements in production and shelf-life in food and cosmetics industries (Ghaderi-Ghahfarokhi et al. 2017; Frewer et al. 2014). Owing to the large surface area, they are expected to be more biologically active than larger sized particles of the same chemical composition (Wang et al. 2016a). This offers several perspectives for food applications. For instance, nanoparticles could be utilized as bioactive compounds in functional foods. Bioactive compounds found naturally in certain foods have physiological benefits and might help to reduce the risk of certain diseases such as cancer (Ghorani and Tucker 2015; Kim et al. 2016). By reducing particle size, nanotechnology contributes to improve the properties of bioactive compounds such as solubility, prolonged residence time and delivery properties in the gastrointestinal tract and efficient absorption through cells (Fabiano et al. 2015). More importantly, functional nanomaterials have opened up new frontiers in materials science and engineering to be an enabling technology for creating high-performance energy conversion and storage devices (Gao et al. 2013; Dai et al. 2012).

The energy issue is one of the most significant topics during the twenty-first century (Pfenninger et al. 2014). It is estimated that the world will need to double its energy supply by 2050 (Armaroli and Balzani 2007; Cook et al. 2010). An overdependence on the non-renewable fossil fuels poses not only ecological problems but also serious and continuous impacts on the global society and economy (Asif and Muneer 2007; Omer 2008). Increasing energy demand, reduction of fossil fuel reserves, and environmental pollution have promoted the research of efficient and low-emission energy conversion devices (Bromberg et al. 2001). The importance of developing new types of energy is evident from the fact that global energy consumption is accelerating at an alarming rate due to the rapid economic growth on the global scale, the increase of the world population, and the increasing dependence of mankind on energy equipment (Dincer 2000). For this purpose, advanced technologies for both energy conversion (e.g., solar cells and fuel cells) and storage (e.g., supercapacitors and batteries) have received extensive research around the world. Nanotechnology has opened up new areas in materials science and engineering to meet this challenge (Arico et al. 2005). As with all other devices, the performance of energy-related devices strongly depends on the characteristics of the materials they utilize. Recent development in materials science, particularly nanomaterials, has facilitated the research and development of energy technologies. Comparing to traditional energy materials, nanomaterials possess unique properties useful for enhancing the energy conversion and storage performances (Arico et al. 2005; Zhang et al. 2013). Nanomaterials have attracted much attention in various energy devices including fuel cells, solar cells, light-emitting diodes, sensors, lithium-ion batteries, supercapacitors, thermoelectric devices, and memory devices (Zhao et al. 2015; Wang et al. 2015; Sun et al. 2015a; Xing et al. 2016; Song et al. 2015; Zhu et al. 2014; Devi et al. 2015; Su et al. 2014; Wu et al. 2015; Peng et al. 2014; Chen and Dai 2014; Ortega et al. 2017; Tan et al. 2015). As a very promising catalyst for oxygen reduction reactions (ORR), nanomaterials are effective alternatives to Pt-based electrocatalysts in fuel cell systems (Zhang et al. 2015a; Ganesan et al. 2015). In recent years, nanomaterials have also been actively researched as electrode materials in lithium-ion batteries and electrochemical supercapacitors (Hu et al. 2015; Mondal et al. 2015; Lu et al. 2017). The wide application of nanomaterials benefits from the progress in the synthesis of novel nanostructured materials with different sizes and various topographies. As the size of nanomaterials is reduced to the nanometer scale, new chemical and physical properties have emerged due to the well-known quantum size effects (Zalfani et al. 2016). In addition, nanomaterials can provide a larger specific surface area, which is advantageous to energy devices compared with their bulk masses, as the reaction/interaction between the device and the interaction medium can be significantly enhanced (Lü et al. 2014; Gao et al. 2013). As a result, people have made tremendous efforts to exploit the unique properties of nanomaterials and have made tremendous progress in developing high-performance energy conversion and storage devices as shown in Fig. 1.1.

In the past two decades, there have been many scientific efforts devoted to the design and preparation of nanomaterials with controlled morphology and tailored anisotropic nanostructures such as one-dimensional (1D) nanowires (NWs), nanorods (NRs), and nanotubes (NTs), two-dimensional (2D) nanoplates, and three-dimensional (3D) hierarchical structures have exhibited new fundamental (Shen et al. 2017; Qu et al. 2018; Wu et al. 2017; Guo et al. 2017; Lee et al. 2017; Xin et al. 2018). 0D structures have a short diffusion length and a minimal surface area, so agglomeration tends to occur during the cycle. 1D nanomaterials have attracted wide attention from the academic and industrial world due to their low cost, controllable size, and large-scale manufacturing capabilities (Liu et al. 2014; Ma et al. 2016; Wang et al. 2018). These nanomaterials have a unique, versatile, tunable structure with a nano interface, a high surface to volume ratio, and a large surface area, which can improve the performance of energy devices (Lin et al. 2017; Wei et al. 2017; Li et al. 2018). 1D structure has a fast electron transport in one dimension and a short ion diffusion length in the radial direction, but the static structure and fixed size limit the non-adjustable specific surface area and porosity properties (Mao et al. 2018). In contrast, 2D nanomaterials with high aspect ratios are very attractive materials for energy storage applications due to their unique electronic, mechanical and optical properties, quantum confinement, large surface area, and surface orientation properties (Wu et al. 2014; Peng et al. 2017). With the thickness of atoms or molecules and the infinite plane length, 2D nanomaterials have different atomic structures from their bulk counterparts including atomic arrangements, chemical valences, coordination numbers, and bond length differences. In addition, their more exposed internal atoms inevitably induce the formation of various defects, which will have a non-negligible effect on their chemical and physical properties. In fact, 2D nanomaterials have shown fascinating properties in the energy storage field including good mechanical flexibility, short ion diffusion length, and a large exposed surface electrochemical process (Pomerantseva and Gogotsi 2017; Tan et al. 2017). In addition, 2D nanomaterials have been extensively studied as active or supporting materials for various energy storage applications such as lithium ions, sodium ions, lithium sulfur, and metal-air batteries (Ji et al. 2016; Agubra et al. 2016; Ji et al. 2011). In general, 3D structures have the following advantages: (1) The large specific surface area of nanoplatelets can be maintained because the restacking of nanoplatelets is effectively suppressed. Therefore, a large number of electrochemically active sites are exposed to the electrolyte so that sufficient electrochemical reactions can be performed. (2) In the preparation of 3D structures, the use of nanosized 2D nanoplatelets as building blocks can result in a large number of pores or channels, thereby effectively reducing

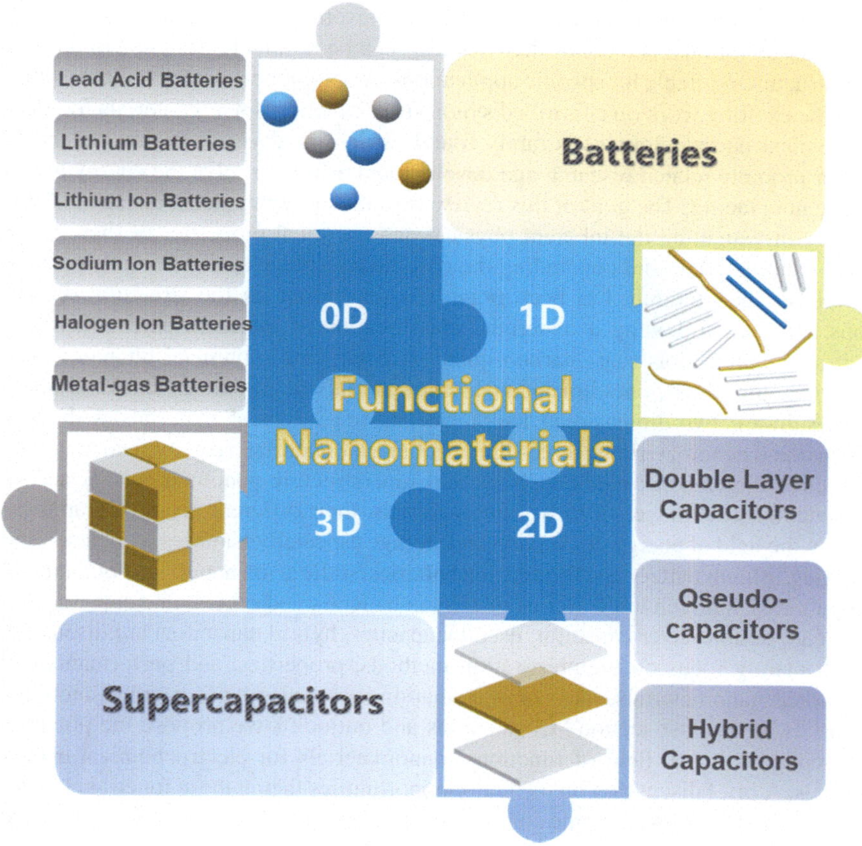

Fig. 1.1 Functional nanostructured materials for various high-performance energy conversion and storage devices Reprinted from Ref. Shen et al. (2017), copyright 2017, with permission from WILEY–VCH; reprinted from Ref. Qu et al. (2018), copyright 2018, with permission from The Royal Society of Chemistry; reprinted from Ref. Wu et al. (2017), copyright 2017, with permission from American Chemical Society; reprinted from Ref. Guo et al. (2017), copyright 2017, with permission from ACS Applied Materials and Interfaces; reprinted from Ref. Lee et al. (2017), copyright 2017, with permission from The Royal Society of Chemistry; reprinted from Ref. Xin et al. (2018), copyright 2018, with permission from The Royal Society of Chemistry; reprinted from Ref. Liu et al. (2014), copyright 2014, with permission from Macmillan Publishers Limited; reprinted from Ref. Ma et al. (2016), copyright 2016, with permission from Elsevier; reprinted from Ref. Wang et al. (2018), copyright 2018, with permission from The Royal Society of Chemistry

the distance of ions and mass transport. (3) Electron transfer is significantly enhanced since the 3D structure is usually built directly on a conductive substrate or hybridized with a conductive material such as a carbonaceous material. Notably, compared with 2D nanomaterials, 3D architectures have better processability due to the inhibited aggregation (Zhang et al. 2017; Choi et al. 2012, 2015; Wang et al. 2016b; Chen et al. 2014; Zhao et al. 2013).

It is very important to understand the basic anisotropic growth process of nanomaterials, so rational and controlled synthesis can be rationally designed to prepare nanostructures suitable for specific applications. A comprehensive book that summarizes the exciting work on controlled synthesis of functional nanomaterials for electrochemical energy storages is rarely found, which, however, is highly needed to further promote related research and development efforts to solve the energy issues we are now facing. The goal of this review is to illustrate the recent advance of this field by investigating the inherent physical and chemical properties of these functional nanomaterials and concluding the specific advantages and potentials of these materials. The literature has been organized depending on the structural dimensions and then following a materials' classification of the nano-objects involves such as quantum dots (e.g., carbon quantum dots, metal-nonmetal quantum dots, other quantum dots), one-dimensional nanomaterials (e.g., one-dimensional metal oxide/sulfide), two-dimensional nanomaterials (e.g., typical materials with two-dimensional nanomaterials), three-dimensional nanomaterials (e.g., typical materials with three-dimensional nanomaterials) and superstructure nanomaterials. Based on the potential advantages of these nanomaterials, their different promising applications in the field of energy conversion and storage especially batteries (e.g., lead-acid batteries, lithium battery and lithium-ion batteries, sodium ion battery and other metal ion batteries, halogen ion batteries metal-gas batteries and others) and supercapacitors (e.g., double-layer capacitor, qseudocapacitor, hybrid capacitor) are discussed, which mainly focus on the preparation methods, properties, and performances of functional nanomaterials, and some formation mechanisms of specific functional materials. In the last section "Conclusions and outlook", we propose the potential developments in the field of functional nanomaterials for electrochemical energy storages. A brief discussion on the major opportunities facing these functional nanomaterials will be showed. Some concluding remarks will try to determine what could be the next challenges of this fascinating research area.

We hope that this book will constitute a useful tool for the non-specialized readers who want to get an overview of the current trends related to functional nanomaterials for electrochemical energy storages, or for experts who want to look for a precise entry in a particular domain of application. Because of the explosion of publications in this exciting and emerging field, we do not claim that this book includes all of the published work (especially the most recently published work). We apologize to the authors of many outstanding research papers, that owing to the large activity in this field, we have unintentionally left out.

References

Abe H, Liu J, Ariga K (2016) Catalytic nanoarchitectonics for environmentally compatible energy generation. Mater Today 19(1):12–18

Agubra VA, Zuniga L, Flores D, Villareal J, Alcoutlabi M (2016) Composite nanofibers as advanced materials for Li-ion, Li-O_2 and Li-S batteries. Electrochim Acta 192:529–550

Arico AS, Bruce P, Scrosati B, Tarascon J-M, Van Schalkwijk W (2005) Nanostructured materials for advanced energy conversion and storage devices. Nat Mater 4(5):366

Armaroli N, Balzani V (2007) The future of energy supply: challenges and opportunities. Angew Chem Int Ed 46(1–2):52–66

Asif M, Muneer T (2007) Energy supply, its demand and security issues for developed and emerging economies. Renew Sustain Energy Rev 11(7):1388–1413

Barkalina N, Charalambous C, Jones C, Coward K (2014) Nanotechnology in reproductive medicine: emerging applications of nanomaterials. Nanomed Nanotech Biol Med 10(5):e921–e938

Biju V (2014) Chemical modifications and bioconjugate reactions of nanomaterials for sensing, imaging, drug delivery and therapy. Chem Soc Rev 43(3):744–764

Boverhof DR, Bramante CM, Butala JH, Clancy SF, Lafranconi M, West J, Gordon SC (2015) Comparative assessment of nanomaterial definitions and safety evaluation considerations. Regul Toxicol Pharmacol 73(1):137–150

Bromberg L, Cohn D, Rabinovich A, Heywood J (2001) Emissions reductions using hydrogen from plasmatron fuel converters. Int J Hydrogen Energy 26(10):1115–1121

Candelaria SL, Shao Y, Zhou W, Li X, Xiao J, Zhang J-G, Wang Y, Liu J, Li J, Cao G (2012) Nanostructured carbon for energy storage and conversion. Nano Energy 1(2):195–220

Carrow JK, Gaharwar AK (2015) Bioinspired polymeric nanocomposites for regenerative medicine. Macromol Chem Phys 216(3):248–264

Catherine L, Olivier P (2017) Gold nanoparticles for physics, chemistry and biology

Chen T, Dai L (2014) Flexible supercapacitors based on carbon nanomaterials. J Mater Chem A 2(28):10756–10775

Chen Z, Yuan Y, Zhou H, Wang X, Gan Z, Wang F, Lu Y (2014) 3D nanocomposite architectures from carbon-nanotube-threaded nanocrystals for high-performance electrochemical energy storage. Adv Mater 26(2):339–345

Choi B, Chang S, Lee Y, Bae J, Kim H, Huh Y (2012) 3D heterostructured architectures of Co3O4 nanoparticles deposited on porous graphene surfaces for high performance of lithium ion batteries. Nanoscale 4(19):5924–5930

Choi SH, Ko YN, Lee JK, Kang YC (2015) 3D MoS_2–graphene microspheres consisting of multiple nanospheres with superior sodium ion storage properties. Adv Func Mater 25(12):1780–1788

Choi S, Lee H, Ghaffari R, Hyeon T, Kim DH (2016) Recent advances in flexible and stretchable bio-electronic devices integrated with nanomaterials. Adv Mater 28(22):4203–4218

Commission E (2011) Commission Recommendation of 18 October 2011 on the definition of nanomaterial. Official Journal of the European Communities: Legis

Cook TR, Dogutan DK, Reece SY, Surendranath Y, Teets TS, Nocera DG (2010) Solar energy supply and storage for the legacy and nonlegacy worlds. Chem Rev 110(11):6474–6502

Dai L, Chang DW, Baek JB, Lu W (2012) Carbon nanomaterials for advanced energy conversion and storage. Small 8(8):1130–1166

Devi RV, Doble M, Verma RS (2015) Nanomaterials for early detection of cancer biomarker with special emphasis on gold nanoparticles in immunoassays/sensors. Biosens Bioelectron 68:688–698

Dincer I (2000) Renewable energy and sustainable development: a crucial review. Renew Sustain Energy Rev 4(2):157–175

Dutta A, Datta J (2014) Energy efficient role of Ni/NiO in PdNi nano catalyst used in alkaline DEFC. J Mater Chem A 2(9):3237–3250

Fabiano A, Chetoni P, Zambito Y (2015) Mucoadhesive nano-sized supramolecular assemblies for improved pre-corneal drug residence time. Drug Dev Ind Pharm 41(12):2069–2076

Feynman R (2018) There's plenty of room at the bottom. In: Feynman and computation

Frewer L, Gupta N, George S, Fischer A, Giles E, Coles D (2014) Consumer attitudes towards nanotechnologies applied to food production. Trends Food Sci Technol 40(2):211–225

Gai S, Yang G, Yang P, He F, Lin J, Jin D, Xing B (2018) Recent advances in functional nanomaterials for light–triggered cancer therapy. Nano Today 19:146–187

Ganesan P, Prabu M, Sanetuntikul J, Shanmugam S (2015) Cobalt sulfide nanoparticles grown on nitrogen and sulfur codoped graphene oxide: an efficient electrocatalyst for oxygen reduction and evolution reactions. ACS Catalysis 5(6):3625–3637

Gao M-R, Xu Y-F, Jiang J, Yu S-H (2013) Nanostructured metal chalcogenides: synthesis, modification, and applications in energy conversion and storage devices. Chem Soc Rev 42(7):2986–3017

Gawande MB, Goswami A, Felpin F-X, Asefa T, Huang X, Silva R, Zou X, Zboril R, Varma RS (2016) Cu and Cu-based nanoparticles: synthesis and applications in catalysis. Chem Rev 116(6):3722–3811

Ghaderi-Ghahfarokhi M, Barzegar M, Sahari M, Gavlighi HA, Gardini F (2017) Chitosan-cinnamon essential oil nano-formulation: application as a novel additive for controlled release and shelf life extension of beef patties. Int J Biol Macromol 102:19–28

Ghorani B, Tucker N (2015) Fundamentals of electrospinning as a novel delivery vehicle for bioactive compounds in food nanotechnology. Food Hydrocolloids 51:227–240

Guo R, Guo Y, Duan H, Li H, Liu H (2017) Synthesis of orthorhombic perovskite-type $ZnSnO_3$ Single-crystal nanoplates and their application in energy harvesting. ACS Appl Mater Interf 9(9):8271–8279

Hatto P (2011) ISO consensus definitions relevant to nanomaterials and nanotechnologies. In: 4th annual nano safety for success dialogue ISO TC 229and BSI NTI/1 Nano. https://doi.org/10.5772/58672technologiesstandardizationcommittees29thand30thMarch

Hochbaum AI, Yang P (2009) Semiconductor nanowires for energy conversion. Chem Rev 110(1):527–546

Hu X, Zhang W, Liu X, Mei Y, Huang Y (2015) Nanostructured Mo-based electrode materials for electrochemical energy storage. Chem Soc Rev 44(8):2376–2404

Jain KK, Jain KK (2017) The handbook of nanomedicine. Springer, Berlin

Ji L, Lin Z, Alcoutlabi M, Zhang X (2011) Recent developments in nanostructured anode materials for rechargeable lithium-ion batteries. Energy Environ Sci 4(8):2682–2699

Ji L, Meduri P, Agubra V, Xiao X, Alcoutlabi M (2016) Graphene-based nanocomposites for energy storage. Adv Energy Mater 6(16):1502159

Kagan CR, Fernandez LE, Gogotsi Y, Hammond PT, Hersam MC, Nel AE, Penner RM, Willson CG, Weiss PS (2016) Nano Day: celebrating the next decade of nanoscience and nanotechnology 10(10):9093–9103

Kim H-G, Bae J-H, Jastrzebski Z, Cherkas A, Heo B-G, Gorinstein S, Ku Y-G (2016) Binding, antioxidant and anti-proliferative properties of bioactive compounds of sweet paprika (*Capsicum annuum* L.). Plant Foods Hum Nutr 71(2):129–136

Kumar S, Ahlawat W, Kumar R, Dilbaghi N (2015) Graphene, carbon nanotubes, zinc oxide and gold as elite nanomaterials for fabrication of biosensors for healthcare. Biosens Bioelectron 70:498–503

Lee DJ, Yu S-H, Lee HS, Jin A, Lee J, Lee JE, Sung Y-E, Hyeon T (2017) Facile synthesis of metal hydroxide nanoplates and their application as lithium-ion battery anodes. J Mater Chem A 5(18):8744–8751

Lin N, Huang J, Dufresne A (2012) Preparation, properties and applications of polysaccharide nanocrystals in advanced functional nanomaterials: a review. Nanoscale 4(11):3274–3294

Lin D, Liu Y, Cui Y (2017) Reviving the lithium metal anode for high-energy batteries. Nat Nanotechnol 12(3):194

Liu N, Lu Z, Zhao J, Mcdowell MT, Lee HW, Zhao W, Cui Y (2014) A pomegranate-inspired nanoscale design for large-volume-change lithium battery anodes. Nat Nanotechnol 9(3):187–192

Li J, Cheng F, Huang H, Li L, Zhu J-J (2015) Nanomaterial-based activatable imaging probes: from design to biological applications. Chem Soc Rev 44(21):7855–7880

Li D, Wang Y, Sun Y, Lu Y, Chen S, Wang B, Zhang H, Xia Y, Yang D (2018) Turning gelidium amansii residue into nitrogen-doped carbon nanofiber aerogel for enhanced multiple energy storage. Carbon 137:31–40

Lu X, Wang C, Favier F, Pinna N (2017) Electrospun nanomaterials for supercapacitor electrodes: designed architectures and electrochemical performance. Adv Energy Mater 7:1601301

Lü Y, Zhan W, He Y, Wang Y, Kong X, Kuang Q, Xie Z, Zheng L (2014) MOF-templated synthesis of porous Co_3O_4 concave nanocubes with high specific surface area and their gas sensing properties. ACS Appl Mater Interf 6(6):4186–4195

Ma X, Luo W, Yan M, He L, Mai L (2016) In situ characterization of electrochemical processes in one dimensional nanomaterials for energy storages devices. Nano Energy 24:165–188

Mao J, Zhou T, Zheng Y, Gao H, Liu HK, Guo Z (2018) Two-dimensional nanostructures for sodium-ion battery anodes. J Mater Chem A 6(8):3284–3303

Manoharan M (2008) Research on the frontiers of materials science: the impact of nanotechnology on new material development. Technol Soc 30(3–4):401–404

Min SY, Kim TS, Lee Y, Cho H, Xu W, Lee TW (2015) Organic nanowire fabrication and device applications. Small 11(1):45–62

Mondal S, Rana U, Malik S (2015) Graphene quantum dot-doped polyaniline nanofiber as high performance supercapacitor electrode materials. Chem Commun 51(62):12365–12368

Murphy CJ (2002) Nanocubes and nanoboxes. Science 298(5601):2139–2141

Nazir S, Hussain T, Ayub A, Rashid U, MacRobert AJ (2014) Nanomaterials in combating cancer: therapeutic applications and developments. Nanomed Nanotechnol Biol Med 10(1):19–34

Omer AM (2008) Energy, environment and sustainable development. Renew Sustain Energy Rev 12(9):2265–2300

Ortega S, Ibáñez M, Liu Y, Zhang Y, Kovalenko MV, Cadavid D, Cabot A (2017) Bottom-up engineering of thermoelectric nanomaterials and devices from solution-processed nanoparticle building blocks. Chem Soc Rev 46(12):3510–3528

Ouyang S, Hu X, Zhou Q (2015) Envelopment–internalization synergistic effects and metabolic mechanisms of graphene oxide on single-cell chlorella vulgaris are dependent on the nanomaterial particle size. ACS Appl Mater Interf 7(32):18104–18112

Peng X, Peng L, Wu C, Xie Y (2014) Two dimensional nanomaterials for flexible supercapacitors. Chem Soc Rev 43(10):3303–3323

Peng L, Fang Z, Zhu Y, Yan C, Yu G (2017) Holey 2D nanomaterials for electrochemical energy storage. Adv Energy Mater. https://doi.org/10.1002/aenm.201702179

Perreault F, De Faria AF, Elimelech M (2015) Environmental applications of graphene-based nanomaterials. Chem Soc Rev 44(16):5861–5896

Pfenninger S, Hawkes A, Keirstead J (2014) Energy systems modeling for twenty-first century energy challenges. Renew Sustain Energy Rev 33:74–86

Pomerantseva E, Gogotsi Y (2017) Two-dimensional heterostructures for energy storage. Nat Energy. https://doi.org/10.1038/nenergy.2017.89

Qu C, Zhang L, Meng W, Liang Z, Zhu B, Dang D, Dai S, Zhao B, Tabassum H, Gao S (2018) MOF-Derived α-NiS nanorods on graphene as an electrode for high-energy-density supercapacitors. J Mater Chem A 6:4003–4012

Rengan AK, Bukhari AB, Pradhan A, Malhotra R, Banerjee R, Srivastava R, De A (2015) In vivo analysis of biodegradable liposome gold nanoparticles as efficient agents for photothermal therapy of cancer. Nano Lett 15(2):842–848

Sampaio M, Bacsa R, Benyounes A, Axet R, Serp P, Silva CG, Silva A, Faria JL (2015) Synergistic effect between carbon nanomaterials and ZnO for photocatalytic water decontamination. J Catal 331:172–180

Shao A, Xie Y, Zhu S, Guo Z, Zhu S, Guo J, Shi P, James TD, Tian H, Zhu WH (2015) Far-red and near-IR AIE-active fluorescent organic nanoprobes with enhanced tumor-targeting efficacy: shape-specific effects. Angew Chem 127(25):7383–7388

Shen L, Lv H, Chen S, Kopold P, van Aken PA, Wu X, Maier J, Yu Y (2017) Carbon nanowires: peapod-like Li_3VO_4/N-doped carbon nanowires with pseudocapacitive properties as advanced materials for high-energy lithium-ion capacitors. Adv Mater. https://doi.org/10.1002/adma.201770196

Si Y, Chen M, Wu L (2016) Syntheses and biomedical applications of hollow micro-/nano-spheres with large-through-holes. Chem Soc Rev 45(3):690–714

Sichert JA, Tong Y, Mutz N, Vollmer M, Fischer S, Milowska KZ, García Cortadella R, Nickel B, Cardenas-Daw C, Stolarczyk JK (2015) Quantum size effect in organometal halide perovskite nanoplatelets. Nano Lett 15 (10):6521–6527

Sobon G (2015) Mode-locking of fiber lasers using novel two-dimensional nanomaterials: graphene and topological insulators. Photonics Res 3(2):A56–A63

Song J, Li J, Li X, Xu L, Dong Y, Zeng H (2015) Quantum dot light-emitting diodes based on inorganic perovskite cesium lead halides ($CsPbX_3$). Adv Mater 27(44):7162–7167

Stark WJ, Stoessel PR, Wohlleben W, Hafner A (2015) Industrial applications of nanoparticles. Chem Soc Rev 44(16):5793–5805

Su X, Wu Q, Li J, Xiao X, Lott A, Lu W, Sheldon BW, Wu J (2014) Silicon-based nanomaterials for lithium-ion batteries: a review. Adv Energy Mater. https://doi.org/10.1002/aenm.201300882

Sun H, Deng J, Qiu L, Fang X, Peng H (2015a) Recent progress in solar cells based on one-dimensional nanomaterials. Energy Environ Sci 8(4):1139–1159

Sun Y, Sills RB, Hu X, Seh ZW, Xiao X, Xu H, Luo W, Jin H, Xin Y, Li T (2015b) A bamboo-inspired nanostructure design for flexible, foldable, and twistable energy storage devices. Nano Lett 15(6):3899–3906

Tan C, Liu Z, Huang W, Zhang H (2015) Non-volatile resistive memory devices based on solution-processed ultrathin two-dimensional nanomaterials. Chem Soc Rev 44(9):2615–2628

Tan C, Cao X, Wu X-J, He Q, Yang J, Zhang X, Chen J, Zhao W, Han S, Nam G-H, Sindoro M, Zhang H (2017) Recent advances in ultrathin two-dimensional nanomaterials. Chem Rev 117(9):6225–6331

Tanaka K, Chujo Y (2015) Recent progress of optical functional nanomaterials based on organoboron complexes with β-diketonate, ketoiminate and diiminate. NPG Asia Mater 7(11):e223

Tao X, Wang J, Ying Z, Cai Q, Zheng G, Gan Y, Huang H, Xia Y, Liang C, Zhang W (2014) Strong Sulfur binding with conducting magnéli-phase Ti_nO_{2n-1} nanomaterials for improving lithium-sulfur batteries. Nano Lett 14(9):5288–5294

Wang Y-J, Zhao N, Fang B, Li H, Bi XT, Wang H (2015) Carbon-supported Pt-based alloy electrocatalysts for the oxygen reduction reaction in polymer electrolyte membrane fuel cells: particle size, shape, and composition manipulation and their impact to activity. Chem Rev 115(9):3433–3467

Wang P, Lombi E, Zhao F-J, Kopittke PM (2016a) Nanotechnology: a new opportunity in plant sciences. Trends Plant Sci 21(8):699–712

Wang J, Zhang X, Wei Q, Lv H, Tian Y, Tong Z, Liu X, Hao J, Qu H, Zhao J (2016b) 3D self-supported nanopine forest-like Co_3O_4@$CoMoO_4$ core–shell architectures for high-energy solid state supercapacitors. Nano Energy 19:222–233

Wang L, Xiong Q, Xiao F, Duan H (2017) 2D nanomaterials based electrochemical biosensors for cancer diagnosis. Biosens Bioelectron 89:136–151

Wang C, Kaneti YV, Bando Y, Lin J, Liu C, Li J, Yamauchi Y (2018) Metal–organic framework-derived one-dimensional porous or hollow carbon-based nanofibers for energy storage and conversion. Mater Horiz 5:394–407

Wei Q, Xiong F, Tan S, Huang L, Lan EH, Dunn B, Mai L (2017) Energy storage: porous one-dimensional nanomaterials: design, fabrication and applications in electrochemical energy storage. Adv Mater. https://doi.org/10.1002/adma.201602300

Wu XL, Jiang LY, Cao FF, Guo YG, Wan LJ (2009) $LiFePO_4$ nanoparticles embedded in a nanoporous carbon matrix: superior cathode material for electrochemical energy-storage devices. Adv Mater 21(25–26):2710–2714

Wu LC, Chen YJ, Mao ML, Li QH, Zhang M (2014) Facile Synthesis of spike-piece-structured Ni(OH)2 interlayer nanoplates on nickel foam as advanced pseudocapacitive materials for energy storage. ACS Appl Mater Interf 6(7):5168–5174

Wu S, Xu R, Lu M, Ge R, Iocozzia J, Han C, Jiang B, Lin Z (2015) Graphene-containing nanomaterials for lithium-ion batteries. Adv Energy Mater. https://doi.org/10.1002/aenm.201500400

Wu Q, Yang L, Wang X, Hu Z (2017) From carbon-based nanotubes to nanocages for advanced energy conversion and storage. Acc Chem Res 50(2):435–444

Xin YC, Perumal A, Bruno A, Yantara N, Veldhuis S, Martinezsarti L, Chandran B, Chirvony VS, Lo SZA, So J (2018) Self-assembled hierarchical nanostructured perovskites enable highly efficient LEDs via energy cascade. Energy Environ Sci 11:1770–1778

Xing J, Yan F, Zhao Y, Chen S, Yu H, Zhang Q, Zeng R, Demir HV, Sun X, Huan A (2016) High-efficiency light-emitting diodes of organometal halide perovskite amorphous nanoparticles. ACS Nano 10(7):6623–6630

Zalfani M, van der Schueren B, Mahdouani M, Bourguiga R, Yu W-B, Wu M, Deparis O, Li Y, Su B-L (2016) ZnO quantum dots decorated 3DOM TiO_2 nanocomposites: symbiose of quantum size effects and photonic structure for highly enhanced photocatalytic degradation of organic pollutants. Appl Catal B 199:187–198

Zhang H (2015) Ultrathin two-dimensional nanomaterials. ACS nano 9(10):9451–9469

Zhang A, Lieber CM (2015) Nano-bioelectronics. Chem Rev 116(1):215–257

Zhang Q, Uchaker E, Candelaria SL, Cao G (2013) Nanomaterials for energy conversion and storage. Chem Soc Rev 42(7):3127–3171

Zhang J, Zhao Z, Xia Z, Dai L (2015a) A metal-free bifunctional electrocatalyst for oxygen reduction and oxygen evolution reactions. Nat Nanotechnol 10(5):444

Zhang X, Wang K, Liu M, Zhang X, Tao L, Chen Y, Wei Y (2015b) Polymeric AIE-based nanoprobes for biomedical applications: recent advances and perspectives. Nanoscale 7(27):11486–11508

Zhang H, Yun Q, Lu Q, Zhang X, Tan C (2017) Three-dimensional architectures constructed from transition metal dichalcogenide nanomaterials for electrochemical energy storage and conversion. Angew Chem. https://doi.org/10.1002/anie.201706426

Zhao B, Jiang S, Su C, Cai R, Ran R, Tadé MO, Shao Z (2013) A 3D porous architecture composed of TiO_2 nanotubes connected with a carbon nanofiber matrix for fast energy storage. J Mater Chem A 1(39):12310–12320

Zhao Z, Li M, Zhang L, Dai L, Xia Z (2015) Design principles for heteroatom-doped carbon nanomaterials as highly efficient catalysts for fuel cells and metal–air batteries. Adv Mater 27(43):6834–6840

Zhu C, Yang G, Li H, Du D, Lin Y (2014) Electrochemical sensors and biosensors based on nanomaterials and nanostructures. Anal Chem 87(1):230–249

Chapter 2
Synthesis of Quantum Dots

When these bulk 2D materials are converted into forms with lateral dimensions generally smaller than 100 nm (typically < 10 nm), quantum dots (QDs) could be produced resulting from the strong quantum confinement. The rising graphene quantum dots (GQDs) and carbon dots (C-dots) have attracted considerable attention as a result of their tremendous potentials in application of biomedicine, on account of their small size and their excellent performance in terms of photoluminescence properties, physicochemistry, photostability, and biocompatibility. The preparation of C-dots and GQDs could be roughly divided into two categories: "top-down" method and "bottom-up" method.

2.1 Carbonaceous Quantum Dots

The great success achieved so far in graphene materials is triggering immense enthusiasm for exploring two-dimensional (2D) layered inorganic materials such as hexagonal boron nitride (h-BN), (Li et al. 2015; Lei et al. 2015; Bonaccorso et al. 2015) transition metal dichalcogenides (TMDCs), (Kormányos et al. 2014; Zhou et al. 2016) graphitic carbon nitride (g-C_3N_4), (Abdolmohammad-Zadeh and Rahimpour 2016; Wang et al. 2014a) germanene (Wei et al. 2013) and silicene, (Xu et al. 2018) to meet new application requirements (Deng et al. 2016; Rao et al. 2013; Huang et al. 2014; Miró et al. 2014; Rim et al. 2016; Wang et al. 2017). When these bulk 2D materials are converted into forms with lateral dimensions generally smaller than 100 nm (typically < 10 nm), quantum dots (QDs) could be produced resulting from the strong quantum confinement (Buzaglo et al. 2016; Liu et al. 2013). As early as 1988, molybdenum disulfide (MoS_2) and tungsten disulfide (WS_2) nanoclusters with particle sizes of 10–35 Å were made via cleavage of the van der Waals layers of 2D bulk materials through penetrating solvent molecules (Peterson et al. 1988). This stimulated researchers to explore new types of nanoclusters of graphene,

H. Pang et al., *Synthesis of Functional Nanomaterials for Electrochemical Energy Storage*,
https://doi.org/10.1007/978-981-13-7372-5_2

(León and Pacheco 2011) h-BN, (Beheshtian et al. 2013) layered double hydroxides (LDHs), (Tokudome et al. 2016) molybdenum diselenide ($MoSe_2$) and tungsten diselenide (WSe_2), etc. (Huang and Kelley 2000). Such semiconductor nanoclusters represent clear quantum confinement effects, and thus are also classified as "QDs" (Mainwaring et al. 2006). The as-prepared QDs are usually regarded as zero dimensional (0D), however, they are just smaller forms of their native layered forms in terms of the lateral length dimension. Their 2D lattices in the bulk form can be still maintained by a certain degree when they are exfoliated into the nanoscale. Therefore, herein, "2D-QDs" is reasonably used as an abbreviation for "QDs derived from 2D inorganic materials" (Wang et al. 2016).

The burgeoning GQDs and C-dots have attained great attention due to their enormous development prospect with respect to biomedical applications. These illuminant carbon nanocrystals supply optical sensing and biological imaging with unprecedented opportunities. Owing to their small dimension and great biocompatibility, they could also function as effective carriers in the application of drug delivery when allowing isochronous visual control of releasing dynamics. What's more, their distinctive catalytic and physicochemical performance promise various applications in field of biomedicine. There have been already several preeminent review articles in regard to syntheses, performance, and application prospect of C-dots (Baker and Baker 2010) and GQDs (Zhang et al. 2012b).

Meanwhile, preparation of C-dots and GQDs may be ordinarily split into "top-down" method and "bottom-up" method. The former involves making carbonaceous materials cut or broken down through some approaches on basis of chemistry, electrochemistry, or physics. The latter is actualized by making small organic molecules thermally decomposed and carbonized or by making small aromatic molecules gradually chemically fused (Zheng et al. 2015).

2.1.1 Top-Down Method

Acidic Oxidation. Some methods have been extensively used to make C-dots and GQDs exfoliated from carbon fiber, carbon nanotube, (Tao et al. 2012) graphene oxide (GO), (Wang et al. 2011) soot, (Liu et al. 2007) coal, (Ye et al. 2013) carbon black, (Xia and Zheng 2012) and activated carbon, (Qiao et al. 2009) such as strong acid treatment. These methods are suitable for large-scale processable production from easily accessible low-cost carbon source. When using these methods, it is unavoidable to introduce oxygenated groups with a negative charge on the resultant C-dots, which makes them defective in graphitic structure and hydrophilic. One problem often met is that it is difficult to make excess oxidizing agent (e.g., HNO_3) completely removed.

Hydrothermal or Solvothermal Synthesis. Hydrothermal synthesis quintessentially uses reduced GO (rGO) sheets working as the precursors with a thermal method, and the precursors are heretofore treated with oxidizer (e.g., HNO_3, O_3), introducing epoxy groups onto the carbon lattice and ascertaining the cutting sites. In hydrothermal circumstances, GQDs are ultimately synthesized via a method with chemical cutting and deoxidization in alkali medium (e.g., NaOH, NH_3) (Pan et al. 2010, 2012; Zhu et al. 2011; Shen et al. 2011). On the basis of GO sheets, Zhu et al. illustrated the first sonication combined with solvent thermal preparation of GQDs, with DMF acting as solvent. Because of superior edge influence and quantum restriction, GQDs own pronounced features of QDs and graphene (Chen et al. 2014; Dinari et al. 2015). Consequently, GQDs, with their simple synthesis routes and great electrochemical performance (Low et al. 2013; Wang et al. 2014b). Chao et al. synthesized the graphene foam (GF) that are supported GQD and anchored it on VO_2 array electrode, which was called GVG (Chao et al. 2015). The formation of the GVG is shown in Fig. 2.1b. The growth mechanism includes the self-assembly and crystallographic orientation processes. In the solvothermal reaction process, the VOC_2O_4 nucleates on the GF surface, and the particles bond with each other, reducing the overall energy. Afterwards, a belt structure produced in the light of the strong anisotropy of monoclinic VO_2 crystal. Eventually, two single belts combine together in order to produce the intersected nanobelts structure according to the oriented attachment mechanism. Park et al. (2016) reported that introducing the GQDs into the cathode dramatically improved sulfur/sulfide utilization, realizing high performance. Additionally, the GQDs induced the integrity of sulfur-carbon electrode composite's structure through oxygen-enriched functional groups. The hierarchical architecture made charge transfer fast when reducing the wastage of lithium polysulfides, on account of the physicochemical performance of GQDs. The mechanism through which great cycling and rate properties are obtained was completely studied by the analysis of capacity and voltage profiles. GQDs could be fabricated by an improved Hummers' method, and then GQD-S and GQD-S/Carbon Black (CB) could be obtained through the hydrothermal processes with simple hybrid. The TEM and scanning TEM images of nanosized sulfur on the GQD electrodes in the Li_2S_8 catholyte are shown in Fig. 2.1b–e. GQDs tightly covered the nanosized sulfur particles' surface, which was demonstrated through lattice fringes in corresponding with (111) planes.

Electrochemical Exfoliation. GQDs and C-Dots can be prepared by electrochemical cutting of carbon precursors like carbon nanotubes, rGO film, graphite rods, and 3-dimensional (3D) CVD-grown graphene. Meanwhile, it has been suggested that OH· radical and O· radical formed on the basis of oxidation of water at the anode function as electrochemical "scissors" for the sake of the releasion of GQDs or C-dots (Lu et al. 2009). Some corrosion processes could be started near the edges and sped up at defect sites. When using organic solvents, Ananthanarayanan et al. ascribed the desquamation process to the capability of the electrolytic anion to insert between graphene layers and the electrical stress (Ananthanarayanan et al. 2014). Functional groups or heteroatoms can be adhibited or doped on the synthetic carbon nanodots lying on the used electrolyte (Li et al. 2012c). The abovementioned electrochemical strategy is uncomplicated (mainly one step) and usually of high yield.

Physical Routes. Microwave extraction provides rapid and even heating for the reaction medium, then considerably enhancing the reaction rate and making the product yields and quality better (Li et al. 2012a) Ultrasound may generate alternant high-pressure and low-pressure waves in fluid, resulting in the formation and severe

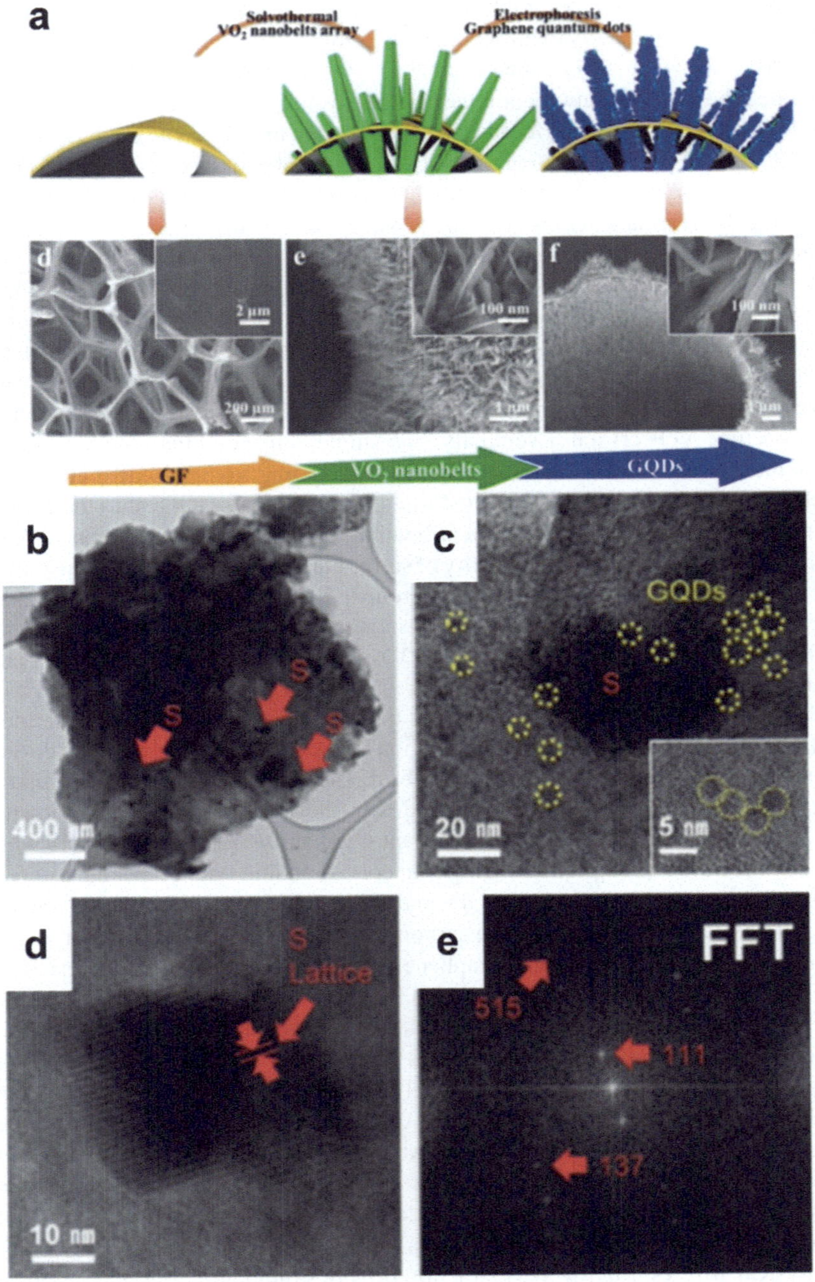
a
Solvothermal
VO2 nanobelts array
Electrophoresis
Graphene quantum dots
d
2 μm
200 μm
e
100 nm
1 μm
f
100 nm
1 μm
GF
VO2 nanobelts
GQDs
b
S
S
S
400 nm
c
GQDs
S
20 nm
5 nm
d
S
Lattice
10 nm
e
FFT
515
111
137

◂**Fig. 2.1** Synthesis procedures of GF supported GQDs-coated VO_2 nanobelts array. TEM and Scanning TEM images of nanosized S on GQDs electrode in Li_2S_8 cathode electrolyte when cycling 20 cycles in the state of charged: **a** low and high magnification images of nanosulfur in GQD. Lumps represent nanosulfur-containing GQDs electrodes, and the small black particles are nanosulfur, **b** GQDs coated on nanosulfur particles, **d** High-resolution transmission electron microscope (HRTEM) image displaying the lattice fringes of the nanosulfur as well as the GQDs, **e** The fast Fourier transform (FFT) of the initial HRTEM image is c in the middle of the filtered image. The two bright spots indicate sulfur particles, and the other spots represent the GQDs lattice plane. **a** Reprinted from Ref. Chao et al. (2015), copyright 2014, with permission from American Chemical Society. **b–e** Reprinted from Ref. Park et al. (2016), copyright 2016, with permission from Nature

collapse of little vacuum bubbles, which produce intense hydrodynamic shear forces and high-speed liquid jets in order to decompose the carbon with layer structure into GQDs (Tan et al. 2012). In recent, Prasad et al. prepared GQD-like quantum dots through the ultrasonic desquamation of polythiophene in DMF (Prasad et al. 2014). Fascinatingly, these QDs don't need to photobleach under consecutive laser irradiation. In the meantime, laser irradiation with high power can ablate carbon materials to obtain C-dots. Nevertheless, this method needs sophisticated equipment.

2.1.2 *Bottom-Up Method*

Stepwise Organic Synthesis. GQDs with well-defined monodispersed structures can be synthesized by solution chemistry approach, despite with low-throughput and trouble to avoid aggregation aroused by π-π interaction. For example, Yan et al. revealed that aryl groups oxidative condensation of polyphenylene dendritic precursor via stepwise solution chemistry brought about melt graphene moieties and ultimate formation of GQDs including 168, 132, and 170 carbon atoms that are conjugated (Yan et al. 2010b) Covalent interaction between 2′, 4, 6′-trialkyl phenyl groups and the edges of graphene-based materials stabilizes GQDs in liquid. More recently, homogeneous GQDs (with different dimensions and different colors) could be produced, with unsubstituted hexa-peri-hexabenzocoronene acting as precursor (Liu et al. 2011).

Pyrolysis or Carbonization of Organic Precursors. It has been extensively reported that C-dots and GQDs can be synthetized via thermally decompose or carbonize small organic molecules. With small organic molecules heated above their smelting point, they condensate, nucleate, and subsequently form larger C-dots or GQDs. The used precursors contain organic salts (e.g., diethylene glycolammonium citrate or octadecylammonium citrate), (Bourlinos et al. 2008) coffee grounds, (Hsu et al. 2012) glycerol, (Lai et al. 2012) L-glutamic acid, (Wu et al. 2013) ascorbic acid, (Jia et al. 2012) citric acid, (Dong et al. 2012a; Ju and Chen 2014) and ethylenediaminetetraacetic acid disodium salt (EDTA-2Na) (Deng et al. 2013). Other than simple combustion, plasma (Wang et al. 2012; Kim and Suh 2014) and microwave (Yin et al. 2013) can be used. These approaches are uncomplicated, cost-efficient, extensible, and allow heteroatoms to naturally inherit from the precursors.

2.1.3 Synthetic or Post-synthetic Strategies

2.1.3.1 Size and Shape Control

As described by Sk and his co-workers, PL transmission of a GQD is determined by its size in large part, and its shape to some extent (Sk et al. 2014). Stepwise organic synthesis can control both shape and size of synthetic GQDs precisely. Lu et al. prepared well-defined GQDs by mean of cage-opening of C60 molecules with ruthenium catalysis (Yan et al. 2010b). The appearance of the GQDs could be tailored to various shapes, such as triangular, parallelogram, trapezoid, hexagon, mushroom, and so on, with the different density of carbon clusters and the change in annealing temperature. C-dots with well defined and controllable size have been achieved by templated pyrolysis on the basis of organic precursors (Lu et al. 2011). While a co-polymer serves as soft template to provide C-dots with restrictions, the structure of mesoporous silica with order that functions as hard template prevents aggregation. C-dots prepared upon the foundation of single-chain polymeric nanoparticles as well have narrow distribution of size (Liu et al. 2011). Post-synthesis separation techniques like dialysis, gel electrophoresis, (Zhu et al. 2013) ultra-filtration, (Xia and Zheng 2012) column chromatography, (Zheng et al. 2013) or anion-exchange high-performance liquid chromatography (Li et al. 2010) can be used in order to reduce the dimensional changes. For instance, polyacryalamide gel electrophoresis (PAGE) already has been applied to split as-prepared C-dots up into fluorescent bands of nine with the emission peak range of 415 nm (violet) to 615 nm (orange-red), which indicates the advantage of gel electrophoresis to obtain multicolor C-dots with monodisperse (Vinci et al. 2013).

2.1.3.2 Surface Engineering

Controlling Oxidation. On the one side, oxygenated functional groups upon C-dots make them hydrophilic and supply further functionalization with convenient chemical handles. On the other side, these groups serve as surface emissive traps, causing the decrease of the PL efficiency. Hence, the characters of these nanodots could be tailored by the control of the degree of their oxidation. Besides, it has been illustrated that reducing oxygenated GQDs can improve quantum yield (QY) while oxidizing GQDs makes the emission red-shift. For instance, the green microwave-synthesized GQDs with $NaBH_4$ reducing made the emission shift to blue and a onefold increase in the QY.

Polymer Passivation. Polymer passivation with surface emissive traps is usually used to make the QY of GQDs and C-dots better. The most frequently used polymer is polyethylene glycol (PEG), even though others have been reported as well such as branched polyethyleneimine (BPEI) (Liu et al. 2007; Wang et al. 2013; Liu et al. 2012). A layer of PEG1500N can make the QY of C-dots more than 75%. In contrast research, Shen et al. prepared bare GQDs and GQDs that were passivated with PEG. The result indicated that the QY of the modified GQDs was one time higher than the former (Dong et al. 2012b).

Surface passivation. Surface passivation makes the synthesis process complicated and the particle sizes increased whereby placing restrictions on applications. When it comes to attaching Chemical Moieties, various chemical groups have been adhered to GQDs and C-dots in the synthesis process or after synthesis such as diamine thiol, alkylamine, and hydrazide (Shen et al. 2011). The electron-donating groups often improve QY by avoiding non-radiative associativity and usually cause evident wavelength shift. For instance, GQDs synthesized by green oxidation become blue when substituting alkylamine for carboxyl. In Tetsuka et al.'s work, it was illustrated that the emission wavelength of GQDs might be widely adjusted (blue to yellow) by regulating the degree of functionalization of amine (Zhu et al. 2012).

2.1.3.3 Heteroatom Doping

Heteroatom doping (by far the most common is, nitrogen doping) could be applied to fine-tune or attain new PL and the other physicochemical performance of C-dots and GQDs (Tetsuka et al. 2012; Luo et al. 2013). Heteroatoms could be retained from precursors in the process of synthesis. Wei et al. used the Maillard reaction between glucose and amino acids for the sake of the systematic preparation of a train of N-doped C-dots accompanied with high QY (about 69.1%) and PL tuned through N-doping level (shorter emission wavelength along with greater N-doping through basic amino acids) (Wang et al. 2014b). In addition, it has been clearly observed that N-doping (principally pyrrolic, potentially pyridinic, but not graphitic doping configuration) on GQDs enhances QY and leads to blue shift in emission because of the electron-withdrawing capability of nitrogen atoms. A number of researches also present that N-doping has the ability to provide up-conversion property for GQDs (Wei et al. 2014). Some other elements [e.g., Si, (Li et al. 2012b) P, (Qian et al. 2014) S, (Prasad et al. 2013; Kwon et al. 2013) and B (Fan 2014)] have as well been doped into C-dots and GQDs, in order to change PL properties or obtain greater catalytic performance. S/N co-doped C-dots and GQDs are able to achieve QY as great as 73 and 71%, severally (Dong et al. 2013).

2.2 Metal/Metal Oxides Quantum Dots

2.2.1 *Microwave Irradiation*

Recent advances in technology have now made microwave irradiation a novel synthetic method for the preparation of nanosized materials since the chemical reaction rate of microwave heating is as much as 1000 times faster than the conventional heating (Gawande et al. 2014). The use of microwave irradiation has become increasingly popular in inorganic synthesis, especially synthesis of metal oxides. Recently, nanostructured metal oxides such as Fe_2O_3 (Qiu et al. 2011), NiO (Song and Lian 2010), ZnO (Yang et al. 2013), CuO (Lu and Wang 2011), and Co_3O_4 (Yan et al. 2010a) have been successfully prepared by a microwave-assisted process.

Zhou et al. (2014) shown a unique method to synthesis a Co_3O_4 QDs/graphene compound by a microwave irradiation process. Graphene oxide (GO) sheets as precursors were synthesized by chemical exfoliation of flake graphite powder by an improved Hummers process. Graphene nanosheet was prepared via the thermal reduction of graphene oxide (GO) sheets in air atmosphere at 200 °C for 30 min then in Ar atmosphere at 450 °C for 3 h. Co_3O_4 quantum dots/graphene composite was synthesized by a facile microwave irradiation process. In a typical synthesis, 69 mg of grapheme nanosheets were dispersed in the varied solvent of 20 mL of deionized water and 30 mL of ethyl alcohol under sonication for 30 min. Then 1 g of $Co(CH_3COO)_2 \cdot 4H_2O$ was dissolved in the above solution and 5 mL of 25% $NH_3 \cdot H_2O$ was put in. The mixture was sonicated for about 10 min to form a homogeneous fuscous slurry. The suspension was finally moved into a microwave oven (XH MC-1) with reflux equipment and was heated at 81 °C for 5 min by microwave irradiation with a power of 500 W. The black products were collected by several rinse-centrifugation cycles with distilled water and absolute ethanol and were dried overnight in vacuum at 60 °C for further characterization.

2.2.2 *Heat Reduction*

Germanium QDs (Ge QDs) embedded in an N-doping GN matrix with a sponge-like architecture (Ge/GN sponge) was synthesized by freeze-drying and then a thermal reduction procedure (Fig. 2.2a–c). In a typical fabrication procedure, 0.2 g GeO_2 was put into 5 mL distilled water. After adding 1 mL ethylenediamine (EDA), the suspension came to be clear under the thermal situations. Then, some graphite oxide (GO) was out into the solution. Dilute the mixture to 20 mL with distilled water and continue stirring for 20 min. The mixture was then freeze-dried for one day. To finish, Ge/GN sponge was prepared by calcining the solid precursor in flowing H_2/AR (7 vol% H_2) atmosphere at 650 °C for 2 h. Pure GN was also synthesized via the same process but no GeO_2 was added. Ge QDs were well distributed in N-doped GN interconnects, about 2–5 nm. Based on the whole memory influence and network

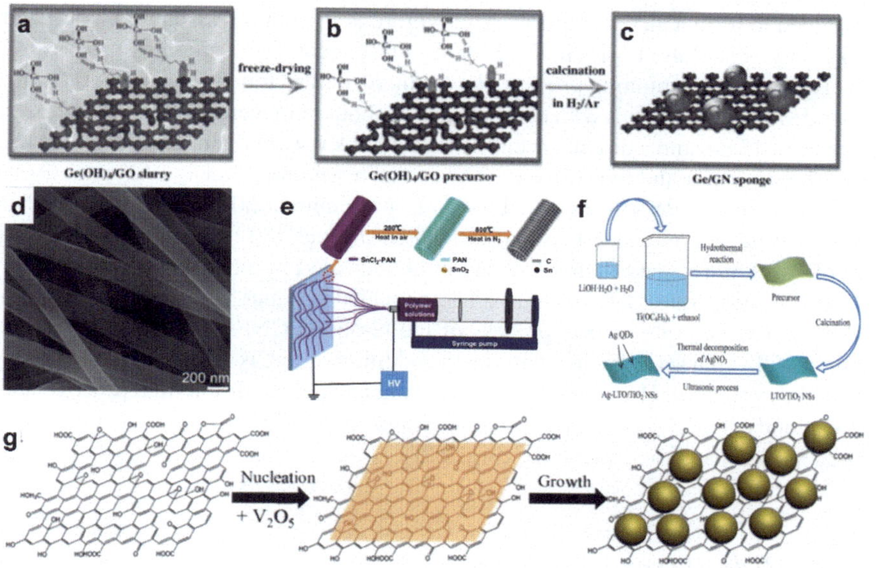

Fig. 2.2 **a–c** Diagrammatic sketch of the two-step procedure for the synthesis of the Ge/GN sponge. **d** TEM of Sn QDs@CNFs. **e** Schematic illustration of electrospinning process of Sn QDs@CNFs. **f** Fabrication procedure of the Ag-LTO/TiO_2 NS composite materials. **g** Diagrammatic sketch synthesis methods of VQDG. The black network indicates the reduced graphene oxide, the orange transparent plate represents the V_2O_5 sol and the blackish yellow particles are V_2O_5 QD. **a–c** Reprinted from Ref. Chao et al. (2014), copyright 2014, with permission from WILEY–VCH. **d, e** Reprinted from Ref. Zhang et al. (2014), copyright 2014, with permission from Elsevier. **f** Reprinted from Ref. Ge et al. (2016), copyright 2016, with permission from The Royal Society of Chemistry. **g** Reprinted from Ref. Han et al. (2013), copyright 2013, with permission from Elsevier

structure of highly conductive N-doped GN as the framework of Ge QDs, it is proved that Ge/GN sponge has good charge transfer and conductivity, good rate performance, and cycle stability (Qin et al. 2014). Because of the hydrophobicity of carbon surface, it is difficult to add active Sn with high capacity into these 1D carbon constructions. Surfactants and/or templates always need to be uniformly dispersed, which inevitably increases production costs and reduces electronic conductivity (Zhang et al. 2012a). Hence, it is desirable to synthesize uniform dispersion Sn QDs within conductive CNFs supports that not only retains the high capacity of the nanosized material but also shows excellent cycling performance by avoiding excessive subreactions and aggregation of Sn. Zhang's group (Zhang et al. 2014) reported a facile, direct method to prepare Sn QDs embedded in N-doped CNFs via an electrospinning method as well as subsequent annealing step in nitrogen atmosphere (Fig. 2.2d, e). The short diffusion path for both electrons and ions provided via the ultra-small Sn particles advance enhanced the rate property. Typically, 0.8 g of polyacrylonitrile (PAN, Mw = 150,000, Sigma) was dissolved in 10 mL of *N*,*N*-dimethylformamide (DMF, Tianjin Chemicals, 99.0%), and then 0.28 g of Tin(II) chloride dehydrate ($SnCl_2 \cdot 2H_2O$)

was added to form a mixed solution. After vigorous stirring for 4 h at 60 °C, buff and sticky polymer solution was obtained for the subsequent electrospinning process. As for a typical electrospinning process, the spinneret had an inner diameter of 0.6 mm. Grounded aluminum strips with parallel gaps of about 1 cm were used as collectors. A distance of 15 cm and a direct current voltage of 18 kV were maintained between the tip of the spinneret and the collector. The as-electrospinning $SnCl_2$/PAN nanofibers (denoted as $SnCl_2$-PAN) were dried for 24 h in vacuum at 60 °C. The dried $SnCl_2$-PAN were further annealed in an air-circulated oven at 280 °C for 2 h at a heating rate of 2 1C min^{-1}, and then 62 Zhang et al. annealed in a N2 flow at 800 1C for 1 h at a heating rate of 5 1C min1. The annealed product was designated as Sn QDs@CNFs. The preparation process of the N-doped CNFs was the same as the method of Sn QDs@CNFs, just without the addition of $SnCl_2 \cdot 2H_2O$. Ag-LTO/TiO_2 NSs were synthesized by Ge et al. (2016) through a facile hydrothermal process, and subsequently, $AgNO_3$ was then calcinated and thermally decomposed; ultrasonic treatment was used to disperse Ag (Fig. 2.2f).

2.2.3 *Hydrothermal Reaction*

Due to the feature of high purity in the production of particles, good dispersion, good and controllable crystal shape and low production cost, hydrothermal synthesis is also an effective method (Zubair et al. 2017). Han and co-workers (2013) published a two-step solution phase synthesis procedure to obtain the V_2O_5 QD/graphene hybrid (VQDG) nanocomposite by controlling the nucleation and growth procedures (Fig. 2.2 g). A two-step solution phase process was used to synthesize the VQDG with the as-prepared rGO suspension (~0.1 mg mL^{-1}). The vanadium sol and aniline were mixed in a 100 mL beaker and stirred at room temperature for 30 min with the molar ratio of 1:0.03. Next, 13 mL rGO suspension was added into the mixture and stirred at 80 °C for 24 h. The sample was then transferred into a 100 mL autoclave and heated at 180 °C for 48 h. After washed and dried at 80 °C for 24 h, a blackish yellow powder was obtained.

2.3 Nonmetallic Quantum Dots

2.3.1 *Mechanical Crushing Method*

The mechanical crushing method is a direct method to grind bulk to nanometer scale. Rangasamy et al. exhibited graphene/Si-CuO quantum dots (Gr/Si-CuO QD) layered structure. Si nanopowder was prepared by mechanical crushing process. In order to synthesize Si quantum dots (QD), the crushed Si powder (0.2 g) was mixed with (100 mL) DI water: (10 mL) HF solution followed by sonication for

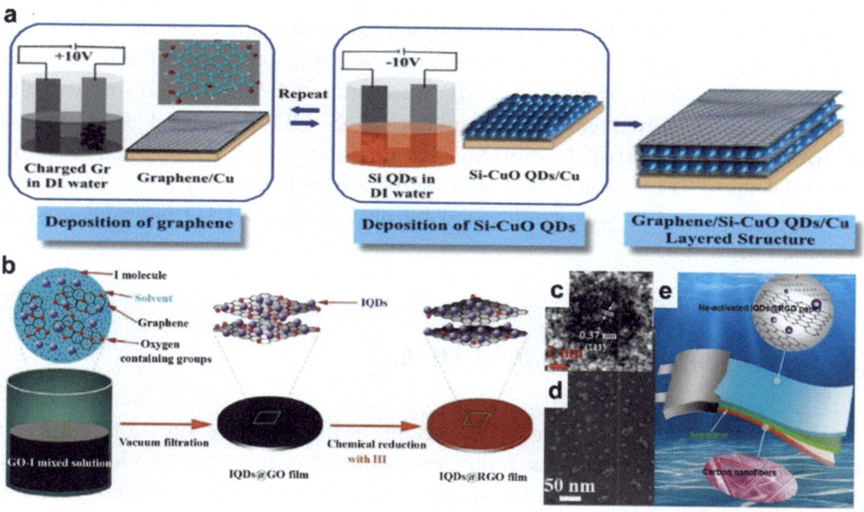

Fig. 2.3 **a** Fabrication of multilayer Gr/Si-CuO QD. **b** Diagrammatic sketch of the preparation of IQDs@RGO cathode. **c** High-resolution TEM images of IQDs@RGO cathodes. **d** High magnification FE-SEM image of the IQDs@RGO cathode. **e** Diagrammatic sketch of the flexible Na-ion full-cell. **a** Reprinted from Ref. Rangasamy et al. (2014), copyright 2014, with permission from Elsevier. **b–e** Reprinted from Ref. Gong et al. (2016), copyright 2016, with permission from Wiley–VCH

2 h. The Si QD (0.15 g) solution was then used for second electrophoresis with Cu electrodes by applying a voltage of 10 V for 60 s (Rangasamy et al. 2014). By repeating cathodic depositions and subsequent drying process, Si-CuO thin film on Cu substrate was fabricated. The Gr/Si-CuO layered thin films have been synthesized by alternating electrophoretic technique as depicted above (Fig. 2.3a). The layered Gr/Si-CuO QD samples were dried in a vacuum oven at 60 °C for 24 h (as-prepared). Further reduction process was applied by annealing at 400 °C for 30 min under Ar atmosphere with 50 mTorr (base Pressure 10^{-7} mTorr) in a vacuum chemical vapor deposition (CVD) system which leads to the formation of interlayer Cu_3Si in Si-CuO QD.

2.3.2 *Oxidant-Triggered Exfoliation Method*

The reasonable surface engineering of electrostatically active nanostructures is very ideal, because it can not only ensure the energy storage of high surface control, but also maintain the integrity of the structure, so as to achieve a long-term and high-speed cycle. A kind of spontaneous mild oxidant (H_2O_2) initiated exfoliation process was used to prepare subminiature MoS_2 quantum dots. Typically, 200 mg of pristine MoS_2 flake were distributed in a solution of H_2O_2 (30 wt% aqueous solution) and NMP (v/v = 1:1) and stirred at 35 °C for 10 h. The mixture was centrifuged for 30 min at 5000 rpm and the upper half of the supernatant was collected. After centrifugation for 2–4 times, the mixture was transparent and the ultra-small MoS_2 quantum dots distributed in NMP were prepared. In order to improve the electrochemical performance, the subminiature MoS_2 QDs were used as the surface sensitizer (Xu et al. 2016). Herein Xu's group (Song et al. 2017) tried to bring MoS_2 QDs and LTO together through a facile assembly technique, the $Li_4Ti_5O_{12}$ (LTO)/MoS_2 composite was synthesized by anchoring 2D LTO nanosheets with ultra-small MoS_2 QDs.

2.3.3 *Concentration Gradient Method*

Due to the abundance of iodine in the ocean, iodine is a low-cost alternative to renewable energy storage devices. Iodine easily dissolves the solution. Gong and co-workers synthesized free-standing iodine QDs (IQDs) decorated on reduced graphene oxide (IQDs@RGO) by a new and facile "concentration gradient" two-step method (Fig. 2.3b). The FE-SEM and TEM images (Fig. 2.3c, d) reveal an average diameter $\approx$6 nm for the IQDs and demonstrated that IQDs were homogeneously dispersed on and adhered strongly to the RGO (Fig. 2.3e) (Gong et al. 2016).

References

Abdolmohammad-Zadeh H, Rahimpour E (2016) A novel chemosensor based on graphitic carbon nitride quantum dots and potassium ferricyanide chemiluminescence system for Hg (II) ion detection. Sens Actuators B: Chem 225:258–266

Ananthanarayanan A, Wang X, Routh P, Sana B, Lim S, Kim D-H, Lim K-H, Li J, Chen P (2014) Facile synthesis of graphene quantum dots from 3D graphene and their application for Fe^{3+} sensing. Adv Func Mater 24(20):3021–3026

Baker SN, Baker GA (2010) Luminescent carbon nanodots: emergent nanolights. Angew Chem Int Ed 49(38):6726–6744

Beheshtian J, Tabar MB, Bagheri Z, Peyghan AA (2013) Exohedral and endohedral adsorption of alkaline earth cations in BN nanocluster. J Mol Model 19(3):1445–1450

Bonaccorso F, Colombo L, Yu G, Stoller M, Tozzini V, Ferrari AC, Ruoff RS, Pellegrini V (2015) Graphene, related two-dimensional crystals, and hybrid systems for energy conversion and storage. Science 347(6217):1246501

Bourlinos AB, Stassinopoulos A, Anglos A, Anastasiadis SH, Zboril R, Karakassides M, Giannelis EP (2008) Surface functionalized carbogenic quantum dots. In: APS meeting, pp 455–458

Buzaglo M, Shtein M, Regev O (2016) Graphene quantum dots produced by microfluidization. Chem Mater 28(1):21–24

Chao D, Zhu C, Xia X, Liu J, Xiao Z, Jin W, Pei L, Lin J, Hua Z, Shen Z (2015) Graphene quantum dots coated VO_2 arrays for highly durable electrodes for Li and Na ion batteries. Nano Lett 15(1):565

Chen Q, Hu Y, Hu C, Cheng H, Zhang Z, Shao H, Qu L (2014) Graphene quantum dots-three-dimensional graphene composites for high-performance supercapacitors. Phys Chem Chem Phys 16(36):19307–19313

Deng Y, Zhao D, Chen X, Wang F, Song H, Shen D (2013) Long lifetime pure organic phosphorescence based on water soluble carbon dots. Chem Commun 49(51):5751–5753

Deng D, Novoselov K, Fu Q, Zheng N, Tian Z, Bao X (2016) Catalysis with two-dimensional materials and their heterostructures. Nat Nanotechnol 11(3):218

Dinari M, Momeni MM, Goudarzirad M (2015) Nanocomposite films of polyaniline/graphene quantum dots and its supercapacitor properties. Surf Eng 32(7):535–540

Dong Y, Shao J, Chen C, Li H, Wang R, Chi Y, Lin X, Chen G (2012a) Blue luminescent graphene quantum dots and graphene oxide prepared by tuning the carbonization degree of citric acid. Carbon 50(12):4738–4743

Dong Y, Wang R, Li G, Chen C, Chi Y, Chen G (2012b) Polyamine-functionalized carbon quantum dots as fluorescent probes for selective and sensitive detection of copper ions. Anal Chem 84(14):6220–6224

Dong Y, Pang H, Yang HB, Guo C, Shao J, Chi Y, Li CM, Yu T (2013) Carbon-based dots co-doped with nitrogen and sulfur for high quantum yield and excitation-independent emission. Angew Chem Int Ed Engl 52(30):7800–7804

Fan L (2014) Surrounding media sensitive photoluminescence of boron-doped graphene quantum dots. Electrochem Soc 70:149–156

Gawande MB, Shelke SN, Zboril R, Varma RS (2014) Microwave-assisted chemistry: synthetic applications for rapid assembly of nanomaterials and organics. Acc Chem Res 47(4):1338–1348

Ge H, Cui L, Zhang B, Ma T-Y, Song X-M (2016) Ag quantum dots promoted $Li_4Ti_5O_{12}/TiO_2$ nanosheets with ultrahigh reversible capacity and super rate performance for power lithium-ion batteries. J Mater Chem A 4(43):16886–16895

Gong D, Wang B, Zhu J, Podila R, Rao AM, Yu X, Xu Z, Lu B (2016) An iodine quantum dots based rechargeable sodium-iodine battery. Adv Energy Mater 7(3):1601885

Han C, Yan M, Mai L, Tian X, Xu L, Xu X, An Q, Zhao Y, Ma X, Xie J (2013) V_2O_5 quantum dots/graphene hybrid nanocomposite with stable cyclability for advanced lithium batteries. Nano Energy 2(5):916–922

Hsu P-C, Shih Z-Y, Lee C-H, Chang H-T (2012) Synthesis and analytical applications of photoluminescent carbon nanodots. Green Chem 14(4):917–920

Huang JM, Kelley DF (2000) Synthesis and characterization of $MoSe_2$ and WSe_2 nanoclusters. Chem Mater 12(10):2825–2828

Huang X, Tan C, Yin Z, Zhang H (2014) 25th anniversary article: hybrid nanostructures based on two-dimensional nanomaterials. Adv Mater 26(14):2185–2204

Jia X, Li J, Wang E (2012) One-pot green synthesis of optically pH-sensitive carbon dots with upconversion luminescence. Nanoscale 4(18):5572–5575

Ju J, Chen W (2014) Synthesis of highly fluorescent nitrogen-doped graphene quantum dots for sensitive, label-free detection of Fe (III) in aqueous media. Biosens Bioelectron 58(10):219–225

Kim J, Suh JS (2014) Size-controllable and low-cost fabrication of graphene quantum dots using thermal plasma jet. ACS Nano 8(5):4190–4196

Kormányos A, Zólyomi V, Drummond ND, Burkard G (2014) Spin-orbit coupling, quantum dots, and qubits in monolayer transition metal dichalcogenides. Phys Rev X 4(1):011034

Kwon W, Lim J, Lee J, Park T, Rhee SW (2013) Sulfur-incorporated carbon quantum dots with a strong long-wavelength absorption band. J Mater Chem C 1(10):2002–2008

Lai C-W, Hsiao Y-H, Peng Y-K, Chou P-T (2012) Facile synthesis of highly emissive carbon dots from pyrolysis of glycerol; gram scale production of carbon dots/$mSiO_2$ for cell imaging and drug release. J Mater Chem 22(29):14403–14409

Lei Z, Xu S, Wan J, Wu P (2015) Facile preparation and multifunctional applications of boron nitride quantum dots. Nanoscale 7(45):18902–18907

León A, Pacheco M (2011) Electronic and dynamics properties of a molecular wire of graphane nanoclusters. Phys Lett A 375(47):4190–4197

Liu H, Ye T, Mao C (2007) Fluorescent carbon nanoparticles derived from candle soot. Angew Chem Int Ed Engl 46(34):6473–6475

Liu R, Wu D, Feng X, Müllen K (2011) Bottom-up fabrication of photoluminescent graphene quantum dots with uniform morphology. J Am Chem Soc 133(39):15221–15223
Liu C, Zhang P, Zhai X, Tian F, Li W, Yang J, Liu Y, Wang H, Wang W, Liu W (2012) Nanocarrier for gene delivery and bioimaging based on carbon dots with PEI-passivation enhanced fluorescence. Biomaterials 33(13):3604–3613
Liu F, Jang M-H, Ha HD, Kim J-H, Cho Y-H, Seo TS (2013) Facile synthetic method for pristine graphene quantum dots and graphene oxide quantum dots: origin of blue and green luminescence. Adv Mater 25(27):3657–3662
Li H, He X, Kang Z, Huang H, Liu Y, Liu J, Lian S, Tsang CHA, Yang X, Lee ST (2010) Water-soluble fluorescent carbon quantum dots and photocatalyst design. Angew Chem 122(26):4532–4536
Li M, Wu W, Ren W, Cheng HM, Tang N, Zhong W, Du Y (2012b) Synthesis and upconversion luminescence of N-doped graphene quantum dots. Appl Phys Lett 101(10):768
Li Y, Zhao Y, Cheng H, Hu Y, Shi G, Dai L, Qu L (2012c) Nitrogen-doped graphene quantum dots with oxygen-rich functional groups. J Am Chem Soc 134(1):15–18
Li LL, Ji J, Fei R, Wang CZ, Lu Q, Zhang JR, Jiang LP, Zhu JJ (2012a) A facile microwave avenue to electrochemiluminescent two-color graphene quantum dots. Adv Func Mater 22(14):2971–2979
Li H, Tay RY, Tsang SH, Zhen X, Teo EHT (2015) Controllable synthesis of highly luminescent boron nitride quantum dots. Small 11(48):6491–6499
Low CTJ, Walsh FC, Chakrabarti MH, Hashim MA, Hussain MA (2013) Electrochemical approaches to the production of graphene flakes and their potential applications. Carbon 54(4):1–21
Luo PG, Sahu S, Yang S-T, Sonkar SK, Wang J, Wang H, LeCroy GE, Cao L, Sun Y-P (2013) Carbon "quantum" dots for optical bioimaging. J Mater Chem B 1(16):2116–2127. https://doi.org/10.1039/C3TB00018D
Lu J, Yang JX, Wang J, Lim A, Wang S, Loh KP (2009) One-pot synthesis of fluorescent carbon nanoribbons, nanoparticles, and graphene by the exfoliation of graphite in ionic liquids. ACS Nano 3(8):2367–2375
Lu J, Yeo PS, Gan CK, Wu P, Loh KP (2011) Transforming C60 molecules into graphene quantum dots. Nat Nanotechnol 6(4):247
Lu LQ, Wang Y (2011) Sheet-like and fusiform CuO nanostructures grown on graphene by rapid microwave heating for high Li-ion storage capacities. J Mater Chem 21(44):17916–17921
Mainwaring DE, Let AL, Rix C, Murugaraj P (2006) Titanium sulphide nanoclusters formed within inverse micelles. Solid State Commun 140(7):355–358
Miró P, Audiffred M, Heine T (2014) An atlas of two-dimensional materials. Chem Soc Rev 43(18):6537–6554
Pan D, Zhang J, Li Z, Wu M (2010) Hydrothermal route for cutting graphene sheets into blue-luminescent graphene quantum dots. Adv Mater 22(6):734–738
Pan D, Guo L, Zhang J, Xi C, Xue Q, Huang H, Li J, Zhang Z, Yu W, Chen Z, Li Z, Wu M (2012) Cutting sp2 clusters in graphene sheets into colloidal graphene quantum dots with strong green fluorescence. J Mater Chem 22(8):3314–3318
Park J, Moon J, Kim C, Kang JH, Lim E, Park J, Lee KJ, Yu S, Seo J, Lee J (2016) Graphene quantum dots: structural integrity and oxygen functional groups for high sulfur|[sol]|sulfide utilization in lithium sulfur batteries 8(5):e272
Peterson MW, Nenadovic MT, Rajh T, Herak R, Micic OI, Goral JP, Nozik AJ (1988) ChemInform abstract: quantized colloids produced by dissolution of layered semiconductors in acetonitrile. Cheminform 19(24):1400–1402
Prasad KS, Pallela R, Kim DM, Shim YB (2013) Microwave-assisted one-pot synthesis of metal-free nitrogen and phosphorus dual-doped nanocarbon for electrocatalysis and cell imaging. Part Part Syst Charact 30(6):557–564
Prasad K, Chen Y, Sk M, Than A, Wang Y, Sun H, Lim KH, Dong X, Chen P (2014) Fluorescent quantum dots derived from PEDOT and their applications in optical imaging and sensing. Mater Horizons 1(5):529–534

Qian Z, Shan X, Chai L, Ma J, Chen J, Feng H (2014) Si-Doped carbon quantum dots: a facile and general preparation strategy, bioimaging application, and multifunctional sensor. ACS Appl Mater Interfaces 6(9):6797–6805

Qiao ZA, Wang Y, Gao Y, Li H, Dai T, Liu Y, Huo Q (2009) Commercially activated carbon as the source for producing multicolor photoluminescent carbon dots by chemical oxidation. Chem Commun 46(46):8812–8814

Qin J, Wang X, Cao M, Hu C (2014) Germanium quantum dots embedded in N-Doping graphene matrix with sponge-like architecture for enhanced performance in lithium-ion batteries. Chemistry 20(31):9675–9682

Qiu G, Huang H, Genuino H, Opembe N, Stafford L, Dharmarathna S, Suib SL (2011) Microwave-Assisted hydrothermal synthesis of nanosized α-Fe_2O_3 for catalysts and adsorbents. J Phys Chem C 115(40):19626–19631

Rangasamy B, Hwang JY, Choi W (2014) Multi layered Si–CuO quantum dots wrapped by graphene for high-performance anode material in lithium-ion battery. Carbon 77(10):1065–1072

Rao CNR, Ramakrishna Matte HSS, Maitra U (2013) Graphene analogues of inorganic layered materials. Angew Chem Int Ed 52(50):13162–13185

Rim YS, Bae S-H, Chen H, De Marco N, Yang Y (2016) Recent progress in materials and devices toward printable and flexible sensors. Adv Mater 28(22):4415–4440

Shen J, Zhu Y, Yang X, Zong J, Zhang J, Li C (2011) One-pot hydrothermal synthesis of graphene quantum dots surface-passivated by polyethylene glycol and their photoelectric conversion under near-infrared light. New J Chem 36(1):97–101

Sk MA, Ananthanarayanan A, Huang L, Lim KH, Chen P (2014) Revealing the tunable photoluminescence properties of graphene quantum dots. J Mater Chem C 2(34):6954–6960

Song X, Lian G (2010) Facile synthesis of polycrystalline NiO nanorods assisted by microwave heating. J Am Ceram Soc 91(10):3465–3468

Song S, Heo J, Lee TK, Park S, Walker B, Sang KK, Jin YK (2017) Optically tunable plasmonic two-dimensional Ag quantum dot arrays for optimal light absorption in polymer solar cells. J Phys Chem C 121(33):17569–17576

Tan D, Zhou S, Qiu J (2012) Comment on "upconversion and downconversion fluorescent graphene quantum dots: ultrasonic preparation and photocatalysis." ACS Nano 6(8):1059–1064

Tao H, Yang K, Ma Z, Wan J, Zhang Y, Kang Z, Liu Z (2012) In vivo NIR fluorescence imaging, biodistribution, and toxicology of photoluminescent carbon dots produced from carbon nanotubes and graphite. Small 8(2):281–290

Tetsuka H, Asahi R, Nagoya A, Okamoto K, Tajima I, Ohta R, Okamoto A (2012) Optically tunable amino-functionalized graphene quantum dots. Adv Mater 24(39):5333–5338

Tokudome Y, Morimoto T, Tarutani N, Vaz PD, Nunes CD, Prevot V, Stenning GBG, Takahashi M (2016) Layered double hydroxide nanocluster: aqueous, concentrated, stable, and catalytically-active colloids towards green chemistry. ACS Nano 10(5):5550–5559

Vinci JC, Ferrer IM, Seedhouse SJ, Bourdon AK, Reynard JM, Foster BA, Bright FV, Colón LA (2013) Hidden properties of carbon dots revealed after HPLC fractionation. J Phys Chem Lett 4(2):239–243

Wang Q, Zheng H, Long Y, Zhang L, Gao M, Bai W (2011) Microwave–hydrothermal synthesis of fluorescent carbon dots from graphite oxide. Carbon 49(9):3134–3140

Wang J, Wang CF, Chen S (2012) Amphiphilic Egg-derived carbon dots: rapid plasma fabrication, pyrolysis process, and multicolor printing patterns. Angew Chem Int Ed 51(37):9297–9301

Wang Z, Xia J, Zhou C, Via B, Xia Y, Zhang F, Li Y, Xia L, Tang J (2013) Synthesis of strongly green-photoluminescent graphene quantum dots for drug carrier. Colloids Surf, B 112(3):192–196

Wang W, Jimmy CY, Shen Z, Chan DK, Gu T (2014a) gC 3 N 4 quantum dots: direct synthesis, upconversion properties and photocatalytic application. Chem Commun 50(70):10148–10150

Wang X, Sun G, Routh P, Kim DH, Huang W, Chen P (2014b) Heteroatom-doped graphene materials: syntheses, properties and applications. Chem Soc Rev 43(20):7067–7098

Wang X, Sun G, Li N, Chen P (2016) Quantum dots derived from two-dimensional materials and their applications for catalysis and energy. ChemInform 47(22):2239–2262

Wang L, Xiong Q, Xiao F, Duan H (2017) 2D nanomaterials based electrochemical biosensors for cancer diagnosis. Biosens Bioelectron 89:136–151
Wei Y, Yu H, Li H, Ming H, Pan K, Huang H, Liu Y, Kang Z (2013) Liquid-phase plasma synthesis of silicon quantum dots embedded in carbon matrix for lithium battery anodes. Mater Res Bull 48(10):4072–4077
Wei W, Xu C, Wu L, Wang J, Ren J, Qu X (2014) Non-enzymatic-browning-reaction: a versatile route for production of nitrogen-doped carbon dots with tunable multicolor luminescent display. Sci Rep 4(1):3564
Wu X, Tian F, Wang W, Chen J, Wu M, Zhao JX (2013) Fabrication of highly fluorescent graphene quantum dots using L-glutamic acid for in vitro/in vivo imaging and sensing. J Mater Chem C 1(31):4676–4684
Xia X, Zheng Y (2012) Comment on "one-step and high yield simultaneous preparation of single- and multi-layer graphene quantum dots from CX-72 carbon black". J Mater Chem 22(18):8764–8766
Xu G, Yang L, Wei X, Ding J, Zhong J, Chu PK (2016) MoS_2-Quantum-dot-interspersed $Li_4Ti_5O_{12}$ nanosheets with enhanced performance for Li- and Na-ion batteries. Adv Func Mater 26(19):3349–3358
Xu Y, Wang X, Zhang WL, Lv F, Guo S (2018) Recent progress in two-dimensional inorganic quantum dots. Chem Soc Rev 47(2):586–625
Yang SJ, Nam S, Kim T, Im JH, Jung H, Kang JH, Wi S, Park B, Park CR (2013) Preparation and exceptional lithium anodic performance of porous carbon-coated ZnO quantum dots derived from a metal-organic framework. J Am Chem Soc 135(20):7394–7397
Yan J, Wei T, Qiao W, Shao B, Zhao Q, Zhang L, Fan Z (2010a) Rapid microwave-assisted synthesis of graphene nanosheet/Co 3 O 4 composite for supercapacitors. Electrochim Acta 55(23):6973–6978
Yan X, Cui X, Li LS (2010b) Synthesis of large, stable colloidal graphene quantum dots with tunable size. J Am Chem Soc 132(17):5944
Ye R, Xiang C, Lin J, Peng Z, Huang K, Yan Z, Cook NP, Samuel EL, Hwang CC, Ruan G (2013) Coal as an abundant source of graphene quantum dots. Nat Commun 4(11):2943
Yin JY, Liu HJ, Jiang S, Chen Y, Yao Y (2013) Hyperbranched polymer functionalized carbon dots with multistimuli-responsive property. Acs Macro Lett 2(11):1033–1037
Zhang B, Yu Y, Huang Z, He YB, Jang D, Yoon WS, Mai YW, Kang F, Kim JK (2012a) Exceptional electrochemical performance of freestanding electrospun carbon nanofiber anodes containing ultrafine SnOx particles. Energy Environ Sci 5(12):9895–9902
Zhang Z, Zhang J, Chen N, Qu L (2012b) Graphene quantum dots: an emerging material for energy-related applications and beyond. Energy Environ Sci 5(10):8869–8890
Zhang G, Zhu J, Zeng W, Hou S, Gong F, Li F, Cheng Chao LI, Duan H (2014) Tin quantum dots embedded in nitrogen-doped carbon nanofibers as excellent anode for lithium-ion batteries. Nano Energy 9:61–70
Zheng XT, Than A, Ananthanaraya A, Kim DH, Chen P (2013) Graphene quantum dots as universal fluorophores and their use in revealing regulated trafficking of insulin receptors in adipocytes. ACS Nano 7(7):6278–6286
Zheng XT, Ananthanarayanan A, Luo KQ, Chen P (2015) Glowing graphene quantum dots and carbon dots: properties, syntheses, and biological applications. Small 11(14):1620–1636
Zhou X, Shi J, Liu Y, Su Q, Zhang J, Du G (2014) Microwave irradiation synthesis of Co_3O_4 quantum dots/graphene composite as anode materials for Li-ion battery. Electrochim Acta 143(12):175–179
Zhou K, Zhang Y, Xia Z, Wei W (2016) As-prepared MoS_2 quantum dot as a facile fluorescent probe for long-term tracing of live cells. Nanotechnology 27(27):275101
Zhu S, Zhang J, Qiao C, Tang S, Li Y, Yuan W, Li B, Tian L, Liu F, Hu R (2011) Strongly green-photoluminescent graphene quantum dots for bioimaging applications. Chem Commun 47(24):6858–6860

Zhu S, Zhang J, Tang S, Qiao C, Wang L, Wang H, Liu X, Li B, Li Y, Yu W (2012) Surface chemistry routes to modulate the photoluminescence of graphene quantum dots: from fluorescence mechanism to up-conversion bioimaging applications. Adv Func Mater 22(22):4732–4740

Zhu B, Sun S, Wang Y, Deng S, Qian G, Wang M, Hu A (2013) Preparation of carbon nanodots from single chain polymeric nanoparticles and theoretical investigation of the photoluminescence mechanism. J Mater Chem C 1(3):580–586

Zubair M, Daud M, Mckay G, Shehzad F, Al-Harthi MA (2017) Recent progress in layered double hydroxides (LDH)-containing hybrids as adsorbents for water remediation. Appl Clay Sci 143:279–292

Chapter 3
Synthesis of One-Dimensional Nanomaterials

With the development of nanotechnology, nanostructures have been a focus of research, particularly in the field of electrochemical energy storage. Since the groundbreaking discovery of carbon nanotubes (CNTs) in the 1990s, 1D nanostructures have attracted much research interest because of their remarkable physical/chemical properties and great potential in nanotechnology applications. In particular, 1D nanostructure can provide direct current pathways, shorten the ion diffusion distance, lower the charge–discharge time, increase the electrolyte–electrode contact area, limit mechanical degradation, and accommodate volume expansion. 1D nanostructure has one singular nanostructure, and types of 1D homostructure include nanowires, nanorods, nanotubes, and nanobelt. Furthermore, unlike typical 1D nanostructure, complex 1D nanostructures consist of multiple components. Types of complex 1D nanostructure include core–shell nanostructures, array architectures, branched nanowires, hollow nanostructure, and fiber structures. This synergistic effect between each component endows heterostructured electrodes with better electrical conductivity, greater electrochemical cycle stability and reversibility, faster ion transport, improved mechanical stability, etc. The synthesis of 1D nanostructures includes both top-down and bottom-up approaches. Representative bottom-up routes, including various vapor-based approaches and solution-based synthetic approaches, have been studied extensively. An integrated overview of the synthetic methods of 1D nanomaterials is presented in this section.

3.1 Metal Oxides/Sulfides

Metal oxides and sulfides are often used as active material for electrochemical energy storage equipment because of their large capacity. Numerous works have shown that by taking one step further into fabricating 1D nanoarchitecture or nanocomposites with rational design, much better performance could be achieved. Particular attention

H. Pang et al., *Synthesis of Functional Nanomaterials for Electrochemical Energy Storage*,
https://doi.org/10.1007/978-981-13-7372-5_3

is placed on strategies employed in fabricating, with examples such as 1D core–shell architecture, 1D porous architecture, and composite comprising one or more carbon matrices.

3.1.1 Vapor-Phase Fabrication

3.1.1.1 Chemical Vapor Deposition

Chemical vapor deposition (CVD) is an approach in which a material reacts on special substrate surfaces or vaporized precursors decompose to form typical 1D solid materials. The precursor material is carried via an inert gas, and the reaction occurs in a vacuum chamber. If there are chemical reactions involved in the process of vapor deposition, the approach is called chemical vapor deposition; the converse is physical vapor deposition. Chemical vapor deposition techniques include hot filament, microwave plasma, atmospheric pressure, thermal, photoassisted, plasma-enhanced, and low-temperature chemical vapor deposition.

A universal schematic has been reported to show the primary influencing factors in the catalyst-assisted growth of ZnO nanowires and the formation of vapor–solid materials. The different influencing factors in the growth of ZnO nanowires have been discussed in detail. Various parameters, such as reaction gas and carrier flow, substrate and powder temperature, were presented in the selected shape diagrams (Menzel et al. 2012). Most importantly, we could identify typical parameter ranges for vapor–solid growth, film formation and catalyst-assisted vapor–liquid–solid growth using the shape diagrams. In particular, experiment has also demonstrated the controlled growth of catalyst-assisted and vapor–solid ZnO nanowires in a chemical vapor deposition reactor via the carbothermal reduction process based on either downstream or upstream deposition techniques. In addition, a single-crystalline nanowires have been synthesized by CVD method (Mudusu et al. 2017). The SnS nanomaterials grown at temperatures between 600–700 °C have 1D wire-like morphology. The nanowires have an average diameter of about 12–15 nm with lengths up to several microns. Furthermore, the SnS nanowires consist of smooth and uniform surfaces. TEM analysis reveals that the 1D SnS nanowires consist of single-crystalline cubic crystal construction with a preferential growth direction of <100>.

3.1.1.2 Carbothermal Reduction

Carbothermal reduction is a facile approach for the synthesis of 1D materials, particularly metal oxide-functionalized nanostructures. Various nanomaterials for energy applications, including $LiFePO_4$, MnO_2, and ZnO, have been synthesized by carbothermal reduction. Additionally, this method is compatible with industry because of its low-cost precursor materials. The oxide reduction mechanism by gaseous intermediates, such as CO_2 and CO, follows (3.1) and (3.2):

$$MO\ (s) + CO\ (g) \rightarrow M\ (s) + CO_2(g) \tag{3.1}$$

$$CO_2(g) + C\ (s) \rightarrow 2CO\ (g) \tag{3.2}$$

For instance, metal oxide reduction by gaseous metal carbides can occur through the reaction of carbon and metal vapor, as in (3.3) and (3.4):

$$2MO\ (s) + MC_2(g) \rightarrow 3M\ (s) + 2CO\ (g) \tag{3.3}$$

$$M\ (s) + 2C\ (s) \rightarrow MC_2(g) \tag{3.4}$$

Then, thermal reduction by H_2O and H_2 gaseous intermediates follows, as in (3.5), (3.6), and (3.7):

$$MO\ (s) + H_2(g) \rightarrow M\ (s) + H_2O\ (g) \tag{3.5}$$

$$C\ (s) + H_2O\ (g) \rightarrow H_2(g) + CO\ (g) \tag{3.6}$$

$$MO\ (s) + C\ (s) \rightarrow M\ (s) + CO\ (g) \tag{3.7}$$

Cu_2O nanostructures have received attention in different applications such as energy storage, (Minami et al. 2013; Mittiga et al. 2006) water splitting, (Paracchino et al. 2011) gas sensing, (Deng et al. 2012) solar energy conversion, and the photocatalytic degradation of pollutants (Zheng et al. 2009). The unique aligned nanowire structure can afford inherently efficient architectures with increased surface area for charge transfer processes, minimized hole and electron diffusion lengths, and increased superficial area for charge transfer in charge collecting electrodes (Lewis 2007). Accordingly, vertically aligned Cu_2O nanowires can observably diversify and enhance the portfolio of Cu_2O-based nanostructure applications (Wu et al. 2014). Many synthetic methods for obtaining Cu_2O nanowires have been proposed. Distributed single-crystal Cu_2O nanowire arrays with a diameter of approximately 300 nm and a length of over 20 μm have been synthesized by hydrothermal processes (Tan et al. 2007; Hacialioglu et al. 2012). High-density Cu_2O nanowires (approximately 200 nm) have been synthesized using porous nanotemplates by electrochemical techniques; however, the length was limited to 4 μm due to diffusion-restricted electrolyte transport in the confined nanopores. Cu_2O nanowire arrays have been fabricated by the facile carbothermal reduction of CuO nanowires. CuO nanowires were first obtained by the thermal oxidation of copper foils. The oxidation of 26 μm copper foils at 700 °C for 10 h converted all the substrate to phase-pure cupric oxide. Two significant effects were observed when the transformation was initiated at a temperature of 350 °C. First and foremost, the $CuO \rightarrow Cu_2O$ reduction reaction occurs via the diffusion of O_2 under a reducing atmosphere provided by the existence of carbon monoxide. Additionally, the Cu_2O nanowires were observed as periodic aggregations, which were probably driven by Rayleigh instability during phase transformation.

3.1.1.3 Pulse Laser Deposition

Pulsed laser deposition is a physical vapor deposition approach, which uses a pulsed, high-power laser beam as the heating source. Complex oxide materials have various excellent features, including superconducting, (McCammon 1997) piezoelectric (Eerenstein et al. 2006) and ferroelectric(Wang et al. 2003) properties, and have been widely applied in electronic devices for advanced energy technologies, such as sensors, memories, and nonvolatile micromotors. Lead-zirconium-titanium is the most promising complex oxide material because of its excellent piezoelectric coefficient, high dielectric constant, and large hysteresis. There are various synthetic approaches to this material such as electrophoresis, (Limmer et al. 2010) hydrothermal synthesis (Xu et al. 2005), and template infiltration (Hernandez et al. 2010).

Work on the growth of $PbZr_{0.2}Ti_{0.8}O_3$ subuliform nanowire arrays by the pulsed laser deposition method suggests that stoichiometry can be controlled by a relatively low-temperature process (Chen et al. 2012). Single-crystal $PbZr_{0.2}Ti_{0.8}$ subuliform nanowire arrays were successfully synthesized on a $SrTiO_3$ substrate by a pulsed laser deposition process. The unique tapered morphology of the nanowires was attributed to the excess coating of the $PbZr_{0.2}Ti_{0.8}$ layer by lateral growth of $PbZr_{0.2}Ti_{0.8}$ adatoms during the $PbZr_{0.2}Ti_{0.8}$ nanowire growth process. The growth conditions for $PbZr_{0.2}Ti_{0.8}$ nanowires were studied at different pressures and temperatures. The experiment also demonstrated that the tapered $PbZr_{0.2}Ti_{0.8}$ nanowires underwent a Frank-van der Merwe growth mechanism, after the formation of 3D islands by a Stranski–Krastanov growth mechanism, followed axial growth on the lowest energy [001] crystal face via a vapor–solid mechanism. Nevertheless, under special conditions, such as specific substrate temperatures (<725 and >850 °C) or lower or higher pressures (<200 or >400 mTorr), the formation of $PbZr_{0.2}Ti_{0.8}$ nanowires was suppressed, while the $PbZr_{0.2}Ti_{0.8}$ thin film grown by the layer-by-layer mechanism remained. Most importantly, the direction of $PbZr_{0.2}Ti_{0.8}$ nanowire array growth could be controlled along the (001), (110), and (111) crystal faces of the $SrTiO_3$ substrate.

3.1.1.4 Atomic-Layer Deposition

Atomic-layer deposition is a cycled, self-limited chemical vapor deposition process of separate precursors via separate precursor pulses. This method is usually used to synthesize conformal thin films and to control the film thickness down to the nanometer level (George 2010).

Understanding the evolution and formation of crystalline and amorphous phases by atomic-layer deposition is necessary to create high-quality, multifunctional dielectric coatings/films, and to predict the surface functionalization (Shi et al. 2013). Integrated, atomistic electron-microscopy research of TiO_2 structures at designed growth cycles during atomic-layer deposition has elucidated transformation sequences and different atom arrangement processes during TiO_2 atomic-layer deposition growth. The evolution of TiO_2 structures during atomic-layer deposition was revealed by following the transformation from amorphous layers to amorphous particles to metastable crystallites and finally to stable crystalline forms. These changes were attributed to the Ostwald-Lussac law, which controlled the ratio and sequence of disparate phases of the TiO_2 nanostructures during the relatively high-temperature atomic-layer deposition. Moreover, the crystalline and amorphous mixture enabled specific anisotropic crystal growth at a relatively high temperature, forming TiO_2 nanorods by a vapor-phase oriented attachment mechanism.

3.1.1.5 Vapor–Liquid–Solid

Vapor–liquid–solid refers to a growth mechanism in which the material is directly absorbed via liquid catalysts during crystal growth. In the vapor–liquid–solid method, the 1D nanostructure is developed by solid precipitation from disperse and supersaturated catalyst droplets.

Titanium dioxide is a very promising material for various applications, including, for example, supercapacitors, lithium-ion batteries, dye-sensitized solar cells and water splitting systems, and there are various methods to synthesize TiO_2 nanostructures. However, intentionally controlling the spatial location and range of the length/diameter seems to be unfeasible with other common methods. Therefore, a position- and size-controlled vapor–liquid–solid method for synthesizing nanowires is a solution to overcome the above issues. Nevertheless, creating a TiO_2 single-crystal nanowire by the vapor–liquid–solid mechanism has been a challenging issue because of the difficulty in understanding and controlling complex nanomaterial transport events across three phases. Experiments have demonstrated that the vapor–liquid–solid growth of 1D TiO_2 structures can emerge consistently only within a quite narrow range of material flux (Zhuge et al. 2012). This phenomenon distinctly contrasts that of typical vapor–liquid–solid metal oxides, such as ZnO, In_2O_3, SnO_2, and MgO, whose metal oxide nanowires are easily grown by the vapor–liquid–solid mechanism with much wider ranges of material flux. Moreover, experimental results have shown that rutile TiO_2 nanowires primarily grow along the (0 0 1) direction, which differs from the usual (0 0 1)-oriented growth of TiO_2 nanowires formed by vapor-phase processes. This method, based on the control of material flux, is a reasonable strategy to tailor TiO_2 nanowires using the vapor–liquid–solid mechanism.

The growth of 1D metal oxides can be achieved through electron-beam evaporation processes (Yu and Lee 2014). Most metal oxides are easily resolved by the condensed electron-beam method. As seen in Fig. 3.1a, b the metal oxide source

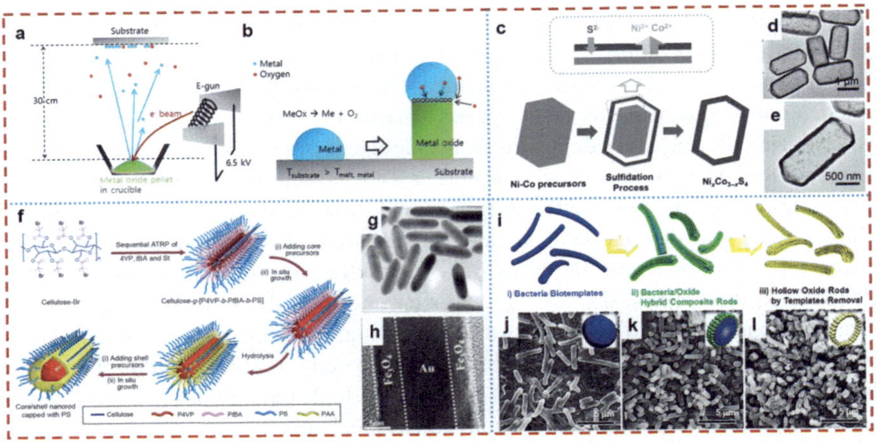

Fig. 3.1 Schematic illustrations of the vapor-based method and templated growth process. **a** Illustrations of electron-beam evaporation process for 1D nanomaterials growth. **b** Growth mechanism of 1D nanowire by vapor–liquid–solid. **c** The formation of 1D $Ni_xCo_{3-x}S_4$ hollow prisms from Ni/Co precursors. **d** TEM images of 1D Ni_2CoS_4 hollow prisms obtained after sulfidation process in ethanol. **e** TEM images of 1D Ni_2CoS_4 hollow prisms annealed in N_2 at 350 °C. **f** Schematic representation of synthetic process of 1D core–shell nanorods by template of cellulose-g-(P4VP-b-PtBA-b-PS) template. **g** TEM images of Au-Fe_3O_4 core–shell nanostructure templated by the cellulose-g-(P4VP-b-PtBA-b-PS) with a scale bar at 200 nm. **h** The HRTEM image shows the Au core and Fe_3O_4 shell. **i** Diagrammatic drawing of the bacteria-templated synthesis of the 1D nanostructures. The FESEM images of **j** the pure bacteria, **k** the 1D bacteria/cobalt oxide rods, and **l** the hollow cobalt oxide rods. The insets show the cross section of sample. **a**, **b** Reprinted from Ref. Yu and Lee (2014), copyright 2014, with permission from Springer Nature. **c**–**e** Reprinted from Ref. Yu et al. (2014), copyright 2014, with permission from WILEY–VCH. **f**–**h** Reprinted from Ref. Pang et al. (2016), copyright 2016, with permission from Science. **i**–**l** Reprinted from Ref. Shim et al. (2011), copyright 2011, with permission from American Chemical Society

is decomposed by a condensed electron beam, consequently creating a nanodot of self-catalytic metal for vapor–liquid–solid growth. 1D metal oxides are grown at a temperature above their melting point to achieve self-catalysis via dissolved oxygen. Furthermore, the morphology of the 1D nanostructures, including uniformity and density, depends on the surface migration energy and surface energy of the substrate.

3.1.2 Templated Growth

The templating technique is significant for the facile construction of promising materials with controlled microstructures/nanostructures and desired functions (Chen et al. 2015). The different hard templates with channel or porous structures such as MCM-41 silica and anodized alumina membranes have been extensively explored for the synthesis of nanotubes and nanorods. At the same time, various soft templates, such as block copolymers and surfactants, primary function as a structure-directing assemble the reacting species. Moreover, these templates offer advantages in terms of designing wire-like porous and sphere-like nanostructures. At present, another type of soft template, i.e., biological templates including DNA, virus particles, bacterium, etc., has been used to produce unique nanostructures and deposit nanoparticles. SnS has been an attractive material for Na-ion batteries. Lou et al. have manufactured SnS nanotubes composed of ultrathin nanosheets by a templating method. In order to boost the electrochemical property, carbon-coated SnS nanotubes have also been designed by adding glucose (He et al. 2017). Significantly, the tube-like C/SnS nanostructures show enhanced sodium storage properties in terms of good rate capability and cycling performance. 1D hollow construction is of great interest to a wide variety of applications. Many efforts have been devoted to developing technologies for the effective synthesis of different hollow constructions. Among these methods, synthesis involving templates has been demonstrated to be the most versatile route to generate different hollow architectures (Lai et al. 2012; Oh and Hyeon 2013). In this regard, many methods have been developed to manufacture hollow nanostructures with anisotropic shapes for noble metals, transition metal sulfides, and oxides, based on various principles, for instance, self-assembly, chemical etching, thermal decomposition, Kirkendall effect, and galvanic replacement, among others (Zhang et al. 2012). Nevertheless, the development of non-spherical hollow constructions is more challenging on account of the less controllable coating around high-curvature surfaces and the deficiency of non-spherical templates. Significantly, a hollow 1D $Ni_xCo_{3-x}S_4$ nanoprisms have been fabricated by sacrificial template process (Yu et al. 2014). The synthetic strategy for 1D $NixCo_{3-x}S_4$ nanoprisms is shown in Fig. 3.1c. Tetragonal nanoprisms of Co/Ni acetate hydroxide precursors with controllable Co/Ni molar ratios were first synthesized and used as the sacrificial templates. After a sulfidation process with thioacetamide in ethanol, the solid precursor could be transformed into the 1D $Ni_xCo_{3-x}S_4$ prisms with a hollow interior. In FESEM observations, the inner cavities of highly uniform 1D prisms are elucidated by the sharp contrast between the edge and the center (Fig. 3.1d). After annealing, polycrystalline hollow prisms were obtained without apparent deformation in appearance, as revealed by TEM (Fig. 3.1e) The specific compositional and 1D structural features are beneficial for electrochemical applications. Additionally, the resultant $Ni_xCo_{3-x}S_4$ nanoprisms manifest a high specific capacitance with enhanced cycling stability, making them potential electrode materials for electrochemical energy storage devices.

3.1.2.1 Block Copolymer Templating

In an alternative route, a series of 1D nanostructures, including core–shell nanorods, nanotubes, and plain nanorods, has been synthesized with precisely controlled compositions and dimensions by using functional block copolymers with narrow molecular weight distributions and specific structures as nanoreactors. The cylindrical nanoreactors enable an ultrahigh degree of control over the shape, surface chemistry, size, properties, and architecture of 1D nanocrystals. Figure 3.1f illustrates the synthetic strategies for 1D core–shell nanorods templated by cellulose-g-(P4VP-b-PtBA-b-PS). Figure 3.1g,h displays a representative SEM image of Au-Fe_3O_4 core–shell nanorods with a Fe_3O_4 shell thickness of approximately 4.6 nm (Pang et al. 2016).

3.1.2.2 Biological Templating

In recent years, the bioinspired and biomimetic synthesis of nanostructures has drawn increasing attention. Bio-nanotechnology is a facile strategy to manufacture different metallic oxide nanomaterials with precise control over their morphology, crystal structure, and chemical composition by means of natural bioassemblies and genetic engineering. The biological templating includes bacillus subtilis/gram-positive bacteria templating, tobacco mosaic virus templating, fungal biomineralization templating, and M13 virus templating.

Bacteria templating is promising for the generation of various inorganic micro/nanostructures, through the appropriate combination of typical chemical techniques without rigorous genetic engineering (Shim et al. 2011). These micro/nanostructures could possibly be used for the large-scale and facile production of functional micro/nanomaterials. A high-yield and simple biomineralization process using Bacillus subtilis, a gram-positive bacterium, as a soft template has been reported to produce Co_3O_4 nanostructures. Figure 3.1i–l schematically illustrates the Bacillus subtilis mediated biosynthetic process of the cobalt oxide nanostructures. The insets depict the cross section of each sample. Rod-type cobalt oxide was synthesized at room temperature via electrostatic interactions between cobalt ions and surface structures of the bacteria in aqueous solution. Porous Co_3O_4 hollow rods were prepared by successive heat treatment at 300 °C. Moreover, the rods showed outstanding electrochemical performance and had a high surface area. This inexpensive, environmentally friendly and facile synthetic method for metal oxides with peculiar nanostructures can be used in a number of practical applications such as supercapacitors, catalysts, sensors, and batteries. Biomineralization is a process utilized by organisms to produce composite structures composed of inorganic and organic materials that usually show exceptional properties. Organic molecules, including polysaccharides, peptides and proteins, indirectly guide nanocrystal growth under different ambient conditions and ultimately determine the functional properties and morphology of the materials (BãUerlein 2003; Weiner et al. 2005). ZnO nanowires with high surface-to-volume ratios are especially attractive (Atanasova et al. 2011). The extraordinary rod-like morphology of the tobacco mosaic virus (TMV) makes it an appropriate scaffold for the manufacture of 1D wire-like structures (Atanasova et al. 2011). Balci et al. successfully synthesized TMV nanorods coated with ZnO and Pd by electroless deposition (Balci et al. 2009). Additionally, these nanocomposites were applied in TMV-based biological and chemical sensing applications, digital device memories, and energy storage devices (Ricky et al. 2006; Nam et al. 2006). Atanasova et al. successfully synthesized TMV-ZnO nanowires by a biotemplating mineralization process using a TMV template at low temperature.

Furthermore, they provided an advanced method for the miniaturization and functionalization of devices using single TMV nanostructures as templates. In particular, genetic modification of TMV provided methods for inducing the deposition of ZnO nanostructures and thus enhancing the field-effect transistor performance (Kadri et al. 2011). Thus, the skillful design of TMV mutants can be applied to the deposition of other inorganic functional nanomaterials.

Novel bio-electrode materials have been successfully synthesized by a fungal manganese biomineralization process for the first time (Li et al. 2016). Further electrochemical tests showed that the carbonized fungal biomass-mineral electrode material had a fundamental specific capacitance in a supercapacitor and a good cycling stability in a lithium-ion battery. Most importantly, the research provided an advanced biotechnological method for the production of sustainable bio-electrochemical materials.

Composite 1D nanowires have been rationally designed and synthesized by an M13 virus-guidance process (Chen et al. 2014). Among the various types of representative organic templates for nanomaterial synthesis, M13 viruses have special properties such as the genetic tunability of helically arranged primary coat proteins and a high aspect ratio structure (diameter of approximately 6 nm and length of approximately 880 nm. Three unique spinel oxide nanowires of $Mn_xCo_{3-x}O_4$ ($x = 0, 1, 2$) were templated from the M13 virus by a two-step reaction. From the heterogeneous nanowires, three-dimensional bio-composite nanowires with a length of 1 μm and a diameter of 50 nm were produced.

3.1.2.3 AAO Templating

Hierarchical manganese dioxide nanowire-nanofibril arrays have been fabricated on an anodized aluminum oxide (AAO) template (Duay et al. 2013). A conformal layer of manganese dioxide nanofibrils was evenly grown on the surface of single manganese dioxide nanowires. The synthetic mechanism of the hierarchical manganese dioxide nanostructures was characterized by electron energy-loss spectroscopy (EELS), electrochemical measurements, electron microscopy, and Raman spectroscopy. The charge storage mechanism of this complex nanomaterial was investigated in different solvents at slow scan rates. In acetonitrile electrolyte, the MnO_2 nanofibrils exhibited capacitance due to cation insertion, while in aqueous electrolyte, the MnO_2 nanofibrils exhibited capacitance due to surface adsorption and double-layer processes. Additionally, the MnO_2 nanofibrils exhibited controllable parameters including nanowire diameter, amount of nanofibril material, and nanowire length.

3.1.2.4 MOFs Templating

Owing to metal–organic frameworks (MOFs) diverse compositional and structural functionalities, which formed by supramolecular assembly of metal ions with organic struts. They have been intensively employed as an emerging class of templates or precursors to construct hollow architectures in the last few years (Zhang et al. 2016; Hu et al. 2016; Cai et al. 2015; Hu and Chen 2014; Zhang et al. 2013; Liu et al. 2016; Wu et al. 2015). Accordingly, hollow structured MOFs have enabled the generation of hollow constructions with complexity in the building blocks and compositions (Chen et al. 2016; Han et al. 2016; Zou et al. 2014). For instance, Zou and co-workers have been studied the novel hollow octahedra composed of carbon stabilized $ZnO/ZnFe_2O_4$ nanoparticles by hollow MOF-5 as both the precursor and self-sacrificing template. A surface-energy-driven process may be responsible for the formation of hollow MOF nanocages. Despite the progress achieved to date, the construction of nanosized 1D hollow MOFs is still relatively less reported because of the limited morphologies of the MOF precursors (Zhang et al. 2014a, b; Dai et al. 2016; Xu et al. 2015; Yang et al. 2015; Avci et al. 2016; Hu et al. 2012; Pang et al. 2013). Notably, Lou et al. reported the synthesis of 1D CoS_2 hollow prisms by MOFs using two-step diffusion-controlled method. Uniform zeolitic imidazolate framework-67 hollow prisms assembled by nanopolyhedra are first synthesized by a transformation method (Yu et al. 2016). After, the zeolitic imidazolate framework-67 building blocks are converted into CoS_2 hollow particles to form the 1D hollow nanoprisms via a sulfidation process with an additional annealing treatment. Furthermore, the as-obtained CoS_2 nanobubble hollow prisms could be used as an electrode material for Li-ion batteries, which show remarkable electrochemical performance.

3.1.3 *Liquid Phase Synthesis*

3.1.3.1 Sol–Gel Method

The sol–gel approach has been generally used to synthesize organic–inorganic hybrid materials or inorganic nanomaterials. Sol–gel methods include the gel of gelatinized colloidal solutions and the sol of colloidal solutions in the liquid phase. In the sol–gel approach, the sol is a colloidal suspension prepared by mixing water with inorganic metal salts or metal organic compounds. The sol undergoes polymerization and hydrolysis reactions, converting the liquid sol into a gel with a 3D network. The gel has a solid skeleton surrounded by an encapsulating liquid phase.

α-MoO_3 nanorods have been fabricated by a facile sol–gel route (Cong et al. 2015). The growth mechanism was controlled by the decomposition rate of citric acid. Single-crystal and single-phase MoO_3 nanorod arrays were grown in stochastic directions from a transparent silica glass substrate with a mean length and diameter of 500 and 10 nm, respectively. Moreover, the decomposition rate of citric acid plays a significant role in the growth of MoO_3 nanorods, determined from examining the molar ratio of citric acid and molybdate in the precursor. Additionally, the timing of citric acid dissociation was controlled by the solvent and by sintering, which also influences the phase transition and growth of nanorods, as revealed by X-ray diffraction and SEM. Additionally, the synthetic parameters of the α-MoO_3 nanorod arrays were optimized to control the density and length. The longest α-MoO_3 nanorod length of 600 nm was achieved by sintering at 673 K for 15 min using a molar ratio of citric acid to Mo of 1.5:0.5 and dimethylacetamide as the solvent. Furthermore, the experiment also demonstrated that factors influencing the phase change and growth of the MoO_3 nanorods are correlated to the dissociation of the citric acid/Mo metal compound in the sol–gel precursor liquor.

3.1.3.2 Hydrothermal/Solvothermal Methods

The hydrothermal/solvothermal route describes a synthetic strategy that relies on the solubility of the metallic oxide material under high pressure and temperature to synthesize 1D crystalline substances (And and Mao† 2007). This synthetic method is usually implemented in autoclaves. A supercritical fluid is formed under elevated pressure and temperature, which increases the solubility of the solid precursors and leads to the precipitation of 1D nanostructures from the excess inorganic precursor. Hydrothermal processes are efficient for the development of high-quality 1D nanocrystals with high orientation and few defects, and we will describe these methods in greater detail in the following sections. Many nanostructures have been prepared by hydrothermal methods, including MnO_2, (Dai et al. 2014; Wang and Li 2002; Kim et al. 2014) $Na_2V_6O_{16} \cdot 3H_2O$, (Jiaguo Yu et al. 2004) $Ag_2V_4O_{11}$, (Sauvage et al. 2010; Shaoyan et al. 2010) VO_2, (Kyung-Hoon et al. 2013; Niu et al. 2014; Ni et al. 2011) $Ag_4V_2O_6$, (Erin et al. 2005; Albrecht et al. 2015) WO_3, (Gao et al. 2013), etc.

The one-step fabrication of hybrid nanostructures with high conductance and surface area remains a challenge (Li et al. 2014). Based on the theoretical Kirkendall effect, Ag/MnO_x nanotubes were produced by the solvothermal reaction between $KMnO_4$ and Ag nanowire templates. The electrochemical performance and morphology of the Ag/MnO_x composite material were tuned by changing the pH level. Tubular MnO_x nanosheets with Ag nanoparticles were formed in a comparatively acidic environment (pH = 0.76), whereas ultrafine Ag nanoparticles entrapped in amorphous manganese dioxide nanotubes were prepared in neutral solution (pH = 7.00) (Fig. 3.2a). Various volume-dependent experiments demonstrated that the Kirkendall effect was involved in the production of these morphologies (Fig. 3.2b). Additionally, the tubular Ag/MnO_x hierarchical nanostructures showed high electrochemical performance and maintained 80% of their initial capacity after 1000 cycles at 1 A g^{-1}. Most importantly, this synthetic mechanism may provide feasible design guidelines for the preparation of other hierarchical 1D nanocomposites.

3.1.3.3 Microwave-Assisted Method

The microwave-assisted method provides heating during the synthetic process and could reduce the reaction time for inorganic and organic materials through control of the reaction kinetics (Cundy and Cox 2003; Bilecka and Niederberger 2010).

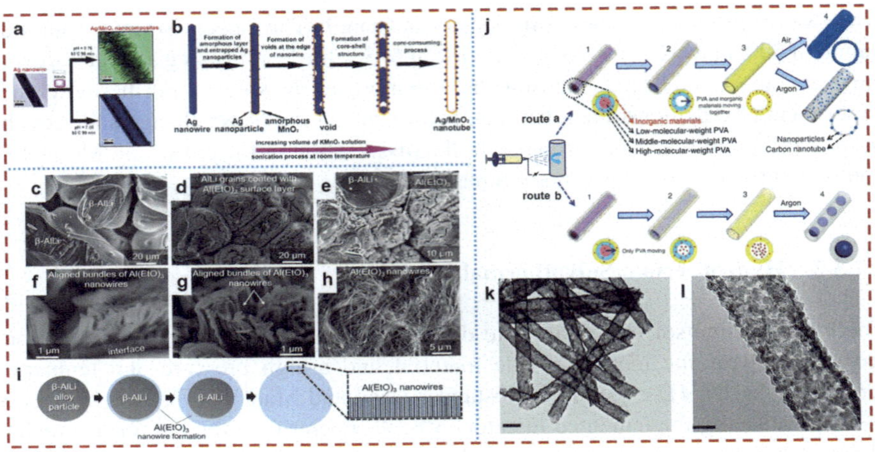

Fig. 3.2 **a** Illustration of the mechanism of solution-based approaches. Schematic sketch of fabrication process from Ag nanowire to Ag/MnO_2 core–shell nanotube and Ag/MnO_x composite nanostructure. TEM images of Ag nanowires (left), Ag/MnO_2 core–shell nanotubes produced at pH 7.00 (upper right) and Ag/MnO_x composite nanosheets prepared at pH 0.76 (low right). **b** Schematic illustration of the evolution processes of Ag/MnO_2 core–shell nanotubes based on the theoretical of Kirkendall effect. **c** SEM images of the initial grain of the as-produced β-AlLi alloy. **d–g** The formation of the $Al(EtO)_3$ nanowires on the β-AlLi particles; and **h** the SEM images of $Al(EtO)_3$ nanowires after the conversion reaction. **i** Schematic of the 1D nanowire forest formation by top-down routine. **j** Schematic sketch illustrating electrostatic spinning and controlled pyrolysis process. **k** TEM image of the multihole $Na_{0.7}Fe_{0.7}Mn_{0.3}O_2$ nanotubes with a scale bar at 200 nm. **l** TEM image of the porous Co_3O_4 nanotubes with scale at 20 nm. **a, b** Reprinted from Ref. Li et al. (2014), copyright 2014, with permission from American Chemical Society. **c–i** Reprinted from Ref. Danni et al. (2017), with permission from Science. **j–l** Reprinted from Ref. Niu et al. (2015), copyright 2015, with permission from Macmillan Publishers Limited

Moreover, microwave provides a progressive heating pattern, which is effective for the synthesis of typical colloidal nanomaterials (Park et al. 2010). The microwave-assisted route can control the shape of a nanomaterial because it stimulates ionic motion, molecular dipolar polarization, faster reaction kinetics, and rapid precursor dissolution.

Tang et al. successfully synthesized vertically aligned ZnO nanorods by both conventional heated water bath and microwave processes at 90 °C on a silicon (100) substrate with a ZnO nanoparticle seed-layer coating (Jie et al. 2015). The optical properties, defects, and morphologies of the ZnO nanorods fabricated by the two processes (pH = 10.07–10.9) were revealed by photoluminescence, X-ray photoelectron spectroscopy (XPS), and scanning electron microscopy (SEM). The experiments demonstrated that the microwave-assisted method produced more symmetrical nanorods with fewer native defects such as zinc vacancies and oxygen interstitials. Thus, microwave-assisted synthesis was shown to be a promising approach for fabricating 1D metal oxide nanostructures.

3.1.3.4 Coprecipitation Method

There are three primary mechanisms of coprecipitation: occlusion, adsorption, and inclusion. Impurities can be adsorbed via weak bonding to the surface of the precipitate, and impurities physically trapped in the crystal lead to occlusions as the crystal grows.

Ma et al. reported the growth mechanism of Ag/Fe_3O_4 core–shell nanowires. They synthesized 1D Ag/Fe_3O_4 core–shell hetero-nanowires by an effective and facile coprecipitation method (Ma et al. 2015). The Ag nanowires acted as a nucleation site for the growth of ferriferous oxide in aqueous solution. Controlling factors affecting the morphology and size of the core–shell nanowires included poly(vinylpyrrolidone) (PVP) concentration, $FeCl_3$/$FeCl_2$ concentration, time, and reaction temperature. The results show that the thickness of the Fe_3O_4 shell could be tuned from 6 to 76 nm and that the morphology could be varied between nanorods and nanospheres. First and foremost, nucleation points were added to the surface of the Ag nanowires by the C=O moieties of PVP. The accumulation of Fe^{2+} and Fe^{3+} on the Ag nanowire surface was promoted by an in situ oxidation reaction between $FeCl_3$/$FeCl_2$ and the Ag nanowire solution. Additionally, the Ag nanowire surface attached to the nucleus of the Fe_3O_4 nanoparticles. Finally, the Fe_3O_4 nanoparticles grew from the Ag nanowire surface. Higher temperature or higher $FeCl_3$/$FeCl_2$ concentration resulted in faster growth or nucleation. Lower temperature and lower concentration led to slower growth or nucleation, causing the formation of Fe_3O_4 nanospheres. Moreover, the Ag-Fe_3O_4 core–shell nanowires exhibited good ferromagnetic and electrical properties at room temperature.

3.1.4 Other Methods

3.1.4.1 Top-Down Routine

1D nanostructures could provide prospects for enhancing the mechanical, electrical, and thermal properties of a broad range of composites and functional materials, nonetheless their synthesis processes are typically expensive and elaborate. Consequently, top-down method has been successfully designed. The method could transform the bulk materials into 1D nanowires under ambient conditions without external stimuli or catalysts (Danni et al. 2017). Figure 3.2i shows the schematic representation of the nanowire formation. The formation process of Al ethoxide nanowires could be seen in Fig. 3.2c–h. The nanowires have been fabricated via minimization of strain energy at the boundary of a chemical. Experimental results show the transformation of multimicrometer-sized particles of magnesium or aluminum alloys into alkoxide nanowires of tunable dimensions, which are converted into 1D oxide nanowires upon heating in air. Particularly, the aluminum oxide nanowires could be used to make separators, which can enhance the rate and safety capabilities of lithium-ion batteries. The approach allows low-cost scalable synthesis of 1D nanomaterials and membranes.

3.1.4.2 Lithographically Patterned Method

Lithographic approaches include dip-pen nanolithography, electron-beam lithography, and focused ion beam lithography and have been exploited to fabricate designed-geometry patterns on typical solid substrates. These techniques have many advantages including the fabrication of well-aligned and large-scale nanometer arrays. Nevertheless, the low throughput and high cost of large-scale nanostructures remain challenges.

Au/MnO_2 core–shell nanowires have been fabricated by a lithographically patterned nanowire electrodeposition process (Yan et al. 2012). The linear Au nanowire core and hemicylindrical MnO_2 shell were prepared on glass. The rectangular cross section of the Au nanowires showed a height, width, and length of approximately 40 nm, 200 nm, and 1–10 mm, respectively. Subsequently, MnO_2 was deposited on the Au nanowires by a potentiostatic electrooxidation process from a manganese ion solution, forming a hemicylindrical and conformal shell structure with a diameter of approximately 50–300 nm. The shell of MnO_2 was mesoporous and in the δ-phase, as revealed by Raman spectroscopy and X-ray diffraction. Moreover, TEM analysis showed that the δ-phase MnO_2 shell was composed of a mesoporous net-like structure with 2 nm fibrils. This mesoporous MnO_2 shell structure had a thickness of 68 ± 3 nm. Moreover, the specific capacitance of the Au/MnO_2 core–shell nanowire array was measured by cyclic voltammetry (CV). The Au/MnO_2 core–shell nanostructure showed a good specific capacitance and stable cycling in aqueous electrolyte. Furthermore, the Au/MnO_2 core–shell nanowires showed hybrid energy storage by the deconvolution of specific capacitance into non-insertion and insertion components. Additionally, a symmetrical capacitor consisting of horizontal, interleaved MnO_2/Au nanowires has been described (Yan et al. 2014). All 750 nanowires in the capacitor had a length of 2.5 mm and consisted of an Au nanowire core (approximately 40×200 nm) and a δ-phase MnO_2 shell (thickness of approximately 60–220 nm). The Au/MnO_2 core–shell nanowires were patterned onto the surface of a glass substrate by lithographically patterned nanowire electrodeposition.

3.1.4.3 Electrospinning

Electrospinning is a large-scale and cost-effective technique for manufacturing 1D nanofibers. Electrospinning uses spinning force and electric fields to eject liquid precursor materials via a fine orifice to typically form fibrous nanostructures (Ren et al. 2015).

At present, a range of practical fabrication techniques, such as hydrothermal synthesis, chemical vapor deposition, physical vapor deposition, electrochemical etching/deposition, electrospinning methods, template-based methods, and laser ablation, has been developed to create specific 1D nanomaterials (Niu et al. 2015). Nevertheless, 1D nanostructures are only realizable for specific materials due to the limited range of appropriate material properties for each synthetic method. Hence, a general technique is required to manufacture nanowires, as well as nanotubes, for a series of inorganic nanomaterials regardless of crystal orientation. Electrospinning methods have been used to manufacture a variety of conductive polymer nanowires and inorganic nanowires. The feasible controlled pyrolysis method and gradient-electrospinning method have been reported with low cost, extensive material diversity, high yield, and good repeatability, providing a promising method to obtain tunable nanotubes for a series of inorganic materials and removing the restriction of crystal growth orientation of different samples. Moreover, a variety of tubular 1D nanomaterials have been synthesized based on this synthetic method including single metal oxide nanotubes (MnO_2, Co_3O_4, CuO and SnO_2), binary metal oxides ($LiCoO_2$, LiV_3O_8, $NiCo_2O_4$, and $LiMn_2O_4$), and multi-element oxides ($LiNi_{1/3}Co_{1/3}Mn_{1/3}O_2$, $Na_{0.7}Fe_{0.7}Mn_{0.3}O_2$, etc.). Transmission electron microscopy (TEM) images of the mesoporous nanotubes are shown in Fig. 3.2k, l. The schematics of the gradient electrospinning and controlled pyrolysis methods are shown in Fig. 3.2j. Following a process of electrostatic spinning, the compound nanowires are straightway placed into a furnace in air at 300 °C. Then, PVA decomposes simultaneously and quickly moves to the outside high molecular weight PVA layer without taking away the desired inorganic materials, leaving inorganic materials in the central zone. The inner inorganic materials and the outer PVA carbonize develop into typical nanoparticles after high-temperature calcining under Ar atmosphere, forming pea-like nanotubes. Accordingly, the electrostatic spinning method could lead to rapid advancements in the progress of 1D nanostructures. Furthermore, these unique pea-like and mesoporous nanotubes, which show outstanding electrochemical performance in supercapacitors, sodium-ion batteries, and lithium-ion batteries, owing to their robust structural stability, high conductivity and large surface area hold great promise not only in electrochemical energy storage devices but also in a number of other frontiers.

3.2 Others

3.2.1 One-Dimensional Graphene

1D graphene nanomaterials are attracting significant attention on account of their excellent thermal, mechanical, and electrical properties, which could lead to a series of significant potential applications. Synthetic methods associated with making 1D carbon nanomaterials could consume a mass of energy and also comparative complex, so a primary challenge is to develop efficient and simple processes to produce them. Xu and co-workers successfully synthesized 1D carbon nanorods

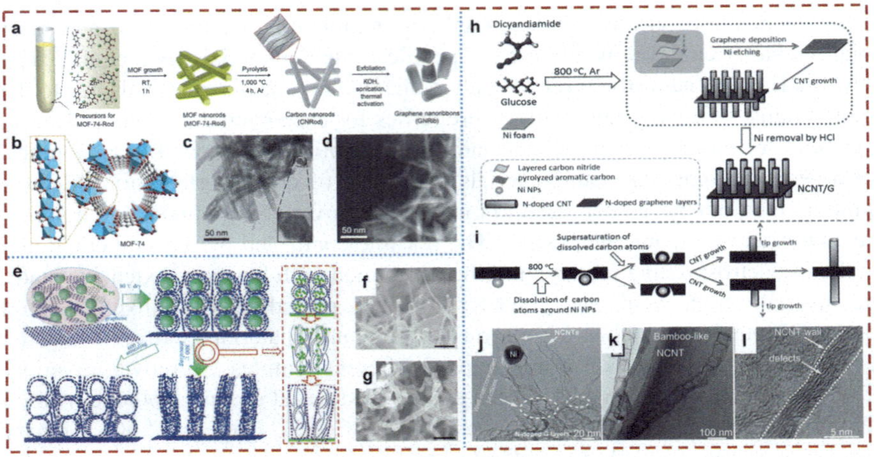

Fig. 3.3 **a** Scheme of synthesis of MOF-74-Rod, carbon nanorods, and 1D graphene nanoribbons. **b** The secondary building unit and 3D crystal structure of MOF-74. **c** TEM images of 1D carbon nanorods. **d** HAADF-STEM images of 1D graphene nanoribbons. **e** Schematic diagrams for the fabrication of 1D graphene nanowires. **f–g** SEM micrographs of 1D graphene nanowires on graphene foam obtained from different suspensions with concentrations of 10 and 20 mg mL^{-1}. **h** Diagrammatic drawing of the N-doped CNT/graphene hybrid structures fabricated by a solid-state growth process. **i** Cross-sectional schematic diagram of the possible tip growth mechanism of 1D CNTs in the N-CNT/graphene hybrid. **j** TEM image of N-doped CNT/graphene hybrid. **k** TEM image of 1D N-doped CNTs in the resulting hybrid. **l** HRTEM image of the N-doped CNTs wall. **a–d** Reprinted from Ref. Pachfule et al. (2016), copyright 2016, with permission from Macmillan Publishers Limited. **e–g** Reprinted from Ref. Liu et al. (2017), copyright 2017, with permission from Elsevier. **h–l** Reprinted from Ref. Ding et al. (2015), copyright 2015, with permission from WILEY–VCH

via thermal transformation of 1D metal–organic frameworks (Pachfule et al. 2016). The strategy for the synthesis of 1D carbon nanorods and graphene nanoribbons is shown in Fig. 3.3a. The synthesis of non-hollow (solid) one-dimensional carbon nanorods with moderate aspect ratio, high surface area is achieved by self-sacrificial and morphology-preserved thermal transformation of MOF-74 (Fig. 3.3b) with a 1D rod-shaped morphology (MOF-74-Rod). TEM images of carbon nanorods and graphene nanoribbons are shown in Fig. 3.3c, d. This is a catalyst-free and self-templated strategy. The solid carbon nanorods could be transformed into graphene nanoribbons by a sonochemical process followed by thermal activation. Significantly, the 1D graphene nanoribbons and 1D carbon nanorods could be used for supercapacitor electrodes. Consequently, this synthetic route is readily scalable and could be used to produce 1D graphene nanoribbons and carbon nanorods on industrial levels. A graphene nanowire on graphene foam has been manufactured by a template strategy. This method involves catalytic process between graphene oxides and polystyrene spheres decomposition products, pyrolysis of polystyrene spheres, and assembly/reduction process of nano-graphene oxides (Liu et al. 2017). The schematic diagram for the fabrication of graphene nanowires is illustrated in Fig. 3.3e.

(1) An aqueous solution with nanosized graphene oxides and polystyrene spheres dropwise added to surface of nickel foam with micro-sized graphene oxides. (2) nanosized graphene oxides and polystyrene spheres coated on the surface of micro-sized graphene oxides after evaporation of water. (3) A solid-state pyrolysis process at 600 °C for the removal of polystyrene spheres template. (4) A solid-state pyrolysis process at 800 °C for the removal of polystyrene spheres template and assembly of nanosized graphene oxides sheets. The block diagram on the right further explained the formation process of the graphene nanowires. What is more, Fig. 3.3f, g shows the SEM images of 1D graphene nanowire on graphene foam obtained from different suspensions with concentrations of 10 and 20 mg mL^{-1}. This graphene nanowire on graphene foam could be used for Na/Li batteries, which shows an excellent rate capability, good reversible capacity, and low discharge-voltage plateau. The outstanding electrochemical performance is attributed to many lateral exposed pores/edges, high graphene crystallinity, hierarchical multidimensional graphene construction, and expansile graphene interlayer distance, which could promote the ion and electron transport. The 1D graphene assembled by graphene sheets provided new opportunities for energy storage equipment of graphene-based assembly in future. Furthermore, a zonal porous graphene has been successfully designed by bottom-up approach (Moreno et al. 2018). This graphene includes an ordered array of pores separated by ribbons, which could be tuned down to the one nanometer range. The chemical composition, density, size, and morphology of the pores are defined with atomic precision by the precise design of the molecular precursors.

3.2.2 Carbon Nanotubes

Carbon nanotubes (CNTs) are allotropes of carbon with a 1D nanostructure. CNTs have unusual properties, which are valuable for optics, nanotechnology, electronics, and other fields of materials science. A variety of techniques have been developed to produce CNTs, including high-pressure carbon monoxide disproportionation(HiPCO), laser ablation, arc discharge, CVD, and so on. In these arc discharge, CVD and laser ablation are batch by batch process. The HiPCO technology is gas-phase continuous process. Most of these treating processes take place in a vacuum or with process gases. The CVD technology is popular, as it has a degree of control over morphology, length, and diameter. Using particulate catalysts, large quantities of 1D nanotubes could be synthesized by these processes. But achieving the repeatability becomes a major problem with CVD growth. Nonetheless, achieving the repeatability becomes a main problem with CVD technology. The HiPCO method has an advantage in continuous growth and catalysis, which makes CNTs more commercially viable. The HiPCO process helps in producing high purity single-walled carbon nanotubes in higher quantity. The HiPCO technology helps in producing high purity single-walled CNTs in higher quantity. The HiPCO reactor operates at a high pressure of 30–50 bar and a high temperature of 900–1100 °C. The carbon source is carbon monoxide and the catalyst is Ni/Fe penta carbonyl. This catalyst acts as the nucleation site for the CNTs to grow. Vertically aligned CNT arrays could be grown by thermal CVD method. A substrate (stainless steel, silicon, quartz, etc.) is coated with a catalytic metal of Ni/Fe/Co layer. In general, the layer is Fe, and is deposited via sputtering to a thickness of 1–5 nm. A 10–50 nm underlayer of alumina is put down on the substrate. This imparts good interfacial and controllable wetting properties. The as-prepared CNTs always have impurities, for example, non-carbonaceous impurities and other forms of carbon (fullerene, amorphous carbon, etc.). The impurities need to be removed to make use of the CNTs in applications. Graphene and CNTs have attracted tremendous attention, because of their unique properties, such as environmental benignity, lower cost, and high electronic conductivity. Some studies have shown that hybrid architectures built from graphene and 1D CNTs show improved properties and synergistic effects as regards electronic devices and energy storage. Furthermore, the chemical properties of sp^2 carbon materials could be further tailored by introducing heteroatom doping, for instance, boron, nitrogen, phosphorus, and sulfur. Consequently, Yu et al. have fabricated an N-doped CNT/graphene hybrid structure by a solid-state growth process (Ding et al. 2015). They use dicyandiamide, glucose, and nickel foam as nitrogen sources, carbon source, and substrate, respectively. The overall synthetic procedure of N-doped 3D CNT/graphene hybrid structures is achieved in a two-step process as shown in Fig. 3.3h, i. The ruffly graphene layers as the support of CNTs can clearly be observed in Fig. 3.3j–l, showing that CNTs are rooted and well interconnected with the in situ produced graphene layers. The nickel foam plays a bifunctional role in providing scaffold for graphene deposition, and offering in situ generated nickel nanoparticles as catalyst for CNT growth on graphene layers without any synthetical catalyst. Conclusively, the hybrid architectures could show stable cyclability and superior rate capability as cathode hosts for lithium-sulfur batteries.

In summary, 1D nanomaterials are widely studied for use in electrochemical energy storage devices because of their relatively short path lengths for ion and electron transport and insertion/extraction, large specific surface areas, and facile strain relaxation during electrochemical cycling. Furthermore, 1D-based nanostructures have attracted much attention from both the research community and the commercial sector worldwide. Much effort has been devoted to improving the electrochemical performance and mechanical properties of 1D nanostructure, and future directions and existing problems are as follows: (1) Nanoscale composite materials may have higher electrochemical conductivity and specific capacitance. Composite nanostructures not only compensate for the disadvantages of the separate components but also incorporate the advantages of all the constituents. Furthermore, the synergistic effects of 1D nanomaterials metal and conducting polymers or carbon nanomaterials can maximize the properties of different components, for example, to further improve the conductivity of the metal oxide/sulfides materials. (2) The results of both theoretical calculations and experimental research have indicated that structural control and interfacial modification can observably enhance the rate performance, cycling ability and specific capacity of 1D metal oxide/sulfides. It is key to design the complex 1D nanostructure (e.g., core–shell nanostructure, array architecture, branched structure, layered structure, porous nanostructures, etc.) based on 1D structure, which can facilitate more direct electron transport pathways and higher surface areas. Accordingly, 1D hetero-nanostructures have played a significant role in the research of energy nanomaterials. (3) For instance, the electrochemical performance can be enhanced by controlling the surface area, pore properties, and morphology. Nanostructures based on metal oxide/sulfides with higher specific surface areas can provide more electro-active sites and make efficient contact with the electrolyte, leading to a higher charge and discharge capacity at large current density. (4) For 1D carbon materials, graphene and CNTs all have remarkable mechanical properties. When they form composite materials with other materials such as metal oxides/sulfides, the volume expansion/contraction of these materials can be effectively restrained. Due to these excellent properties, CNTs or graphene can play a significant role in constructing electrochemical energy storage devices such as stretchable, compressible, bendable, and other devices.

References

Albrecht TA, Sauvage F, Bodenez V, Tarascon JM, Poeppelmeier KR (2015) Room temperature synthesis of the larger power, high silver density cathode material $Ag_4V_2O_6F_2$ for implantable cardioverter defibrillators. Cheminform 40(40):60208–63113

And XC, Mao† SS (2007) Titanium dioxide nanomaterials: synthesis, properties, modifications, and applications. Cheminform 38(41):2891

Atanasova P, Rothenstein D, Schneider JJ, Hoffmann RC, Dilfer S, Eiben S, Wege C, Jeske H, Bill J (2011) Virus-templated synthesis of zno nanostructures and formation of field-effect transistors. Adv Mater 23(42):4918–4922

Avci C, Ariñez-Soriano J, Carné-Sánchez A, Guillerm V, Carbonell C, Imaz I, Maspoch D (2016) Post-synthetic anisotropic wet-chemical etching of colloidal sodalite ZIF crystals. Angew Chem Int Ed Engl 54(48):14417–14421

Balci S, Bittner AM, Schirra M, Thonke K, Sauer R, Hahn K, Kadri A, Wege C, Jeske H, Kern K (2009) Catalytic coating of virus particles with zinc oxide. Electrochim Acta 54(22):5149–5154

BãUerlein E (2003) Biomineralization of unicellular organisms: an unusual membrane biochemistry for the production of inorganic nano- and microstructures. Cheminform 42(6):614–641

Bilecka I, Niederberger M (2010) Microwave chemistry for inorganic nanomaterials synthesis. Nanoscale 2(8):1358–1374

Cai X, Gao W, Ma M, Wu M, Zhang L, Zheng Y, Chen H, Shi J (2015) A prussian blue-based core-shell hollow-structured mesoporous nanoparticle as a smart theranostic agent with ultrahigh pH-responsive longitudinal relaxivity. Adv Mater 27(41):6382–6389

Chen YZ, Liu TH, Chen CY, Liu CH, Chen SY, Wu WW, Wang ZL, He JH, Chu YH, Chueh YL (2012) Taper $PbZr_{(0.2)}Ti_{(0.8)}O_3$ nanowire arrays: from controlled growth by pulsed laser deposition to piezopotential measurements. Acs Nano 6(3):2826–2832

Chen C, Hu X, Wang Z, Xiong X, Hu P, Liu Y, Huang Y (2014) Controllable growth of TiO_2-B nanosheet arrays on carbon nanotubes as a high-rate anode material for lithium-ion batteries. Carbon 69(4):302–310

Chen M, Shen X, Wu Q, Li W, Diao G (2015) Template-assisted synthesis of core–shell α-Fe_2O_3 @TiO_2 nanorods and their photocatalytic property. J Mater Sci 50(11):4083–4094

Chen YM, Yu L, Lou XW (2016) Hierarchical tubular structures composed of Co_3O_4 hollow nanoparticles and carbon nanotubes for lithium storage. Angew Chem Int Ed Engl 55(20):5990–5993

Cong S, Sugahara T, Wei T, Jiu J, Hirose Y, Nagao S, Suganuma K (2015) Growth and extension of one-step sol-gel derived molybdenum trioxide nanorods via controlling citric acid decomposition rate. Cryst Growth Des 15(9):150727040626005

Cundy CS, Cox PA (2003) The hydrothermal synthesis of zeolites: history and development from the earliest days to the present time. Chem Rev 103(3):663

Dai K, Lu L, Liang C, Dai J, Liu Q, Zhang Y, Zhu G, Liu Z (2014) In situ assembly of MnO_2 nanowires/graphene oxide nanosheets composite with high specific capacitance. Electrochim Acta 116(2):111–117

Dai C, Zhang A, Liu M, Guo X, Song C (2016) Hollow ZSM-5 with silicon-rich surface, double shells, and functionalized interior with metallic nanoparticles and carbon nanotubes. Adv Func Mater 25(48):7479–7487

Danni L, Jim B, Alexandre M, Gene B, Gleb Y (2017) Transformation of bulk alloys to oxide nanowires. Science 355(6322):267

Deng S, Tjoa V, Fan HM, Tan HR, Sayle DC, Olivo M, Mhaisalkar S, Wei J, Sow CH (2012) Reduced graphene oxide conjugated Cu_2O nanowire mesocrystals for high-performance NO_2 gas sensor. J Am Chem Soc 134(10):4905–4917

Ding Y-L, Kopold P, Hahn K, van Aken PA, Maier J, Yu Y (2015) Facile solid-state growth of 3D well-interconnected nitrogen-rich carbon nanotube-graphene hybrid architectures for lithium-sulfur batteries. Adv Func Mater 26(7):1112–1119

Duay J, Sherrill SA, Gui Z, Gillette E, Lee SB (2013) Self-limiting electrodeposition of hierarchical MnO_2 and $M(OH)_2/MnO_2$ nanofibril/nanowires: mechanism and supercapacitor properties. ACS Nano 7(2):1200–1214

Eerenstein W, Mathur N, Scott JF (2006) Multiferroic and magnetoelectric materials. Nature 442(7104):759

Erin MS, Heather KI, John TV, Charlotte LS, Kenneth RP (2005) Ag4V2O6F2: An electrochemically active and high silver density phase. Cheminform 36(32):634

Gao L, Wang X, Xie Z, Song W, Wang L, Wu X, Qu F, Chen D, Shen G (2013) High-performance energy-storage devices based on WO_3 nanowire arrays/carbon cloth integrated electrodes. J Mater Chem A 1(24):7167–7173

George SM (2010) Atomic layer deposition: an overview. Chem Rev 110(1):111–131

Hacialioglu S, Meng F, Jin S (2012) Facile and mild solution synthesis of Cu_2O nanowires and nanotubes driven by screw dislocations. Chem Commun 48(8):1174–1176

Han L, Yu XY, Lou XW (2016) Formation of Prussian-Blue-analog nanocages via a direct etching method and their conversion into Ni-Co-mixed oxide for enhanced oxygen evolution. Adv Mater 28(23):4601–4605

Hernandez BA, Chang KS, And ERF, Dorhout PK (2010) Sol−Gel template synthesis and characterization of $BaTiO_3$ and $PbTiO_3$ nanotubes. Cheminform 33(20):480–482

He P, Fang Y, Yu XY, Lou XW (2017) Hierarchical nanotubes constructed by carbon-coated ultrathin SnS nanosheets for fast capacitive sodium storage. Angew Chem Int Ed Engl. https://doi.org/10.1002/ange.201706652

Hu L, Chen Q (2014) Hollow/porous nanostructures derived from nanoscale metal-organic frameworks towards high performance anodes for lithium-ion batteries. Nanoscale 6(3):1236–1257

Hu M, Furukawa S, Ohtani R, Sukegawa H, Nemoto Y, Reboul J, Kitagawa S, Yamauchi Y (2012) Synthesis of Prussian blue nanoparticles with a hollow interior by controlled chemical etching. Angew Chem 124(4):1008–1012

Hu H, Han L, Yu M, Wang Z, Lou XW (2016) Metal–organic-framework-engaged formation of Co nanoparticle-embedded carbon@Co_9S_8 double-shelled nanocages for efficient oxygen reduction. Energy Environ Sci 9(1):107–111

Jiaguo Y, Wingkei H, Wu L, Wang X (2004) A simple and general method for the synthesis of multicomponent $Na_2V_6O_{16}{\cdot}3H_2O$ single-crystal nanobelts. J Am Chem Soc 126(11):3422

Jie T, Chai J, Jian H, Deng L, Xuan SN, Sun L, Venkatesan T, Shen Z, Tay CB, Chua SJ (2015) ZnO nanorods with low intrinsic defects and high optical performance grown by facile microwave-assisted solution method. Acs Appl Mater Interfaces 7(8):4737–4743

Kadri A, Maiss E, Amsharov N, Bittner AM, Balci S, Kern K, Jeske H, Wege C (2011) Engineered Tobacco mosaic virus mutants with distinct physical characteristics in planta and enhanced metallization properties. Virus Res 157(1):35–46

Kim IY, Ha HW, Kim TW, Paik Y, Choy JH, Hwang SJ (2014) Origin of improved electrochemical activity of β-MnO_2 nanorods: effect of the Mn valence in the precursor on the crystal structure and electrode activity of manganates. J Phys Chem C 113(51):21274–21282

Kyung-Hoon L, Young-Woo L, A-Ra K, Cao G, Kyung-Won P (2013) Single-crystalline mesoporous molybdenum nitride nanowires with improved electrochemical properties. J Am Ceramic Soc 96(1):37–39

Lai X, Halpert JE, Wang D (2012) Recent advances in micro-/nano-structured hollow spheres for energy applications: From simple to complex systems. Energy Environ Sci 5(2):5604–5618

Lewis NS (2007) Toward cost-effective solar energy use. Science 315(5813):798–801

Limmer SJ, Seraji S, Forbess MJ, Wu Y, Chou TP, Nguyen C, Cao GZ (2010) Electrophoretic growth of lead zirconate titanate nanorods. Adv Mater 13(16):1269–1272

Liu J, Wu C, Xiao D, Kopold P, Gu L, Van Aken PA, Maier J, Yu Y (2016) MOF-derived hollow Co_9S_8 nanoparticles embedded in graphitic carbon nanocages with superior Li-Ion storage. Small 12(17):2354–2364

Liu X, Chao D, Su D, Liu S, Chen L, Chi C, Lin J, Shen ZX, Zhao J, Mai L (2017) Graphene nanowires anchored to 3D graphene foam via self-assembly for high performance Li and Na ion storage. Nano Energy 37:108–117

Li Y, Fu H, Zhang Y, Wang Z, Li X (2014) Kirkendall effect induced one-step fabrication of tubular Ag/MnOx nanocomposites for supercapacitor application. J Phys Chem C 118(13):6604–6611

Li Q, Liu D, Jia Z, Csetenyi L, Gadd GM (2016) Fungal biomineralization of manganese as a novel source of electrochemical materials. Current Biol Cb 26(7):950–955

Ma J, Wang K, Zhan MS (2015) Growth mechanism, electrical and magnetic properties of Ag-Fe_3O_4 core-shell nanowires. Acs Appl Mater Interfaces 7(29):16027–16039

McCammon C (1997) Perovskite as a possible sink for ferric iron in the lower mantle. Nature 387(6634):694

Menzel A, Subannajui K, Bakhda R, Wang Y, Thomann R, Zacharias M (2012) Tuning the growth mechanism of ZnO nanowires by controlled carrier and reaction gas modulation in thermal CVD. J Phys Chem Lett 3(19):2815–2821

Minami T, Nishi Y, Miyata T (2013) High-efficiency Cu_2O-based heterojunction solar cells fabricated using a Ga_2O_3 thin film as n-type layer. Appl Phys Express 6(4):044101

Mittiga A, Salza E, Sarto F, Tucci M, Vasanthi R (2006) Heterojunction solar cell with 2% efficiency based on a Cu_2O substrate. Appl Phys Lett 88(16):163502

Moreno C, Vilas-Varela M, Kretz B, Garcia-Lekue A, Costache MV, Paradinas M, Panighel M, Ceballos G, Valenzuela SO, PeãA D (2018) Bottom-up synthesis of multifunctional nanoporous graphene. Science 360(6385):199–203

Mudusu D, Nandanapalli KR, Dugasani SR, Kang JW, Park SH, Tu CW (2017) Growth of single-crystalline cubic structured tin (II) sulfide (SnS) nanowires by chemical vapor deposition. RSC Adv 7(66):41452–41459

Nam KT, Kim D-W, Yoo PJ, Chiang C-Y, Meethong N, Hammond PT, Chiang Y-M, Belcher AM (2006) Virus-enabled synthesis and assembly of nanowires for lithium ion battery electrodes. Science 312(5775):885–888

Niu C, Meng J, Han C, Zhao K, Yan M, Mai L (2014) VO2 nanowires assembled into hollow microspheres for high-rate and long-life lithium batteries. Nano Lett 14(5):2873–2878

Niu C, Meng J, Wang X, Han C, Yan M, Zhao K, Xu X, Ren W, Zhao Y, Lin X (2015) General synthesis of complex nanotubes by gradient electrospinning and controlled pyrolysis. Nat Commun 6:7402

Ni J, Jiang W, Yu K, Gao Y, Zhu Z (2011) Hydrothermal synthesis of VO (B) nanostructures and application in aqueous Li-ion battery. Electrochim Acta 56(5):2122–2126

Oh MH, Hyeon T (2013) Galvanic replacement reactions in metal oxide nanocrystals. Science 340(6135):964–968

Pachfule P, Shinde D, Majumder M, Xu Q (2016) Fabrication of carbon nanorods and graphene nanoribbons from a metal–organic framework. Nat Chem 8(7):718–724

Pang M, Cairns AJ, Liu Y, Belmabkhout Y, Zeng HC, Eddaoudi M (2013) Synthesis and integration of Fe-soc-MOF cubes into colloidosomes via a single-step emulsion-based approach. J Am Chem Soc 135(28):10234–10237

Pang X, He Y, Jung J, Lin Z (2016) 1D nanocrystals with precisely controlled dimensions, compositions, and architectures. Science 353(6305):1268

Paracchino A, Laporte V, Sivula K, Grätzel M, Thimsen E (2011) Highly active oxide photocathode for photoelectrochemical water reduction. Nat Mater 10(6):456

Park J, Joo J, Kwon SG, Jang Y, Hyeon T (2010) Synthesis of monodisperse spherical nanocrystals. Angew Chem Int Ed Engl 46(25):4630–4660

Ren W, Zheng Z, Luo Y, Chen W, Niu C, Zhao K, Yan M, Zhang L, Meng J, Mai L (2015) An electrospun hierarchical LiV_3O_8 nanowire-in-network for high-rate and long-life lithium batteries. J Mater Chem A 3(39):19850–19856

Ricky JT, Chunglin T, Liping M, Jianyong O, Cengiz SO, Yang Y (2006) Digital memory device based on tobacco mosaic virus conjugated with nanoparticles. Nat Nanotechnol 1(1):72–77

Sauvage F, Bodenez V, Tarascon JM, Poeppelmeier KR (2010) Room-temperature synthesis leading to nanocrystalline $Ag_{(2)}V_{(4)}O_{(11)}$. J Am Chem Soc 132(19):6778–6782

Shaoyan Z, Weiyang L, Chunsheng L, Jun C (2010) Synthesis, characterization, and electrochemical properties of $Ag_2V_4O_{11}$ and $AgVO_3$ 1-D nano/microstructures. ChemInform 38(18):24855–24863

Shim HW, Jin YH, Seo SD, Lee SH, Kim DW (2011) Highly reversible lithium storage in *Bacillus subtilis*-directed porous Co_3O_4 nanostructures. ACS Nano 5(1):443

Shi J, Li Z, Kvit A, Krylyuk S, Davydov AV, Wang X (2013) Electron microscopy observation of TiO_2 nanocrystal evolution in high-temperature atomic layer deposition. Nano Lett 13(11):5727–5734

Tan Y, Xue X, Peng Q, Zhao H, Wang T, Li Y (2007) Controllable fabrication and electrical performance of single crystalline Cu2O nanowires with high aspect ratios. Nano Lett 7(12):3723–3728

Wang X, Li Y (2002) Selected-control hydrothermal synthesis of alpha- and beta-$MnO_{(2)}$ single crystal nanowires. J Am Chem Soc 124(12):2880–2881

Wang J, Neaton JB, Zheng H, Nagarajan V, Ogale SB, Liu B, Viehland D, Vaithyanathan V, Schlom DG, Waghmare UV, Spaldin NA, Rabe KM, Wuttig M, Ramesh R (2003) Epitaxial BiFeO3 multiferroic thin film heterostructures. Science 299(5613):1719–1722

Weiner S, Sagi I, Addadi L (2005) Choosing the crystallization path less traveled. Science 309(5737):1027–1028

Wu F, Myung Y, Banerjee P (2014) Rayleigh instability driven nodular Cu_2O nanowires via carbothermal reduction of CuO nanowires. Cryst Growth Des 15(4):1588–1595

Wu R, Wang DP, Rui X, Liu B, Zhou K, Law AW, Yan Q, Wei J, Chen Z (2015) In-situ formation of hollow hybrids composed of cobalt sulfides embedded within porous carbon polyhedra/carbon nanotubes for high-performance lithium-ion batteries. Adv Mater 27(19):3038–3044

Xu G, Ren Z, Du P, Weng W, Shen G, Han G (2005) Polymer-assisted hydrothermal synthesis of single-crystalline tetragonal perovskite $PbZr_{0.52}Ti_{0.48}O_3$ nanowires. Adv Mater 17(7):907–910

Xu X, Zhang Z, Xun W (2015) Well-defined metal-organic-framework hollow nanostructures for catalytic reactions involving gases. Adv Mater 27(36):5365–5371

Yan W, Kim JY, Xing W, Donavan KC, Ayvazian T, Penner RM (2012) Lithographically patterned gold/manganese dioxide core/shell nanowires for high capacity, high rate, and high cyclability hybrid electrical energy storage. Chem Mater 24(12):2382–2390

Yan W, Thai ML, Dutta R, Li X, Xing W, Penner RM (2014) A lithographically patterned capacitor with horizontal nanowires of length 2.5 mm. Acs Appl Mater Interfaces 6(7):5018–5025

Yang J, Zhang F, Lu H, Hong X, Jiang H, Wu Y, Li Y (2015) Hollow Zn/Co ZIF particles derived from core-shell ZIF-67@ZIF-8 as selective catalyst for the semi-hydrogenation of acetylene. Angew Chem Int Ed Engl 127(37):11039–11043

Yu HK, Lee J-L (2014) Growth mechanism of metal-oxide nanowires synthesized by electron beam evaporation: a self-catalytic vapor-liquid-solid process. Sci Rep 4:6589

Yu L, Zhang L, Wu HB, Lou XW (2014) Formation of $NixCo_3-xS_4$ hollow nanoprisms with enhanced pseudocapacitive properties. Angew Chem Int Ed 53(14):3711–3714

Yu L, Yang JF, Lou XW (2016) Formation of CoS_2 nanobubble hollow prisms for highly reversible lithium storage. Angew Chem Int Ed 55(43):13422–13426

Zhang G, Yu L, Wu HB, Hoster HE, Lou XW (2012) Formation of $ZnMn_2O_4$ ball-in-ball hollow microspheres as a high-performance anode for lithium-ion batteries. Adv Mater 24(34):4609–4613

Zhang L, Wu HB, Lou XW (2013) Metal–organic-frameworks-derived general formation of hollow structures with high complexity. J Am Chem Soc 135(29):10664–10672

Zhang Z, Chen Y, Xu X, Zhang J, Xiang G, He W, Wang X (2014a) Well-Defined metal–organic framework hollow nanocages & dagger. Angew Chem Int Ed 53(2):429–433

Zhang F, Wei Y, Wu X, Jiang H, Wang W, Li H (2014b) Hollow zeolitic imidazolate framework nanospheres as highly efficient cooperative catalyst for [3+3] cycloaddition reactions. J Am Chem Soc 136(40):13963–13966

Zhang J, Hu H, Li Z, Lou XW (2016) Double-shelled nanocages with cobalt hydroxide inner shell and layered double hydroxides outer shell as high-efficiency polysulfide mediator for lithium-sulfur batteries. Angew Chem 55(12):3982–3986

Zheng Z, Huang B, Wang Z, Guo M, Qin X, Zhang X, Wang P, Dai Y (2009) Crystal faces of Cu_2O and their stabilities in photocatalytic reactions. J Phys Chem C 113(32):14448–14453

Zhuge F, Yanagida T, Nagashima K, Yoshida H, Kanai M, Xu B, Klamchuen A, Meng G, He Y, Rahong S (2012) Fundamental strategy for creating VLS grown TiO_2 single crystalline nanowires. J Phys Chem C 116(45):24367–24372

Zou F, Hu X, Li Z, Qie L, Hu C, Zeng R, Jiang Y, Huang Y (2014) MOF-Derived Porous $ZnO/ZnFe_2O_4/C$ Octahedra with hollow interiors for high-rate lithium-ion batteries. Adv Mater 26(38):6622–6628

Chapter 4
Synthesis of Two-Dimensional (2D) Nanomaterials

Recently, the development of graphene nanosheets has ushered 2D nanomaterials into the limelight for energy storage and conversion devices. These graphene-like nanostructures, including transition metal hydroxides (TMHs), transition metal oxides (TMOs), transition metal dichalcogenides (TMDs), MXene, etc., feature atomic-level thickness, large surface area, tunable electronic properties, remarkable mechanical strength, and unique confined effect. Until now, variety of synthetic methods, novel designs of electrode, and microstructure tuning of these 2D materials have been discussed to achieve high power and energy densities (Zhang 2015; Xu et al. 2013; Tan et al. 2017; Koski and Cui 2013; Mei et al. 2017; Wu et al. 2014; Dong et al. 2017).

4.1 2D Transition Metal Dichalcogenides

4.1.1 MoS_2

2D TMDs refer to a kind of novel materials with layered structure, the unit cell of which is comprised of a transition metal (M = Mo or W) layer sandwiched between two chalcogen (X = S, Se, or Te) layers in the form of MX_2 (Cao et al. 2013; Muller et al. 2015). Molybdenum disulfide (MoS_2), 2D layers which stacked one over other by van der Waals interaction, has desirable properties similar to graphene (Chen et al. 2013; Ting et al. 2016; Moses et al. 2009; ArunKumar et al. 2017). Exfoliation of the TMDs into single or multiple layer sheets was discovered to enhance the chemical properties of TMDs due to the increase of available surface area and tuning of their electronic properties. This allows for potential applications in capacitors, batteries, vapor sensing, and biosensing applications (Pumera et al. 2014; Chia et al. 2015; Su et al. 2016). However, the electrochemical performance of few-layer or monolayer MoS_2 nanosheets is still impeded by their inherent limitations (Wang et al. 2014; Jiang and Zeng 2015). Despite the large capacity, the exfoliated MoS_2 nanosheets often

H. Pang et al., *Synthesis of Functional Nanomaterials for Electrochemical Energy Storage*,
https://doi.org/10.1007/978-981-13-7372-5_4

suffer from low cycling durability and rate capability, which can be attributed to the poor electronic conductivity of MoS_2, undesired large volume change, and restacking of MoS_2 nanosheets during the cycling. To address these issues, various solutions have been developed to facilitate the electrochemical properties of MoS_2 nanosheets by using carbonaceous materials as the conductive matrix. For example, graphene has established itself as a promising matrix to construct MoS_2-graphene composites for facilitating the electrochemical properties of MoS_2 because of its excellent electrical conductivity, superior flexibility, and high electrochemical stability (Chang et al. 2011; Chang and Chen 2011). There has been increasing interest in the use of MoS_2 for the synthesis and application of functional hybrid nanomaterials.

Deng et al. (2017) demonstrate an efficient electrochemical exfoliation approach to simultaneously and scalablely fabricate 2D MoS_2-grapehene (MoS_2-G) hybrid from combined bulk MoS_2-graphite wafer in a simple electrolytic cell. Figure 4.1a schematically illustrates the synthetic procedure of simultaneous and scalable 2D MoS_2-G hybrid. Different weight ratios of bulk MoS_2 and graphite are ground evenly

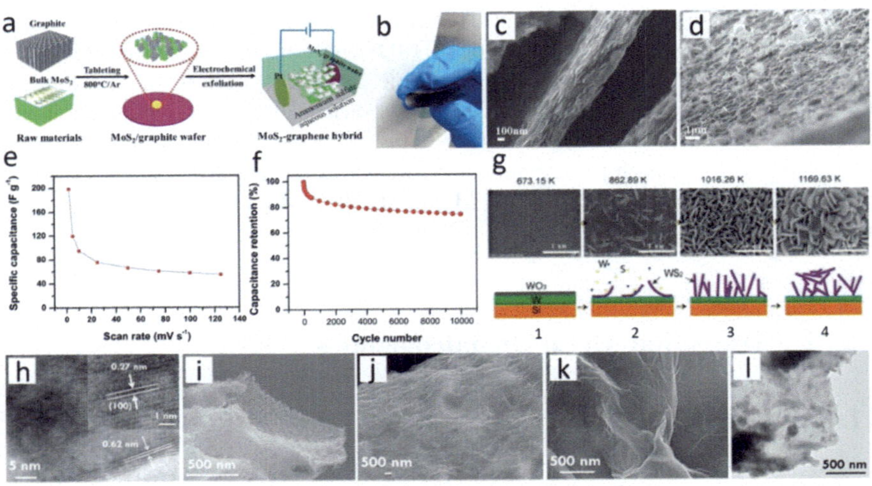

Fig. 4.1 **a** Schematic illustration of the fabrication of 2D MoS_2-G hybrid nanosheets. **b** Optical image of the MoS_2-G film, which can be bent without any damage. **c** Typical SEM image of the MoS_2-G film, showing its layered structure of stacked MoS_2 and graphene nanosheets, whose thickness is about 500 nm. **d** SEM image of the top view of the MoS_2-G film. **e** Specific capacitance value calculated from the CV curves at different scan rates. **f** Specific capacitance retention of the $MoSe_2$ nanosheets as a function of cycle number, measured by charge-discharge at a high current density of 5 A g^{-1} in 0.5 M H_2SO_4 electrolyte. **g** Schematic of the growth process of WS_2 nanoplates, corresponding to the observed SEM images. **h** The HRTEM image of WS_2, the SEM images of **i** WS_2 **j** RGO and **k** WS_2/RGO hybrid, and **l** the TEM image of WS_2/RGO hybrid. **a–d** Reprinted from Ref. Deng et al. (2017), copyright 2017, with permission from Elsevier. **e, f** Reprinted from Ref. Balasingam et al. (2015), copyright 2015, with permission from The Royal Society of Chemistry. **g** Reprinted from Ref. Qian et al. (2016), copyright 2016, with permission from American Chemical Society. **h–l** Reprinted from Ref. Tu et al. (2016), copyright 2016, with permission from Elsevier

and pressed into a wafer with a fixed size by a pressing machine. Subsequently, a thermal treatment is performed at 800 °C under argon atmosphere, and then the resultant wafer is employed as anode and exfoliated in an electrolysis cell with platinum sheets as cathode, ammonium sulfate solution as the electrolyte. At a constant voltage of 5 V, 2D MoS_2-G hybrid is simultaneously and gradually exfoliated from the resultant MoS_2-graphite wafer. Finally, 2D MoS_2-G hybrid can be facilely gained since they are well dispersed in various solvents even after a few days. The resulting products of 2D MoS_2-G hybrid can be directly filtrated to flexible films through a polypropylene separator membrane with a pore size of 0.22 μm (Fig. 4.1b–d).

Xie et al. (2015), by taking MoS_2 as a typical example, established the correlations between the 2D heterointerface and the sodium-ion storage performance of layered metal sulfides/graphene composites. MoS_2/reduced graphene oxide (RGO) nanocomposites were synthesized via a simple hydrothermal method using L-cysteine, phosphomolybdic acid, and GO as precursors. The sheet-on-sheet MoS_2/RGO nanocomposites were formed, which effectively suppresses agglomeration of MoS_2 and RGO. However, in order to realize the functionalities of the 2D MoS_2/RGO heterointerface, it is required to prepare high-quality electrical contact between MoS_2 and RGO. The final MoS_2/RGO composites with sheet-on-sheet structure and controllable heterointerfaces were fabricated by a hydrothermal preparation, and the products underwent a calcination process after that. The in situ hybridization of MoS_2 and RGO plays an important role in constructing MoS_2/RGO heterostructures with enhanced electrochemical performances for Na^+ storage. First, RGO as macromolecular surfactants can effectively stabilize MoS_2 nanosheets with high surface energy, realizing high-dispersive MoS_2 supported on RGO. As a result, more reactive sites are available for the electrode/electrolyte interaction. Second, this soft integration method enables high-quality electrical contact between MoS_2 and RGO, which ensures the functionality of the MoS_2/RGO heterointerface for Na-ion storage performance.

4.1.2 $MoSe_2$

Molybdenum diselenide ($MoSe_2$) structure possesses three atom layers in which molybdenum atom is sandwiched between two selenium atoms, i.e., strong covalent bond characterized the interaction of Se-Mo-Se. On the other hand, interactions between individual layers of $MoSe_2$ are characterized via weak Van der Waals force which is facilitating ion intercalation/deintercalation path during electrochemical charge storage process. In addition, $MoSe_2$ has high theoretical capacity and is embraced of low-cost and abundant elements.

Generally, few-layered $MoSe_2$ nanosheets are synthesized via a chemical vapor deposition (CVD) or rapid thermal processing preparation. However, when these techniques are employed to synthesize a large amount of electrode materials, their shortcomings of high-price and user-hostile cannot be ignored (Bachmatiuk et al. 2014; Mutlu et al. 2014; Xia et al. 2014). Few-layered $MoSe_2$ nanosheets have been large-scale synthesized by Balasingam et al. (2015) via a simple hydrothermal method, and aqueous electrolyte (H_2SO_4) was used to measure their electrochemical capacitance under a symmetric cell configuration with two electrodes. The maximal specific capacitance was 199 F g^{-1}, which was obtained at a scan rate of 2 mV s^{-1}. Furthermore, for the long-term stability test, the capacitance of prepared electrode maintained nearly 75% of its original capacitance after working over 10,000 cycles, which showed it as a suitable material for electrochemical capacitors (Fig. 4.1e, f).

4.1.3 WS_2

Similar to graphene, tungsten disulfide (WS_2) is believed as another promising material for electrochemistry. However, in order to enhance the practicability of this material, a sophisticated synthesis that can control its morphology is required. Fortunately, a facile synthesis method has been proposed by Qian et al. (2016), which could fabricate WS_2 nanostructures with different morphologies. In the synthesis routes, WO_3 and S powders underwent thermal evaporation directly onto Si substrates, which were sputtered with W film. During the process, not only nanostructured W-contained precursors were not needed, but also toxic sulfide gases would not be mentioned. A schematic is presented in Fig. 4.1g. Interestingly, large quantities of horizontally grown WS_2 nanoplates, pure hexagonal, nanoplate formed flowers, and vertically grown nanoplates can be obtained by this preparation, which is realized simply by controlling the distance between the substrate and source powders.

WS_2 is widely considered to be a suitable material for charge accumulation, which thanks to its extensive active sites existing in the spacing between 2D structures and the interspacing of atomic layer, furthermore, the Faradaic reactions generated extra charges on the tungsten centers, which also matter. However, the restacking between the nanosheets and the low electronic conductivity as well as the relative brittle of WS_2 limits its adhibition. To overcome this difficulty, highly conductive carbon materials were applied to enhance the conductivity of the electrode and meanwhile prevent the restacking of WS_2 (Liu et al. 2014; Hu et al. 2013). Tu et al. (2016) reported a supercapacitor electrode based on a 2D hybrid consisting of WS_2 and RGO nanostructures fabricated by using a simple molten salt process. Figure 4.1h presents the HRTEM image for the WS_2 nanomaterial. The SEM images for the WS_2, RGO, and the WS_2/RGO hybrid were shown in Fig. 4.1i–k, respectively. To distinguish the WS_2 and RGO nanostructures more clearly, the TEM image for the WS_2/RGO was obtained as shown in Fig. 4.1l. A surprising high specific capacitance of 1355.67 F g^{-1} was achieved for WS_2/RGO hybrid-based supercapacitor electrode at the scan rate of 1 mV s^{-1}. Considering that WS_2 has the advantage of large charge-accumulating sites that located on the 2D planes, while RGO stands out among other materials for its excellent conductivity and superior connections in the networks structure of WS_2, the synergic effect between them can be the explanation for prominent capacitance of the supercapacitor electrodes. Moreover, the WS_2/RGO-based electrode achieves 98.6% retention of the original capacitance after 5000 cycles, suggesting its outstanding cycling stability. The Coulombic efficiency for the entire measurement is also measured, which is close to 100%. Taking all these results into consideration, the combination of carbon materials and 2D metal sulfide is confirmed to have great potential as storage material, contributing to the relief for the energy issue and development of sustainable society.

4.2 2D Transition Metal Oxides

2D metal oxide nanomaterials have recently received tremendous attention owing to their apparent advantages including high surface to volume ratio and unique chemical and physical properties, which is capable to offer abundant active sites to enhance the adsorption of molecules or ions and improve the transport of ionic species (Zhang 2015). Therefore, these nanomaterials have been applied in multitudinous fields such as energy conversion and storage (Peng et al. 2017; Bao et al. 2015), gas sensors (Kaneti et al. 2013; Kannan et al. 2015), catalysis (Deng et al. 2016; Cheng et al. 2014), and so on. However, it remains challenging to construct 2D transition metal oxide materials nanomaterials with well-defined geometry and functional nanoarchitecture that are tailored for specific applications. Synthetic protocols are needed to create various nanostructures and which can be generalized to many kinds of material types. Additionally, it is urgent to demonstrate the ability to modulate material properties. Recently, many novel methods have been proposed to synthesize 2D TMOs nanomaterials (Xu et al. 2016).

4.2.1 Co_3O_4

Li et al. (2017a) used a self-sacrificing template preparation to synthesize 2D porous Co_3O_4 nanosheets for reversible electrochemical lithium storage. During the calcination process, the Co-precursors are self-linked with each other to form 2D porous Co_3O_4 nanosheets. Figure 4.2a illustrates the formation process of 2D porous Co_3O_4 nanosheets and the advantages for lithium-ion transport. During the calcinations processes, the Co-precursors self-linked with each other to form the 2D porous nanosheets. Interestingly, the 2D nanosheet structure was perfectly inherited in the calcined products (Fig. 4.2b, c), but with porous architectures and rough surfaces in comparison with the Co-precursor. The magnified observations (Fig. 4.2c) further testify that the 2D porous nanosheets structures are constructed by numerous nanoparticles with a narrow size distribution, i.e., most of them are 15–30 nm in diameter. Due to the abundant active surface and the improved charge transport characteristics, 2D porous Co_3O_4 nanosheets exhibit enhanced overall lithium-ion storage properties. Electrochemical measurements display that the 2D porous Co_3O_4 demonstrates an ultrahigh reversible discharge specific capacity of 1000 mA h g^{-1} at

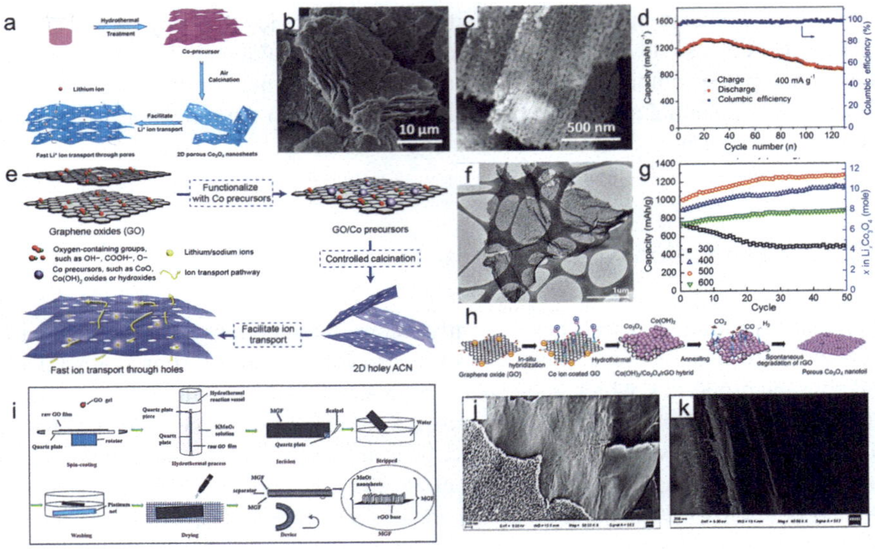

Fig. 4.2 **a** Schematic illustration of the formation process of 2D porous Co_3O_4 nanosheets and the advantages for lithium-ion transport. **b**, **c** SEM images of the as-obtained 2D Co_3O_4 nanosheets at different magnifications. **d** Cycling performance and columbic efficiency of 2D porous Co_3O_4 nanosheets at 400 mA g^{-1}. **e** Schematic illustration of the formation of 2D Holey ACN and their advantages for ion transport. **f** TEM images of porous Co_3O_4 nanofoils. **g** The cycle performance of Co_3O_4 nanofoils anodes at 0.1 °C under 50 cycles. **h** Schematic illustration of the formation process of porous Co_3O_4 nanofoils. **i** The preparation of MGF by spin-coating and hydrothermal process. FESEM images: **j** top view of MGF; **k** cross sections of MGF. **a–d** Reprinted from Ref. Li et al. (2017a; b), copyright 2017, with permission from The Royal Society of Chemistry and the Centre National de la Recherche Scientifique. **e** Reprinted from Ref. Chen et al. (2017a, b), copyright 2017, with permission from American Chemical Society. **f–h** Reprinted from Ref. Eom et al. (2016), copyright 2016, with permission from WILEY-VCH. **i–k** Reprinted from Ref. Li et al. (2016), copyright 2016, with permission from The Royal Society of Chemistry

the current density of 400 mA g^{-1} after 100 cycles and an outstanding rate capability for Li-ion storage (Fig. 4.2d).

Chen et al. (2017b) developed a template-directed strategy to synthesize 2D TMO Co_3O_4 nanosheets with a special holey structure and controllable hole sizes by self-linking of oxide nanoparticles on graphene oxide templates. Moreover, the holey assembly of Co_3O_4 nanoparticles (ACN) is formed by conjugating Co_3O_4 nanocrystals into a free-standing 2D structure. Graphene oxide (GO) has a 2D structure modified with sufficient oxygen-containing functional groups, which promote the growth of metal ion on its surface. During the refluxing process, cobalt ions were anchored on the GO nanosheets through residual functional group and formed the cobalt precursors integrated on GO. During the calcination process, the cobalt precursors self-linked with each other to form the holey ACN due to the thin and highly flexible GO template. The unique structure of the holey ACN satisfies several critical requirements for an ideal lithium and sodium-ion battery electrode (Fig. 4.2e). The as-prepared holey CAN have the hole size of 10 nm and its reversible capacities are 1324 mAh g^{-1} and 566 mAh g^{-1} at the current density of 0.4 A g^{-1} and 0.1 A g^{-1} for lithium and sodium-ion storage, respectively.

Eom et al. (2016) synthesized 2D oxide nanomaterials via imitating a single GO sheet. Unlike the common mixing method for sandwich type or core/shell materials, GO is used for collecting metal ions on the carbon sheet surface and sacrificing to produce a 2D oxide nanofoil consisting of nanocrystals. The negatively charged GO surface adsorbs positively charged Co ions. Then, the mixture of GO and these Co ions forms the hybrid of rGO/$Co(OH)_2$/Co_3O_4 under hydrothermal conditions. Finally, thermal annealing under air condition will leave porous nanofoils of Co_3O_4. Since the reactants are easy to diffuse, the porous shape is extremely advantageous for the electrode material. Moreover, vacancies along the nanocrystal network can offer a short path for mass transmission (Wang et al. 2016). In addition, the Co_3O_4 nanoparticles on the reduced GO (rGO) sheets act as catalysts for the degradation of the carbon backbone during thermal oxidation. Furthermore, due to the existence of oxidation defect regions, rGO can be degraded at a comparatively low temperature compared to the other carbon nanomaterials. The TEM image (Fig. 4.2f) obviously displays porous Co_3O_4 nanofoils, which mimics the original 2D morphology of GO. As shown in Fig. 4.2g, the Co_3O_4 nanofoils show a good reversible capacity of 1279.2 mAh g^{-1} after 50 cycles. According to the conversion mechanism from Co_3O_4 to Li_2O and Co, this capacity far exceeds the theoretical capacity of Co_3O_4. The preparation process of Co_3O_4 nanofoils is clearly shown in Fig. 4.2h.

4.2.2 MnO_x

Among all TMOs, the manganese oxides (MnO_x) like MnO_2, Mn_2O_3 have shown excellent applications in many fields including catalysis, magnetic materials, electronics, supercapacitors, etc. (Umek et al. 2009). In addition, MnO_x attract great interest as promising electrode materials because of their non-toxic nature, abundance, and low cost. Importantly, MnO_x have shown excellent structural flexibility with good chemical and physical properties which express various applications in heterogeneous catalysis, electrocatalysis, supercapacitors, or rechargeable batteries (Han et al. 2008; Cheng et al. 2013; Yan et al. 2010).

MnO_2 as a typical TMO has attracted much attention because of its excellent theoretical specific capacitance and rich redox activity (Shimamoto et al. 2013; Xiao and Cao 2015; Fang et al. 2013; Cai et al. 2016). In order to overcome its poor electrical conductivity and cycling crystal expansion/contraction induced flaking off during repeated cycling processes, conducting carbon materials such as graphene and CNTs were used as supporting materials (Zhao et al. 2015a; Mei and Zhang 2015; Li et al. 2014). Recently, several interesting MnO_2-carbon-based composites have been reported.

Li et al. (2016) present a novel flexible thin MnO_2/graphene film (MGF) electrode based on MnO_2-graphene combination composites as supercapacitor electrodes. In this work, the MnO_2 nanosheets have been planted merely on a side of the graphene film without any binders. It's well known that active materials, current collector and binder are essential parts for fabricating a supercapacitor. In the MGF, the large specific surface provided by the grapheme film makes the usage of the large pseudocapacitance of MnO_2 possible. At the same time, the graphene film also acts as a fast electrotransport current collector without binders. The MnO_2 nanosheets vertically align on the graphene surface forming an open porous structure. Therefore, the MnO_2 surface electrons have a much shorter distance to cross to the current collector. As a result, the interface between the electrolyte and the MnO_2-graphene film electrode increased greatly. Experimental results showed that for the flexible MGF with an excellent capacitance of 280 F g^{-1} at 1 A g^{-1}, after 10,000 cycles the capacitance content is 99%. Further studies showed that a flexible MGF applied symmetric supercapacitor has provided a 77 F g^{-1} capacitance at 1 A g^{-1}, the capacitance content was still at 91% after 10,000 cycles, and the energy density was 10.7 W h kg^{-1} at a power density of 500 W kg^{-1}. Figure 4.2i illustrates the preparation of MGF by spin-coating and hydrothermal process. Figure 4.2j shows the MnO_2 nanosheets vertically grew along the grapheme surface and constructed a porous network structure. The cross section of MGF is showed in Fig. 4.2k.

Mn_2O_3 has presented the good pseudocapacitor electrode materials because of its high specific capacity (Reddy et al. 2010). Furthermore, Mn_2O_3-based electrodes also have low operating voltages, i.e., the average charge and discharge voltages of 1.2 and 0.5 V, respectively (Deng et al. 2014). Over Mn_2O_3 nanoparticles, the nanosheets structures are displayed high surface to volume ratio, big pores, and pore volume, which can accelerate the ions transportation, leading to excellent supercapacitor performance. Li et al. (2017c) present highly dense and uniform Mn_2O_3 nanowalls thin films grown on the Ni foam through hydrothermal process and used as electroactive electrode for the fabrication of electrochemical supercapacitors. From the cyclic voltammetry (CV) results, it has been found that Mn_2O_3 nanowalls-based electrode displays excellent specific capacitance of 480 F g^{-1} at 10 mV s^{-1}. The specific capacitance of 461 F g^{-1} at 0.5 A g^{-1} with excellent recyclability is evaluated by galvanostat discharge-charge measurements.

4.2.3 Fe_3O_4

Among numerous materials advocated as promising anode candidates, Fe_3O_4 attracts extensive interest owing to its nontoxicity, good lithium storage capacity (928 mA h g^{-1}), ecofriendliness, natural abundance, and low cost (Guo et al. 2010; Mitchell et al. 2014). Despite these predominant features, their disadvantages such as sluggish kinetics, agglomeration, severe volume variation (~90%), and low intrinsic electric conductivity during the conversion reaction process, lead to the loss of electrical contact from the current collector and dramatic electrode pulverization, thus resulting in poor cyclic properties (Tuček et al. 2014; Poizot et al. 2010).

To solve the problem of poor rate capability and cycling stability of Fe_3O_4 anode materials in LIBs, Zhang et al. (2016) reported a clean, simple one-step method for synthesis on a large scale of ultrafine Fe_3O_4 nanoparticles (~3 nm) homogeneously embedded in 2D carbon nanonetworks through the thermolysis of iron(III) acetylacetonate ($Fe(C_5H_7O_2)_3$) precursors at 350 °C under vacuum (designated as 2D Fe_3O_4/C nanonetworks). This simple method, based on low-cost precursors, offered a clean and low-cost method to synthesize the Fe_3O_4/carbon composites. Benefitting from the excellent electroconductive nanonetworks, uniform distribution and the ultrafine size of Fe_3O_4 particles, high contact surface area and the efficient protection of the carbon shell, the resultant delivered outstanding cycling properties with excellent Coulombic efficiency (~100%), a large reversible capacity of 1534 mA h g^{-1} at 1 A g^{-1} after 500 cycles, and superior rate capability of 845 mA h g^{-1} at 5 °C. The fabricated 2D Fe_3O_4/carbon nanonetwork can meet almost all the kinetic requirements to achieve fast discharging and charging of ideal electrode materials and long cycle life.

4.2.4 Mixed Oxide

Compared to pure metal oxides, mixed oxides have attracted great interest due to their excellent electronic conductivity. This is a result of the existence of multiple valence states for contributing metals and the synergistic effect of various metallic species (Kaneti et al. 2017). Among various mixed oxides, metal cobaltates (MCo_2O_4, M = Mn, Zn, Ni, etc.) are widely studied for energy storage applications due to their superior electrical conductivity, layered structure, and low cost (Wei et al. 2009).

Zhu et al. (2015) reported the high-quality 2D ultrathin $ZnCo_2O_4$ nanosheets synthesized by microwave-assisted liquid-phase growth and post-annealing procedures. Different from conventional heating methods, the microwave dielectric irradiation features ultrafast, selective, and volumetric heating (Baghbanzadeh et al. 2011). Polar molecules and ions are forced to orderly orientate with the constantly changing electric field caused by microwave activation (Zhu and Chen 2014). The microwave dielectric irradiation is favorable for the 2D anisotropic growth of inorganic nanocrystals. The independent and well-defined nanosheets show ultrathin thickness and a micron-level planar area, indicating a high surface atomic ratio with electronic structure and a special surface, thereby promoting the charge transfer to improve the overall properties of the electrochemical reaction. The ultrathin $ZnCo_2O_4$ nanosheets used as anode materials in LIBs show an excellent reversible lithium storage capacity of 930–980 mAh g^{-1} at 200 mA g^{-1} during 200 cycles with a good cyclic stability and high-rate capability. More importantly, they have expanded the convenient approaches for forming other similar nanosheets, including binary and ternary transition metal oxides (Co_3O_4, NiO, $CuCo_2O_4$, and $NiCo_2O_4$), and to explore more unique performance and promising commercial application offers the possibility.

Kaneti et al. (2018) reported a general template-free method for the fabrication of 2D mesoporous mixed oxide nanosheets, like metal cobaltites (MCo_2O_4, M = Ni, Zn) by the self-deconstruction/reconstruction of highly uniform nanospheres of Co-based metal glycerate into 2D nanosheets of Co-based metal hydroxide/glycerate (Fig. 4.3a). The "self-deconstruction/reconstruction" method has high advantages because the resulting 2D MCo_2O_4 nanosheets show mesoporous characteristics with a narrow pore size distribution and a high surface area of 150–200 m^2 g^{-1}. Moreover, this method also allows the crystallization temperature to obtain a pure MCo_2O_4 phase from the precursor phase. They used 2D mesoporous $NiCo_2O_4$ nanosheets as

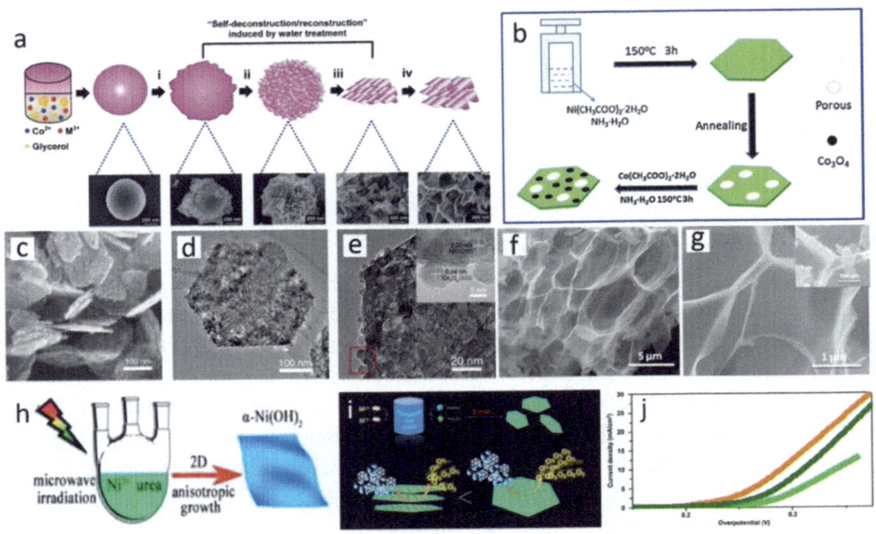

Fig. 4.3 **a** The Schematic illustration showing the proposed general template-free strategy for obtaining mesoporous 2D mixed oxide nanosheets, such as metal cobaltites (MCo_2O_4, M = Ni, Zn) as examples: (i) surface deconstruction of the Co-based metal (M-Co) glycerate nanospheres induced by the "water treatment" process; (ii) surface reconstruction of the M-Co glycerate nanospheres to form hierarchical 3D flower-like nanosheets; (iii) lateral growth of the sheet-like structures assembling the hierarchical 3D structures to form 2D M-Co glycerate/hydroxide nanosheets; (iv) conversion into mesoporous 2D MCo_2O_4 nanosheets via calcination in air at 260 °C. **b** Schematic synthesis of NiO-Co_3O_4 nanocomposites. **c** SEM images of NiO-Co_3O_4 nanocomposite. **d, e** TEM images of NiO-Co_3O_4 nanocomposite. **f, g** FESEM micrograph of mesoporous LNO sample at different magnifications. **h** Schematic illustrating synthesis of α-$Ni(OH)_2$ nanosheets. **i** Schematic illustration for the formation of ultrathin single-unit-cell-thick LDH nanosheets. **j** Linear sweep voltammetry (LSV) curves for the ultrathin single-unit-cell NiFe-LDH nanosheets (orange line); bulk NiFe-LDHs (green line); IrO_2 nanoparticles (blue line). **a** Reprinted from Ref. Kaneti et al. (2018), copyright 2018, with permission from The Royal Society of Chemistry. **b–e** Reprinted from Ref. Zhang et al. (2015), copyright 2015, with permission from Elsevier. **f, g** Reprinted from Ref. Zhang et al. (2017), copyright 2017, with permission from Elsevier. **h** Reprinted from Ref. Zhu et al. (2015), copyright 2014, with permission from Nature Publishing Group. **i, j** Reprinted from Ref. Fan et al. (2015), copyright 2017, with permission from Tsinghua University Press and Springer-Verlag GmbH Germany

representative samples and found that they showed a specific capacitance of 6–20 times and greatly improved capacitance retention compared to $NiCo_2O_4$ nanospheres obtained by directly calcining Ni–Co glycerate nanospheres. This indicates another advantage of the method that is used to enhance the electrochemical properties of mixed oxide products in supercapacitor applications. In addition, the asymmetric supercapacitor (ASC) assembled using the 2D $NiCo_2O_4$ nanosheets//GO shows a superior capacitance retention (91%) at 5 A g^{-1} after 2000 cycles and a maximum energy density (38.53 W h kg^{-1}). It is hoped that this method can be extended to other transition metal elements to produce 2D mixed oxide nanosheets with enhanced electrochemical properties and increased surface area.

NiO and Co_3O_4 have also been selected to construct the hybrid nanocomposite. Zhang et al. (2015) develop a simple hydrothermal reaction to prepare the 2D/0D (NiO–Co_3O_4) hybrid nanocomposite. The scheme of the formation process of NiO–Co_3O_4 nanocomposite is displayed in Fig. 4.3b. Figure 4.3c is the SEM image of NiO–Co_3O_4 nanocomposite, which clearly reveals numerous Co_3O_4 nanoparticles uniformly distributed on the surface of each NiO nanoplate. Quantitative EDS analysis in SEM shows a weight ratio of Co to Ni is about 1:2.8. The hybrid NiO–Co_3O_4 nanostructure in Fig. 4.3d remains the hexagonal structure. As shown in Fig. 4.3e, the high-density Co_3O_4 nanoparticles (3 ~ 5 nm) have been finely anchored on the surface of NiO nanoplate. As an anode material of LIB, the nanocomposite exhibits greatly improved specific capacities and stable cyclability of 633 mA h g^{-1} at 100 mA g^{-1} after 70 cycles, much higher than the corresponding building block alone. The outstanding properties of the NiO/Co_3O_4 composite are ascribed to the synergistic effect of various components and the hybrid structure. This large-scale and cost-efficient synthesis can be extended for the synthesis of TMOs composite for high-performance electrochemical energy storage.

2D perovskite-type metal oxides have the ultrathin characteristics of nanosheets and have a broad application prospect in the field of electrochemical energy storage. The synthesis of 2D perovskite-type $LaNiO_3$ (LNO) nanosheets materials with high conductivity and rich porous structure was firstly reported via a sol-gel method and following heating treatment by Li et al. (2017b). By adjusting the heating temperature and time, the electrochemical properties, crystal structure and morphology of $LaNiO_3$ can be easily adjusted. The FESEM images of the optimized sample at different magnifications were exhibited in Fig. 4.3f, g. As displayed in low-magnification SEM images, a large micron-scale interconnected agglomerates to form an open-pore network structure. It is apparent that the 2D sheet-like morphology is uniform and the average thickness is ~50 nm (insert in Fig. 4.3g). These materials with richly porous morphology and microstructure are very favorable for applications in aspect of energy storage, which are not only beneficial to the electron/ion diffusion and transmission, but also increase the specific surface area of the material, and then provide more reaction sites. When the current density of the optimized sample is 1.0 A g^{-1}, the specific capacitance reaches 139.2 mAh g^{-1}, which has good multiplier performance and good cyclic stability. In addition, the $LaNiO_3$//graphene asymmetric supercapacitor device has a high energy density of 65.8 Wh kg^{-1} at a power density of 1.8 kW kg^{-1} and a specific capacitance retention rate of 92.4% after 10,000 cycles, which is a key step toward practical application.

4.3 2D Transition Metal Hydroxides

TMHs of Co (i.e., $Co(OH)_2$) and Ni (i.e., nickel hydroxide ($Ni(OH)_2$) were mainly found in their 2D nanostructures. For instance, a solid-state asymmetric supercapacitor exhibited a high operation voltage of 1.8 V, which is made up of N-doped graphene as the anode and single layers of β-$Co(OH)_2$ as the cathode (Gao et al. 2015). In addition, the device showed an exceptionally high energy/power density of 98.9 Wh kg^{-1}/17,981 W kg^{-1} and still maintain 93.2% capacitance after 10,000 cycles. The performance similar to LIBS is attributed to the 100% exposure of five atom thick $Co(OH)_2$ monolayers to the surface hydrogen atoms, as the electroactive center of important Faraday redox reactions, as shown in Fig. 4.4b. 2D $Ni(OH)_2$ is another kind of high energy density supercapacitor material with excellent electrochemical properties. Ultrathin, independent α-$Ni(OH)_2$ nanosheets with a thickness of less than 2 nm were fabricated by microwave-assisted liquid-phase growth method (Chen et al. 2014). The growth process of the as-prepared α-$Ni(OH)_2$ nanosheets was shown in Fig. 4.3h. During the process of electrochemical reaction, most of the atoms were exposed to these ultrathin 2D nanostructures. At a current density of 1 A g^{-1}, it shows a maximum specific capacitance of 4172.5 F g^{-1}, and even maintained at 2680 F g^{-1}, at a higher rate of 16 A g^{-1} with 98.5% retention after 2000 cycles.

Layered double hydroxides (LDHs) belong to another two-dimensional (2D) materials which have unique structure consisting of metal hydroxide layers and inorganic/organic gallery anions/molecules (Wang and O'Hare 2012; Fan et al. 2015; Gu et al. 2015). Using the general formula of LDHs to express is $[M^{II}{}_{1-x}M^{III}{}_{x}(OH)_2]^{x+}(A^{n-})_{x/n}\cdot mH_2O$, where divalent metal cations M^{II} (M = Mn, Fe, Co, Ni, Cu, Zn, etc.) and trivalent metal M^{III} (M = Al, Ga, Ti, Cr, Fe, Co, etc.), respectively, and A^{n-} is a charge-balancing anion intercalated between the brucite-like metal hydroxide layers. The easily tailored properties, composition versatility, and low cost of LDHs have led to surging interest in these materials and many applications such as adsorption (Shao et al. 2012), photochemistry (Cho et al. 2014; Zhao et al. 2014a), and electrocatalysis (Liang et al. 2015a).

For instance, the NiFe-LDH system has been investigated as an efficient oxygen evolution reaction (OER) catalyst due to its high activity and stability in basic media; various strategies have been developed to further improve its performance either by hybridizing it with carbon materials or by constructing micro/nanostructures (Gong et al. 2013; Song and Hu 2014). For example, Hu et al. demonstrated that exfoliated LDH nanosheets showed significantly enhanced performance as compared to their bulk phase due to their improved intrinsic catalytic activity and conductivity (Tu et al. 2016). However, the synthesis of exfoliated LDH nanosheets not only required

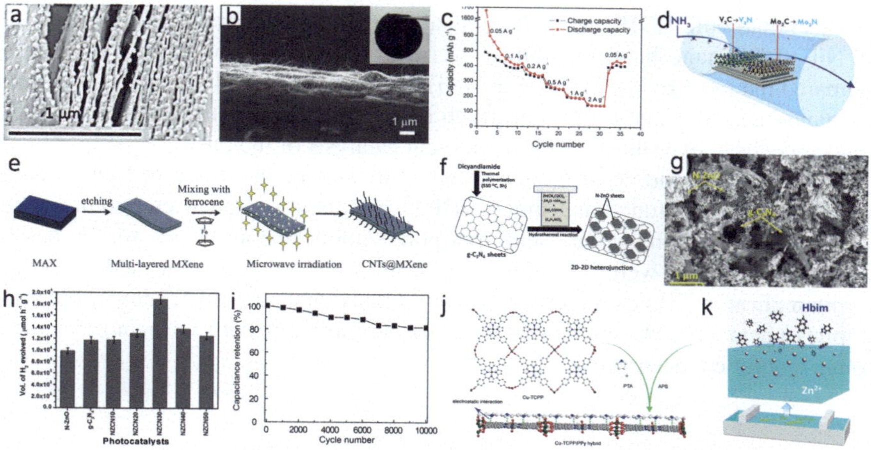

Fig. 4.4 **a** SEM images of PANI-Ti_3C_2. **b** SEM images of the $Ti_3C_2T_x$/CNT composite film. **c** Rate performance of the $Ti_3C_2T_x$/CNT film. **d** Synthesis of 2D transition metal nitrides can be achieved by ammoniation of carbide MXenes (Mo_2CT_x and V_2CT_x) at elevated temperatures. **e** Schematic showing the preparation of CNTs@MXene hybrids. **f** Schematic illustration of hydrothermal synthesis of g-C_3N_4 and N-ZnO, 2D-2D heterojunction. **g** SEM images of NZCN30 2D-2D heterojunction. **h** Volume of H_2 evolved over various photocatalysts. Experimental conditions: 5 mg of catalysts, suspended in 0.3 M of Na_2S, Na_2SO_3 solution under simulated solar light (Xe lamp) for 4 h. **i** Capacitance retention at first 10,000 cycles at 2 A g^{-1}. Electrochemical performance of CAP-2 in an asymmetry supercapacitor cell in 2 M KCl. **j** The proposed formation mechanism of Cu-TCPP@PPy. **k** Schematic illustration for the preparation of large-area zinc benzimidazolate coordination polymer layers at air-water surfaces of a Langmuir trough. **a** Reprinted from Ref. Ren et al. (2018), copyright 2017, with permission from Elsevier. **b, c** Reprinted from Ref. Yu et al. (2018), copyright 2018, with permission from The Royal Society of Chemistry. **d** Reprinted from Ref. Urbankowski et al. (2017), copyright 2017, with permission from The Royal Society of Chemistry. **e** Reprinted from Ref. Zheng et al. (2018), copyright 2018, with permission from The Royal Society of Chemistry. **f–h** Reprinted from Ref. Kumar et al. (2018), copyright 2017, with permission from Elsevier. **i** Reprinted from Ref. Liu et al. (2017), copyright 2017, with permission from American Chemical Society. **j** Reprinted from Ref. Yao et al. (2018), copyright 2017, with permission from Elsevier. **k** Reprinted from Ref. Huang et al. (2018), copyright 2018, with permission from American Chemical Society

complicated multi-step procedures (i.e., hydrothermal process, anion exchange, and exfoliation) but also required two or more days at least to achieve the final product. Additionally, the production yield was relatively low. It is therefore desirable to employ a simple and alternative way to synthesize different kinds of ultrathin atomic-scaled thick LDH nanosheets, which could raise the amount of active edge sites with higher electronic conductivity.

Gao et al. (Song and Hu 2014) report an effective way to obtain ultrathin atomic-thick LDH layers within only 5 min (Fig. 4.3i) (Yu et al. 2015). Representative atomic-thick LDH nanosheets such as CoNi-, NiFe-, CoFe-, and ZnCo-LDHs were constructed as ideal model systems, which showed much higher OER activities (in terms of overpotential, turnover frequency, double-layer capacitance, and stability) than bulk LDHs using the typical co-precipitation method. For example, the NiFe-LDH nanosheets exhibited lower overpotential and higher stability, even compared the most IrO_2 catalyst (Fig. 4.3j). Owing to the rapid synthesis and ultra-thin nanosheets, the unique 2D nanostructures exhibited more surface defects, which could serve as active sites to achieve efficient catalysis of the OER.

A basic understanding of the electronic structures of the different LDHs could be used by density functional theory (DFT) calculations, and it can confirm the higher electronic conductivity and spin-polarization characteristics for Fe-based LDH nanosheets, as well. Therefore, this work not only presents a facile strategy to obtain 2D atomic-thick nanosheets toward highly enhanced OER activity but also supplies a detailed understanding of the electronic structures of LDH nanostructures from a theoretical perspective.

4.4 MXenes

Recently, a new family of 2D early transition metal carbides, nitrides, and carbonitrides, commonly known as MXenes, has been found (Naguib et al. 2011, 2012, 2014). Generally, MXenes are fabricated from MAX phases consisting of layered ternary carbides with the formula $M_{n+1}AX_n$, where M is an early transition metal (Sc, Ti, V, Cr, Zr, Nb, Mo, Hf, or Ta), A is an element from groups 12–16 (Cd, Al, Si, P, S, Ga, Ge, As, In, Sn, Tl, Pb, or S), and X is carbon and/or nitrogen (Ma et al. 2016; Zhao et al. 2012; Ghidiu et al. 2014; Khazaei et al. 2014). The surface-exposed transition metal sheet, MXenes, was synthesized by selecting corrosion element A from MAX phase by strong acid and stripping (Gao et al. 2017). At room temperature, after replacing element A with HF aqueous solution as the etchant, the relatively strong metal bonds of M and A in $M_{n+1}AX_n$ phase are replaced by OH, O or F weak hydrogen bonds (Xiao et al. 2016a). Therefore, the etching and stripping processes usually give MXene samples hydroxyl, oxygen, and fluorine groups. The first principle calculation shows that the electron structure of the generated MXene is obviously different from that of its parent MAX phase. For example, functionalization with OH, O, and F leads to a transition from metal to semiconductor (Lei et al. 2015). MXenes, with both metallic conductivity and hydrophilic behavior, have demonstrated their potential as potential electrode materials for electrochemical energy storage devices, such as lithium-ion batteries (Ren et al. 2016; Chen et al. 2017a), Li-S batteries (Liang et al. 2015b; Zhao et al. 2015c), sodium-ion batteries (Xie et al. 2016; Lian et al. 2017) and supercapacitors (Lukatskaya et al. 2013; Zhao et al. 2014b).

4.4.1 Transition Metal Carbides

Ti_3C_2 is one of the most widely studied members of this family. Particularly, Ti_3C_2 is a promising electrode material for ultracapacitor due to its characteristics of metal conductivity, intercalation of spontaneous ions in solution and capacitance. To further enhance electrochemical performance of Ti_3C_2 electrode materials, one straightforward strategy is the introduction of Ti_3C_2. Conducting polymers, such as polyaniline (PANI), is famous for its large specific surface area (SSA) and relatively high specific capacitance (Liu et al. 2015). In particular, through some methods such as providing faradaic reactions for additional capacity increasing the electric conductivity for faster ion transport, and improving surface wettability for more active sites, nitrogen-containing functional groups in PANI can improve electrochemical performance of Ti_3C_2 (Yan et al. 2010). Multilayered PANI-Ti_3C_2 was synthesized by Ren et al. (2018) via in situ polymerization by low temperature, which could reduce the oxidation of Ti_3C_2 during the polymerization. PANI-Ti_3C_2 images were shown in Fig. 4.4a, homogeneous PANI nanoparticles cover the surfaces of lamellar Ti_3C_2, which results in the increase in SSA of PANI-Ti_3C_2 comparing with Ti_3C_2. When used as supercapacitor electrode material, PANI-Ti_3C_2 showed enhanced performance comparing with Ti_3C_2.

Yu et al. (2018) mixed a delaminated $Ti_3C_2T_x$ MXene with CNTs to prepare a $Ti_3C_2T_x$/CNT self-supporting composite film by a facile vacuum filtration method. Figure 4.4b shows the cross-sectional SEM images of the $Ti_3C_2T_x$/CNT composite film. It is clearly seen that the film still maintains a certain lamellar structure, and carbon nanotubes are interposed between the $Ti_3C_2T_x$ nanoflakes. The thickness of the $Ti_3C_2T_x$/CNT composite film is about 2 μm. Carbon nanotubes look like multilayer fishing nets and wrap the $Ti_3C_2T_x$ nanoflakes, which can effectively suppress the $Ti_3C_2T_x$ nanoflakes from stacking and enhance the electrochemical performance. When used as a binder-free anode, the $Ti_3C_2T_x$/CNT film reveals a high reversible capacity up to 489 mA h g^{-1} together with good cycling performance (Fig. 4.4c). Subsequently, an activated carbon electrode and the $Ti_3C_2T_x$/CNT film as the cathode and the anode, respectively, which delivered a high energy density of 67 Wh kg^{-1} and a good capacity retention of 81.3% after 5000 cycles, to assemble lithium-ion capacitor (LIC). This is the first time that a $Ti_3C_2T_x$/CNT film is utilized as an anode material for LICs.

Zheng et al. (2018) selected Ti_3C_2, Ti_2C, and V_2C as representative MXenes and successfully synthesized CNTs on them by microwave irradiation using carbon fibers (C_f) as the ignitor. The whole preparation process is illustrated in Fig. 4.4e. In such synthesized CNTs@Ti_3C_2, the CNTs serve as spacers preventing the restacking of MXene, and bridge the gap between Ti_3C_2 interlayers, thereby forming a conductive network. On the other hand, Ti_3C_2 presents as a substrate with fine thermal conductivity, high surface area, and catalytically active sites, which facilitate the growth of CNTs. Benefiting from these merits, CNTs@Ti_3C_2 hybrids as anodes of LIBs exhibit excellent electrochemical performance compared with pristine Ti_3C_2 and recently reported MXene composites.

4.4.2 Transition Metal Nitrides

Only a few 2D nitrides including MoN (Xie et al. 2014), GaN (Al Balushi et al. 2016), Ti_4N_3 (Urbankowski et al. 2016), and Ti_2N (Soundiraraju and George 2017) have been reported so far. Current strategies for synthesis of 2D nitrides are mostly limited to exfoliation of layered materials, while a gas-phase synthesis of GaN monolayers under graphene also exists (Al Balushi et al. 2016). Several of those studies, however, lacked focus on upscaling the methods and yielding 2D metal nitrides which could be easily be implemented into devices.

Heat treatment in ammonia (ammoniation) has been used on metal oxides to either N-dope them or transform oxides to 3D nitrides. Although ammoniation of 2D precursors such as GaSe to yield the 2D metal nitride GaN has been reported, the extent of the route of synthesizing other potential 2D metal nitrides via ammoniation has not been fully realized (Sreedhara et al. 2014). TiO_2 can be N-doped by nitridation at 600 °C (Sakthivel and Kisch 2003; Liu et al. 2009), and recently reported nitridation attempts of Ti_3C_2 have yielded only N-doped 2D carbides. Ti_3C_2 was doped with nitrogen atoms by ammoniation at temperatures up to 700 °C in which N comprised up to 20.7 at % of the product (Wen et al. 2017). It was only recently when synthesis of 2D metal nitrides via ammoniation was reported, including 2D MoN via the ammoniation of MoO_3-coated NaCl (Xiao et al. 2017). Urbankowski et al. (2017) report the first transformation of Mo_2CT_x and V_2CT_x carbide MXenes into 2D metal nitrides via ammoniation at 600 °C. To synthesize nitride MXenes, multilayer powders and delaminated films of Mo_2CT_x and V_2CT_x were treated at 600 °C for 1 h at a heating rate of 10 °C h^{-1} in an ammonia, NH_3, atmosphere and cooled at the same rate. The ammonia flow rate through the reactor was approximately 300 cm^3 min^{-1}. During the reaction, nitridation occurs by the replacement of C atoms, in the Mo_2CT_x and V_2CT_x, with N atoms. A schematic of the procedure is displayed in Fig. 4.4d. It is important to note that unlike previously reported 2D MoN and V_2N via salt-templated synthesis with 2D metal oxide precursors (Xiao et al. 2016a, b), the 2D metal nitrides reported here, synthesized with carbide MXene precursors, have different crystal structures than those previously reported. Their results demonstrate that Mo_2N retains the structure of MXene and V_2C transforms into a mixed layered structure of trigonal V_2N and cubic VN.

4.4.3 Transition Metal Carbonitrides

Zhi et al. (2017) developed a one-step and environmental benign ion-thermal route to synthesize 2D hierarchical Mo_2C/C-N hybrid using $(NH_4)_6Mo_7O_{24}4H_2O$ as a molybdenum source, dicyanodiamine ($C_2H_4N_4$) as a carbon and nitrogen source, and NH_4Cl as a 2D structural oriented and ion-thermal agent. As expected, the as-synthesized 2D hierarchical Mo_2C/C-N hybrid is composed of well-dispersed Mo_2C nanocrystals decorated on nitrogen-doped carbon (C-N) frameworks and possesses small grain size and high porosity. This unique structure can reduce the aggregation of Mo_2C nanocrystals, increase the active site, and enhance the interaction between Mo_2C nanocrystals and C-N frameworks, making it exhibit superior hydrogen evolution reaction (HER) activity in both acidic and alkaline media. With regard to large-scale production, the synthetic strategy is facile and inexpensive and can also be extended to fabricate various 2D hierarchical nanomaterials for electrocatalytic applications.

Kumar et al. (2018) reported the preparation of 2D nanosheets heterojunction composed of N-doped ZnO nanosheets (anchored) on graphitic carbon nitride (g-C_3N_4) nanosheets for enhanced photocatalytic hydrogen evolution. The procedure is schematically illustrated in Fig. 4.4f. The N-ZnO nanosheets are loaded over the g-C3N4 nanosheets and they are anchored perfectly over its surface to form 2D-2D heterojunction with intimate face-to-face contact. It is worth mentioning here that such type of 2D-2D heterojunctions can effectively improve the specific surface area and provide more active sites to adsorb reactant species which promote the separation of photogenerated charge carriers significantly to improve the photocatalytic activity (Fig. 4.4g). The optimal heterojunction photocatalyst with 30 wt% of g-C_3N_4 nanosheets (NZCN30) show hydrogen evolution rate of 18,836 μmol h^{-1} $g_{cat}{}^{-1}$ in presence of Na_2S and Na_2SO_3 as sacrificial agents under simulated solar light irradiation (Fig. 4.4h). Supported well by photoluminescence and photoelectrochemical investigations, the enhanced photocatalytic performance of NZCN30 heterojunction reveals the minimum recombination rate and high photoinduced current density, respectively. Moreover, by high face-to-face contact surface area for separation of photogenerated charge carriers in space which facilitates their transfer for H_2 generation, the existence of 2D-2D interfacial contact plays a major role in enhanced H_2 evolution. This work expands the space for the progress of 2D-2D heterojunctions for diverse applications.

4.5 Polymer

2D polymer has properties well fitted for energy storage applications on account of its combination of porosities and layered structure, providing ion diffusion routes through the 2D planes and 1D channels. In addition, if the polymer has an aromatic conjugated scaffold, it mediates charge transfer. The key challenge of 2D polymer synthesis is how to confine chain evolution in two dimensions.

Liu et al. (2017b) reported a C–C coupling based 2D conjugated polymer via endogenous polymerization of phenazine-based monomer carried out under catalyst- and solvent-free condition (Liu et al. 2017a) They extend the endogenous polymerization strategy to different phenazine-based monomers 2-TBTBP and 3-TBQP that prepack into single crystals with different topologies. These are converted by endogenous polymerization into π-conjugated, nanoporous aromatic polymers (CAPs). The products of polymerizing 2-TBTBP and 3-TBQP are denoted as CAP-1 and CAP-2, respectively. Monomers 3-TBQP are densely packed into a quasi-2-D structure, which is useful in confining C–C coupling reaction in two dimensions during the endogenous solid-state polymerization. The synthesized 2D polymer shows distinct lamellar structure with highly regular pore size in the range of 0.7–1.0 nm, which is advantageous for supercapacitive energy storage. When employed as positive electrode in an asymmetric supercapacitor, it provides a specific capacitance of 233 F g^{-1} at a current density of 1.0 A g^{-1} in 2 M KCl aqueous electrolyte. It shows an excellent cycle life with ~80% capacity retention over 10,000 cycles (Fig. 4.4i). These results show that endogenous polymerization is a facile access to 2-D polymer with well-defined pores and conjugated aromatic structure, which can be used as green organic electrodes in supercapacitor applications.

Yao et al. (2018) reported a strategy of growth of conductive polymer (PPy) on 2D coordination polymers (CPs) nanosheets to enhance their capacitive performances (Fig. 4.4j). On one hand, the PPy coating layer can form conductive pathways on the surfaces of 2D CPs; on the other hand, the tight encapsulation of PPy can buffer the interfacial stress resulted from the volume variation during charging and discharging, making the hybrid a potential candidate for high-performance capacitive material. Moreover, the thickness of PPy can also influence the electrochemical performance by affecting the electrical conductivity, ion diffusion resistance, and synergistic cooperation of the composites (Ji et al. 2015). Besides, this strategy is considered to be applicable for growth of other conductive polymers (Such as Polyaniline and Polythiophene) on other 2D CPs (or coordination supramolecular networks). As proof of concept, they choose a typical 2D CP (Cu-TCPP, TCPP = 5, 10, 15, 20-tetrakis(4-carboxyphenyl)porphyrin) as the 2D platform, onto which PPy is grown by a simple in situ chemical oxidation method. In this kind of 2D CPs nanosheets, one TCPP ligand is linked by four Cu paddlewheel metal nodes, i.e., $Cu_2(COO)_4$, to form a 2D layered sheet (Zhao et al. 2015b). The strong π-stacked conjugation and the heteroatom of the TCPP may enhance both the electron transport and the interaction with conductive polymers.

Huang et al. (2018) proposed a synthetic strategy to grow a large area and transferable Zn-coordination 2D polymer, where OH functional groups and the amphoterism of zinc hydroxides play an active role in the separation/selection using the charge exclusion principle. Figure 4.4k depicts the method of synthesizing large-area and free-standing 2D polymeric layers based on zinc(II) benzimidazolate complex on the water surface of a Langmuir trough. The charge barrier feature has been integrated into the Li-S battery to mitigate the polysulfide shuttle effects by the electrostatic shield, largely promoting the Li-S capacity and cycle performance. The synthesis is simple, scalable, and cost-saving, and it is anticipated that this type of functional 2D coordination polymer can soon be adopted for practical use.

References

Al Balushi ZY, Wang K, Ghosh RK, Vilá RA, Eichfeld SM, Caldwell JD, Qin X, Lin Y-C, DeSario PA, Stone G (2016) Two-dimensional gallium nitride realized via graphene encapsulation. Nat Mater 15(11):1166

ArunKumar S, Jegathish V, Soundharya R, JesyAlka M, Arul C, Sathyanarayanan S, Mayavan S (2017) Evaluating the performance of MoS_2 based materials for corrosion protection of mild steel in an aggressive chloride environment. RSC Adv 7(28):17332–17335

Bachmatiuk A, Abelin RF, Quang HT, Trzebicka B, Eckert J, Rummeli MH (2014) Chemical vapor deposition of twisted bilayer and few-layer $MoSe_2$ over SiO_x substrates. Nanotechnology 25(36):365603

Baghbanzadeh M, Carbone L, Cozzoli PD, Kappe CO (2011) Microwave-assisted synthesis of colloidal inorganic nanocrystals. Angew Chem Int Ed 50(48):11312–11359

Balasingam SK, Lee JS, Jun Y (2015) Few-layered $MoSe_2$ nanosheets as an advanced electrode material for supercapacitors. Dalton Trans 44(35):15491–15498

Bao J, Zhang X, Fan B, Zhang J, Zhou M, Yang W, Hu X, Wang H, Pan B, Xie Y (2015) Ultrathin spinel-structured nanosheets rich in oxygen deficiencies for enhanced electrocatalytic water oxidation. Angew Chem 127(25):7507–7512

Fang C, Zhang D, Cai S, Zhang L, Huang L, Li H, Maitarad P, Shi L, Gao R, Zhang J (2013) Low-temperature selective catalytic reduction of NO with NH_3 over nanoflaky MnOx on carbon nanotubes in situ prepared via a chemical bath deposition route. Nanoscale 5(19):9199–9207

Cai S, Hu H, Li H, Shi L, Zhang D (2016) Design of multi-shell Fe_2O_3@MnOx@CNTs for the selective catalytic reduction of NO with NH3: improvement of catalytic activity and SO_2 tolerance. Nanoscale 8(6):3588

Cao L, Yang S, Gao W, Liu Z, Gong Y, Ma L, Shi G, Lei S, Zhang Y, Zhang S (2013) Direct laser-patterned micro-supercapacitors from paintable MoS_2 films. Small 9(17):2905–2910

Chang K, Chen W (2011) In situ synthesis of MoS_2/graphene nanosheet composites with extraordinarily high electrochemical performance for lithium ion batteries. Chem Commun 47(14):4252–4254

Chang K, Chen W, Ma L, Li H, Li H, Huang F, Xu Z, Zhang Q, Lee J (2011) Graphene-like MoS_2/amorphous carbon composites with high capacity and excellent stability as anode materials for lithium ion batteries. J Mater Chem 21(17):6251–6257

Chen YY, Dong M, Wang J, Jiao H (2013) Mechanisms and energies of water gas shift reaction on Fe-, Co-, and Ni-promoted MoS_2 catalysts. Cheminform 44(9):25368–25375

Chen CL, Qi J, Tao J, Zuckermann RN, Deyoreo JJ (2014) Tuning calcite morphology and growth acceleration by a rational design of highly stable protein-mimetics. Sci Rep 4:6266

Chen C, Boota M, Xie X, Zhao M, Anasori B, Ren CE, Miao L, Jiang J, Gogotsi Y (2017a) Charge transfer induced polymerization of EDOT confined between 2D titanium carbide layers. J Mater Chem A 5(11)

Chen D, Peng L, Yuan Y, Zhu Y, Fang Z, Yan C, Chen G, Shahbazian-Yassar R, Lu J, Amine K (2017b) Two-dimensional holey Co_3O_4 nanosheets for high-rate alkali-ion batteries: from rational synthesis to in situ probing. Nano Lett 17(6):3907

Cheng XB, Tian GL, Liu XF, Nie JQ, Zhao MQ, Huang JQ, Zhu W, Hu L, Zhang Q, Wei F (2013) Robust growth of herringbone carbon nanofibers on layered double hydroxide derived catalysts and their applications as anodes for Li-ion batteries. Carbon 62(62):393–404

Cheng H, Kamegawa T, Mori K, Yamashita H (2014) Surfactant-free nonaqueous synthesis of plasmonic molybdenum oxide nanosheets with enhanced catalytic activity for hydrogen generation from ammonia borane under visible light. Angew Chem Int Ed 53(11):2910–2914

Chia X, Eng AY, Ambrosi A, Tan SM, Pumera M (2015) Electrochemistry of nanostructured layered transition-metal dichalcogenides. Chem Rev 115(21):11941–11966

Cho S, Jang J, Park YB, Kim JY, Magesh G, Kim JH, Seol M, Yong K, Lee K, Lee JS (2014) An exceptionally facile method to produce layered double hydroxides on a conducting substrate and their application for solar water splitting without an external bias. Energy Environ Sci 7(7):2301–2307

Deng Y, Wan L, Xie Y, Qin X, Chen G (2014) Recent advances in Mn-based oxides as anode materials for lithium ion batteries. Rsc Adv 4(45):23914–23935

Deng D, Novoselov KS, Fu Q, Zheng N, Tian Z, Bao X (2016) Catalysis with two-dimensional materials and their heterostructures. Nat Nanotechnol 11(3):218

Deng Y, Ding L, Liu Q, Zhan L, Wang Y, Yang S (2017) Two-dimensional MoS_2-graphene hybrid nanosheets for high gravimetric and volumetric lithium storage. Appl Surf Sci 437:384–389

Dong Y, Wu ZS, Ren W, Cheng HM, Bao X (2017) Graphene: a promising 2D material for electrochemical energy storage. Sci Bull 10:724–740

Eom W, Kim A, Park H, Kim H, Han TH (2016) Graphene-mimicking 2D porous Co_3O_4 nanofoils for lithium battery applications. Adv Func Mater 26(42):7605–7613

Fan G, Li F, Evans DG, Duan X (2015) Catalytic applications of layered double hydroxides: recent advances and perspectives. Chem Soc Rev 45(50):7040–7066

Gao S, Sun Y, Lei F, Liang L, Liu J, Bi W, Pan B, Xie Y (2015) Ultrahigh energy density realized by a single-layer β-$Co(OH)_2$ all-solid-state asymmetric supercapacitor. ChemInform 46(8):12789

Gao G, O'Mullane AP, Du A (2017) 2D MXenes: a new family of promising catalysts for the hydrogen evolution reaction. Acs Catal 7(1):494–500

Ghidiu M, Lukatskaya MR, Zhao MQ, Gogotsi Y, Barsoum MW (2014) Conductive two-dimensional titanium carbide 'clay' with high volumetric capacitance. Nature 516(7529):78–81

Gong M, Li Y, Wang H, Liang Y, Wu JZ, Zhou J, Wang J, Regier T, Wei F, Dai H (2013) An advanced Ni-Fe layered double hydroxide electrocatalyst for water oxidation. J Am Chem Soc 135(23):8452–8455

Gu Z, Atherton JJ, Xu ZP (2015) Hierarchical layered double hydroxide nanocomposites: structure, synthesis and applications. Chem Commun 51(15):3024–3036

Guo YG, Hu JS, Wan LJ (2010) Nanostructured materials for electrochemical energy conversion and storage devices. Adv Mater 20(15):2878–2887

Han YF, Chen L, Ramesh K, Widjaja E, Chilukoti S, Surjami IK, Chen J (2008) Kinetic and spectroscopic study of methane combustion over α-Mn_2O_3 nanocrystal catalysts. J Catal 253(2):261–268

Hu B, Qin X, Asiri AM, Alamry KA, Al-Youbi AO, Sun X (2013) WS_2 nanoparticles–encapsulated amorphous carbon tubes: a novel electrode material for supercapacitors with a high rate capability. Electrochem Commun 28(1):75–78

Huang J-K, Li M, Wan Y, Dey S, Ostwal M, Zhang D, Yang C-W, Su C-J, Jeng U-S, Ming J (2018) Functional two-dimensional coordination polymeric layer as a charge barrier in Li–S batteries. ACS Nano 12(1):836–843

Ji W, Ji J, Cui X, Chen J, Liu D, Deng H, Fu Q (2015) Polypyrrole encapsulation on flower-like porous NiO for advanced high-performance supercapacitors. Chem Commun 51(36):7669–7672

Jiang L, Zeng H (2015) Comment on "Strongly luminescent monolayered MoS_2 prepared by effective ultrasound exfoliation." Nanoscale 7(10):4580–4583

Kaneti YV, Yue J, Jiang X, Yu A (2013) Controllable synthesis of ZnO nanoflakes with exposed $(10\bar{1}0)$ for enhanced gas sensing performance. J Phys Chem C 117(25):13153–13162

Kaneti YV, Tang J, Salunkhe RR, Jiang X, Yu A, Wu CW, Yamauchi Y (2017) Nanoarchitectured design of porous materials and nanocomposites from metal-organic frameworks. Adv Mater 29(12):1604898

Kaneti Y, Salunkhe R, Ni LWS, Young C, Jiang X, He Y, Kang Y, Sugahara Y, Yamauchi Y (2018) General template-free strategy for fabricating mesoporous two-dimensional mixed oxide nanosheets via self-deconstruction/reconstruction of monodispersed metal glycerate nanospheres. J Mater Chem A 6:5971–5983

Kannan PK, Late DJ, Morgan H, Rout CS (2015) Recent developments in 2D layered inorganic nanomaterials for sensing. Nanoscale 7(32):13293

Khazaei M, Arai M, Sasaki T, Estili M, Sakka Y (2014) Two-dimensional molybdenum carbides: potential thermoelectric materials of the MXene family. Phys Chem Chem Phys 16(17):7841–7849

Koski KJ, Cui Y (2013) The new skinny in two-dimensional nanomaterials. ACS Nano 7(5):3739–3743

Kumar et al. (2018) Two dimensional N-doped ZnO-graphitic carbon nitride nanosheets heterojunctions with enhanced photocatalytic hydrogen evolution, https://doi.org/10.1016/j.ijhydene.2017.09.113

Lei JC, Zhang X, Zhou Z (2015) Recent advances in MXene: preparation, properties, and applications. 物理学前沿(英文版) 10(3):276–286

Li L, Hu ZA, An N, Yang YY, Li ZM, Wu HY (2014) Facile synthesis of MnO_2/CNTs composite for supercapacitor electrodes with long cycle stability. J Phys Chem C 118(40):22865–22872

Li Z, An Y, Hu Z, An N, Zhang Y, Guo B, Zhang Z, Yang Y, Wu H (2016) Preparation of a two-dimensional flexible MnO_2/graphene thin film and its application in a supercapacitor. J Mater Chem A 4(27):10618–10626

Li L, Jiang G, Sun R, Cao B (2017a) Two-dimensional porous Co_3O_4 nanosheets for high-performance lithium ion batteries. New J Chem 41(24):15283–15288

Li Z, Zhang W, Wang H, Yang B (2017b) Two-dimensional perovskite $LaNiO_3$ nanosheets with hierarchical porous structure for high-rate capacitive energy storage. Electrochim Acta 258:561–570

Li ZY, Akhtar MS, Bui PTM, Yang OB (2017c) Predominance of two dimensional (2D) Mn_2O_3 nanowalls thin film for high performance electrochemical supercapacitors. Chem Eng J 330:1240–1247

Lian P, Dong Y, Wu Z-S, Zheng S, Wang X, Sen W, Sun C, Qin J, Shi X, Bao X (2017) Alkalized Ti_3C_2 MXene nanoribbons with expanded interlayer spacing for high-capacity sodium and potassium ion batteries. Nano Energy 40:1–8

Liang X, Garsuch A, Nazar LF (2015a) Sulfur cathodes based on conductive MXene nanosheets for high-performance lithium-sulfur batteries. Angew Chem 127(13):3979–3983

Liang H, Meng F, Cabán-Acevedo M, Li L, Forticaux A, Xiu L, Wang Z, Song J (2015b) Hydrothermal continuous flow synthesis and exfoliation of NiCo layered double hydroxide nanosheets for enhanced oxygen evolution catalysis. Nano Lett 15(2):1421–1427

Liu G, Wang L, Sun C, Yan X, Wang X, Chen Z, Smith SC, Cheng H-M, Lu GQ (2009) Band-to-band visible-light photon excitation and photoactivity induced by homogeneous nitrogen doping in layered titanates. Chem Mater 21(7):1266–1274

Liu Y, Wang W, Huang H, Gu L, Wang Y, Peng X (2014) The highly enhanced performance of lamellar WS2 nanosheet electrodes upon intercalation of single-walled carbon nanotubes for supercapacitors and lithium ions batteries. Chem Commun 50(34):4485–4488

Liu Y, Ma Y, Guang S, Ke F, Xu H (2015) Polyaniline-graphene composites with a three-dimensional array-based nanostructure for high-performance supercapacitors. Carbon 83:79–89

Liu W, Luo X, Bao Y, Liu YP, Ning G-H, Abdelwahab I, Li L, Nai CT, Hu ZG, Zhao D (2017a) A two-dimensional conjugated aromatic polymer via C-C coupling reaction. Nat Chem 9(6):563

Liu W, Ulaganathan M, Abdelwahab I, Luo X, Chen Z, Rong Tan SJ, Wang X, Liu Y, Geng D, Bao Y (2017b) Two-dimensional polymer synthesized via solid-state polymerization for high-performance supercapacitors. ACS Nano 12(1):852–860

Lukatskaya MR, Mashtalir O, Ren CE, Dall'Agnese Y, Rozier P, Taberna PL, Naguib M, Simon P, Barsoum MW, Gogotsi Y (2013) Cation intercalation and high volumetric capacitance of two-dimensional titanium carbide. Science 341(6153):1502–1505

Ma TY, Cao JL, Jaroniec M, Qiao SZ (2016) Interacting carbon nitride and titanium carbide nanosheets for high-performance oxygen evolution. Angew Chem 128(3):1150–1154

Mei J, Zhang L (2015) Facile and economic synthesis of nitrogen doped graphene/manganese dioxide composites in aqueous solution for energy storage devices. Mater Lett 143:163–166

Mei J, Liao T, Kou L, Sun Z (2017) Two-dimensional metal oxide nanomaterials for next-generation rechargeable batteries. Adv Mater 29(48):1700176

Mitchell E, Gupta RK, Mensahdarkwa K, Kumar D, Ramasamy K, Gupta BK, Kahol P (2014) Facile synthesis and morphogenesis of superparamagnetic iron oxide nanoparticles for high-performance supercapacitor applications. New J Chem 38(9):4344–4350

Moses PG, Mortensen JJ, Lundqvist BI, Nørskov JK (2009) Density functional study of the adsorption and van der Waals binding of aromatic and conjugated compounds on the basal plane of MoS_2. J Chem Phys 130(10):104709

Muller GA, Cook J, Kim HS, Tolbert SH, Dunn B (2015) High performance pseudocapacitor based on 2D layered metal chalcogenide nanocrystals. Nano Lett 15(3):1911

Mutlu Z, Wickramaratne D, Bay HH, Favors ZJ, Ozkan M, Lake R, Ozkan CS (2014) Synthesis, characterization, and electronic structure of few-layer $MoSe_2$ granular films. Phys Status Solidi (a) 211(12):2671–2676

Naguib M (2014) Two-dimensional materials: 25th anniversary article: MXenes: a new family of two-dimensional materials. Adv Mater 26(7):982

Naguib M, Kurtoglu M, Presser V, Lu J, Niu J, Heon M, Hultman L, Gogotsi Y, Barsoum MW (2011) Two-dimensional nanocrystals produced by exfoliation of Ti3 AlC2. Adv Mater 23(37):4248–4253

Naguib M, Mashtalir O, Carle J, Presser V, Lu J, Hultman L, Gogotsi Y, Barsoum MW (2012) Two-dimensional transition metal carbides. ACS Nano 6(2):1322

Peng L, Xiong P, Ma L, Yuan Y, Zhu Y, Chen D, Luo X, Lu J, Amine K, Yu G (2017) Holey two-dimensional transition metal oxide nanosheets for efficient energy storage. Nat Commun 8:15139

Poizot P, Laruelle S, Grugeon S, Dupont L, Tarascon J (2010) Nano-sized transition-metal oxides as negative-electrode materials forlithium-ion batteries. ChemInform 32(3):496–499

Pumera M, Sofer Z, Ambrosi A (2014) Layered transition metal dichalcogenides for electrochemical energy generation and storage. J Mater Chem A 2(24):8981–8987

Qian J, Peng Z, Wang P, Fu X (2016) Bulk fabrication of WS_2 nanoplates: investigation on the morphology evolution and electrochemical performance. ACS Appl Mater Interfaces 8(26):16876

Reddy ALM, Shaijumon MM, Gowda SR, Ajayan PM (2010) Multisegmented Au-MnO_2/carbon nanotube hybrid coaxial arrays for high-power supercapacitor applications. J Phys Chem C 117(42):658–663

Ren CE, Zhao MQ, Makaryan T, Halim J, Boota M, Kota S, Anasori B, Barsoum MW, Gogotsi Y (2016) Back cover: porous two-dimensional transition metal carbide (MXene) flakes for high-performance li-ion storage. Chemelectrochem 3(5):847

Ren Y, Zhu J, Wang L, Liu H, Liu Y, Wu W, Wang F (2018) Synthesis of polyaniline nanoparticles deposited on two-dimensional titanium carbide for high-performance supercapacitors. Mater Lett 214:84–87

Sakthivel S, Kisch H (2003) Photocatalytic and photoelectrochemical properties of nitrogen-doped titanium dioxide. ChemPhysChem 4(5):487–490

Shao M, Ning F, Zhao J, Wei M, Evans DG, Duan X (2012) Preparation of Fe_3O_4@SiO_2@layered double hydroxide core-shell microspheres for magnetic separation of proteins. J Am Chem Soc 134(2):1071

Shimamoto K, Tadanaga K, Tatsumisago M (2013) All-solid-state electrochemical capacitors using MnO_2/carbon nanotube composite electrode. Electrochim Acta 109(11):651–655

Song F, Hu X (2014) Exfoliation of layered double hydroxides for enhanced oxygen evolution catalysis. Nat Commun 5:4477

Soundiraraju B, George BK (2017) Two-dimensional titanium nitride (Ti_2N) MXene: synthesis, characterization, and potential application as surface-enhanced Raman scattering substrate. ACS Nano 11(9):8892–8900

Sreedhara M, Vasu K, Rao C (2014) Synthesis and characterization of few-layer nanosheets of GaN and other metal nitrides. Zeitschrift für anorganische und allgemeine Chemie 640(14):2737–2741

Su S, Sun H, Cao W, Chao J, Peng H, Zuo X, Yuwen L, Fan C, Wang L (2016) Dual-target electrochemical biosensing based on DNA structural switching on gold nanoparticle-decorated MoS_2 nanosheets. ACS Appl Mater Interfaces 8(11):6826

Tan C, Cao X, Wu XJ, He Q, Yang J, Zhang X, Chen J, Zhao W, Han S, Nam GH (2017) Recent advances in ultrathin two-dimensional nanomaterials. Chem Rev 117(9):6225

Ting LRL, Deng Y, Ma L, Zhang YJ, Peterson AA, Yeo BS (2016) Catalytic activities of sulfur atoms in amorphous molybdenum sulfide for the electrochemical hydrogen evolution reaction. ACS Catal 6(2):861–867

Tu CC, Lin LY, Xiao BC, Chen YS (2016) Highly efficient supercapacitor electrode with two-dimensional tungsten disulfide and reduced graphene oxide hybrid nanosheets. J Power Sources 320:78–85

Tuček J, Kemp KC, Kim KS, Zbořil R (2014) Iron-oxide-supported nanocarbon in lithium-ion batteries, medical, catalytic, and environmental applications. ACS Nano 8(8):7571–7612

Umek P, Gloter A, Pregelj M, Dominko R, Jagodič M, Jagličić Z, Zimina A, Brzhezinskaya M, Potočnik A, Filipič C (2009) Synthesis of 3D hierarchical self-assembled microstructures formed from α-MnO_2 nanotubes and their conducting and magnetic properties. J Phys Chem C 113(33):14798–14803

Urbankowski P, Anasori B, Makaryan T, Er D, Kota S, Walsh PL, Zhao M, Shenoy VB, Barsoum MW, Gogotsi Y (2016) Synthesis of two-dimensional titanium nitride Ti_4N_3 (MXene). Nanoscale 8(22):11385–11391

Urbankowski P, Anasori B, Hantanasirisakul K, Yang L, Zhang L, Haines B, May SJ, Billinge SJ, Gogotsi Y (2017) 2D molybdenum and vanadium nitrides synthesized by ammoniation of 2D transition metal carbides (MXenes). Nanoscale 9(45):17722–17730

Zhang W, Li X, Liang J, Tang K, Zhu Y, Qian Y (2016) One-step thermolysis synthesis of two-dimensional ultrafine Fe_3O_4 particles/carbon nanonetworks for high-performance lithium-ion batteries. Nanoscale 8(8):4733–4741

Wang Q, O'Hare D (2012) Recent advances in the synthesis and application of layered double hydroxide (LDH) nanosheets. Chem Rev 112(7):4124–4155

Wang PP, Sun H, Ji Y, Li W, Wang X (2014) Three-dimensional assembly of single-layered $MoS_{(2)}$. Adv Mater 26(6):964–969

Wang T, Shi L, Tang J, Malgras V, Asahina S, Liu G, Zhang H, Meng X, Chang K, He J (2016) A Co_3O_4-embedded porous ZnO rhombic dodecahedron prepared using zeolitic imidazolate frameworks as precursors for CO_2 photoreduction. Nanoscale 8(12):6712

Wei T-Y, Chen C-H, Chien H-C, Lu S-Y, Hu C-C (2009) A cost-effective supercapacitor material of ultrahigh specific capacitances: spinel nickel cobaltite aerogels from an epoxide-driven sol-gel process. Adv Mater 22(3):347–351

Wen Y, Rufford TE, Chen X, Li N, Lyu M, Dai L, Wang L (2017) Nitrogen-doped $Ti_3C_2T_x$ MXene electrodes for high-performance supercapacitors. Nano Energy 38:368–376

Wu ZS, Feng X, Cheng HM (2014) Recent advances in graphene-based planar micro-supercapacitors for on-chip energy storage. Natl Sci Rev 1(2):277–292

Xia J, Huang X, Liu LZ, Wang M, Wang L, Huang B, Zhu DD, Li JJ, Gu CZ, Meng XM (2014) CVD synthesis of large-area, highly crystalline $MoSe_2$ atomic layers on diverse substrates and application to photodetectors. Nanoscale 6(15):8949–8955

Xiao Y, Cao M (2015) Carbon-anchored MnO nanosheets as an anode for high-rate and long-life lithium-ion batteries. Acs Appl Mater Interfaces 7(23):12840–12849

Xiao Y, Hwang J-Y, Sun Y-K (2016a) Transition metal carbide-based materials: synthesis and applications in electrochemical energy storage. J Mater Chem A 4(27):10379–10393

Xiao X, Song H, Lin S, Zhou Y, Zhan X, Hu Z, Zhang Q, Sun J, Yang B, Li T (2016b) Scalable salt-templated synthesis of two-dimensional transition metal oxides. Nat Commun 7:11296

Xiao X, Yu H, Jin H, Wu M, Fang Y, Sun J, Hu Z, Li T, Wu J, Huang L (2017) Salt-templated synthesis of 2D metallic MoN and other nitrides. ACS Nano 11(2):2180–2186

Xie J, Li S, Zhang X, Zhang J, Wang R, Zhang H, Pan B, Xie Y (2014) Atomically-thin molybdenum nitride nanosheets with exposed active surface sites for efficient hydrogen evolution. Chem Sci 5(12):4615–4620

Xie X, Ao Z, Su D, Zhang J, Wang G (2015) MoS_2/graphene composite anodes with enhanced performance for sodium-ion batteries: the role of the two-dimensional heterointerface. Adv Func Mater 25(9):1393–1403

Xie X, Zhao MQ, Anasori B, Maleski K, Ren CE, Li J, Byles BW, Pomerantseva E, Wang G, Gogotsi Y (2016) Porous heterostructured MXene/carbon nanotube composite paper with high volumetric capacity for sodium-based energy storage devices. Nano Energy 26:513–523

Xu M, Liang T, Shi M, Chen H (2013) Graphene-like two-dimensional materials. Chem Rev 113(5):3766–3798

Xu X, Song H, Lin S, Ying Z, Zhan X, Hu Z, Qi Z, Sun J, Bo Y, Li T (2016) Scalable salt-templated synthesis of two-dimensional transition metal oxides. Nat Commun 7:11296

Yan J, Fan Z, Wei T, Qian W, Zhang M, Wei F (2010) Fast and reversible surface redox reaction of graphene–MnO composites as supercapacitor electrodes. Carbon 48(13):3825–3833

Yao H, Zhang F, Zhang G, Luo H, Liu L, Shen M, Yang Y (2018) A novel two-dimensional coordination polymer-polypyrrole hybrid material as a high-performance electrode for flexible supercapacitor. Chem Eng J 334:2547–2557

Yu J, Martin BR, Clearfield A, Luo Z, Sun L (2015) One-step direct synthesis of layered double hydroxide single-layer nanosheets. Nanoscale 7(21):9448

Yu P, Cao G, Yi S, Zhang X, Li C, Sun X, Wang K, Ma Y (2018) Binder-free 2D titanium carbide (MXene)/carbon nanotube composites for high-performance lithium-ion capacitors. Nanoscale 10(13)

Zhang H (2015) Ultrathin two-dimensional nanomaterials. ACS Nano 9(10):9451–9469

Zhang Y, Zhuo Q, Lv X, Ma Y, Zhong J, Sun X (2015) NiO-Co_3O_4 nanoplate composite as efficient anode in Li-ion battery. Electrochim Acta 178:590–596

Zhao Y, Watanabe K, Hashimoto K (2012) Self-supporting oxygen reduction electrocatalysts made from a nitrogen-rich network polymer. J Am Chem Soc 134(48):19528–19531

Zhao Y, Li B, Wang Q, Gao W, Wang CJ, Wei M, Evans DG, Duan X, O'Hare D (2014a) NiTi-Layered double hydroxides nanosheets as efficient photocatalysts for oxygen evolution from water using visible light. Chem Sci 5(3):951–958

Zhao M-Q, Ren CE, Ling Z, Lukatskaya MR, Zhang C, Van Aken KL, Barsoum MW, Gogotsi Y (2014b) Flexible MXene/carbon nanotube composite paper with high volumetric capacitance. Adv Mater 27(2):339–345

Zhao G, Huang X, Wang X, Connor P, Li J, Zhang S, Irvine JS (2015a) Synthesis and lithium-storage properties of MnO/reduced graphene oxide composites derived from graphene oxide plus the transformation of Mn(VI) to Mn(II) by the reducing power of graphene oxide. J Mater Chem A 3(1):297–303

Zhao M, Wang Y, Ma Q, Huang Y, Zhang X, Ping J, Zhang Z, Lu Q, Yu Y, Xu H (2015b) Ultrathin 2D metal–organic framework nanosheets. Adv Mater 27(45):7372–7378

Zhao MQ, Sedran M, Ling Z, Lukatskaya MR, Mashtalir O, Ghidiu M, Dyatkin B, Tallman DJ, Djenizian T, Barsoum MW (2015c) Synthesis of carbon/sulfur nanolaminates by electrochemical extraction of titanium from Ti_2SC. J Electroanal Chem 127(16):4892–4896

Zheng W, Zhang P, Chen J, Tian WB, Zhang YM, Sun ZM (2018) In situ synthesis of CNTs@Ti_3C_2 hybrid structures by microwave irradiation for high-performance anodes in lithium ion batteries. J Mater Chem A 6(8):3543–3551

Zhi G, Liu B, Wang S, Huo L, Gao Z, Zhang J (2017) A novel two-dimensional hierarchical Mo_2C/CN hybrid fabricated via ionothermal route as a robust electrocatalyst for hydrogen evolution reaction. Electrochim Acta 258:970–978

Zhu YJ, Chen F (2014) Microwave-assisted preparation of inorganic nanostructures in liquid phase. Chem Rev 114(12):6462–6555

Zhu Y, Cao C, Zhang J, Xu X (2015) Two-dimensional ultrathin $ZnCo_2O_4$ nanosheets: general formation and lithium storage application. J Mater Chem A 3(18):9556–9564

Chapter 5
Synthesis of Three-Dimensional Nanomaterials

Initially, the research for nanomaterial focuses on the methods of synthesis (Zhang et al. 2017). As further development, new technology and method have been springing up. Faced with various preparation methods, it is difficult to unify classification especially for the synthesis of 3D nanomaterials. In this chapter, it is intended to utilize some classic synthetic methods as examples to demonstrate the methodology and regulation. Herein, we classify it as 7 sections such as Chemical Precipitation Method, Sol-Gel Method, Hydrothermal Method, Solvothermal Method, Thermal Decomposition Method, Microemulsion Method, and Chemical Vapor Deposition Method.

5.1 Chemical Precipitation Method

Due to low cost, simplification and short synthesis cycle, Chemical Precipitation Method, a widespread way, stands out in synthesis of nanomaterial. Hence, without any surfactants and templates, Pang et al. successfully obtained $NH_4CoPO_4 \cdot H_2O$ nano/micro-material (the sketch is shown in Fig. 5.1) (Pang et al. 2012). Ammonium chloride or ammonium phosphate was used as ammonium ion, cobalt chloride, and ammonium phosphate was used as cobalt ion and phosphate anion, respectively. They concluded that viscosity of solvent exerts effects on the ion diffusion rate and architecture growth or assemble. For instance, applying water-polyethylene glycol mixture, they identified that the whole microflower was assembled by nanoplates jointed at short rectangular edges. If compared to the common solvents (e.g., water or ethanol), glycerine is more viscous. The glycerine solvent with high viscosity may slow down the rate of ion diffusion, thus controlling the growth or assembling of particles. When using a water-polyethylene glycol mixed solvent, the morphology of $NH_4CoPO_4 \cdot H_2O$ (M4) is microflowers assembled by many nanoplates, with relatively uniform diameters of 3–5 mm, joined at their short rectangular edges. However, when we used pure water as solvent, hierarchical architecture morphology

H. Pang et al., *Synthesis of Functional Nanomaterials for Electrochemical Energy Storage*,
https://doi.org/10.1007/978-981-13-7372-5_5

Fig. 5.1 **a** The crystal architecture of $NH_4CoPO_4 \cdot H_2O$ unit cell; **b** The simple scheme of the $NH_4CoPO_4 \cdot H_2O$ nano/micromaterials' growth; **c** The possible charge/discharge process. Reprinted from Ref. Pang et al. (2012), copyright 2012, with permission from The Royal Society of Chemistry. SEM images of **d** MHCF and **e–g** MnOx-MHCF at different magnifications (TEM images of MnOx-MHCF at the top-right corner of **f**). Reprinted from Ref. Zhang et al. (2016a, b), copyright 2016, with permission from Wiley

$NH_4CoPO_4 \cdot H_2O$ of sizes in the range 20–30 mm (M5, shown in Fig. 5.1d) can be obtained.

This work, based on Chemical Precipitation Method, explored the influence from different solvents, making provision for the new generation of supercapacitor nanomaterials.

After that, in order to pursue the diversity, Du et al. combined the conventional $CoNi_2S_4$ (abbreviated as CNS) nanoparticles and graphene (GR) via the Chemical Precipitation approach (Du et al. 2014). According to a certain weight ratio (wt%) of synthesized GR (GR-CNS = 5: 100), as-synthesized CNS nanoparticles and GR sheets were mixed in 20 mL of absolute ethanol at room temperature. Subsequently, the mixture was ultrasonicated for 30 min until there was no obvious particulate matter. After 24 h of magnetic stirring, the turbid liquid was centrifuged and dried in a vacuum oven. With 1%, 3%, 5%, 10%, and 30% loaded amount of GR, the $CoNi_2S_4$/graphene nanocomposite (CNS@1%GR, CNS@3%GR, CNS@5%GR, CNS@10%GR, and CNS@30%GR) were set to make a comparison.

In the field of electronic devices and chemical sensors, it is undeniable that polypyrrole (PPy), a type of organic polymer, has gained popularity recently due to its great stability and controllable electrochemical conductivity. Besides, its various advantages have stimulated the application in supercapacitor electrode. Therefore, through a mild condition with $FeCl_3$ as oxidant, Zhao et al. succeeded synthesizing polypyrrole nanowires in order to further research its property in supercapacitor (Zhao et al. 2016). In the representative procedure, MO (0.3 g) was dissolved in 200 mL deionized water in a round bottom flask. PPy (0.70 mL) was then added to form the mixture A. Ferric chloride hexahydrate (2.7 g) was dissolved separately in 23 mL deionized water to prepare an oxidant solution B was added dropwise to the mixture A in a 300 mL beaker. The polymerization proceeded at 25 °C and was stirred for 24 h. The resultant samples were collected by filtration and then washed with Soxhlet extractor, water, and ethanol were used as medium, respectively, to remove the oxidant, MO, and oligomers. Finally, the PPy powder was dried in a vacuum oven at 60 °C for 12 h.

Meanwhile, Li et al. prepared 3D flower-like Ni/Co-LDHs microspheres and explored the synthetic mechanism of the LDHs microspheres (Li et al. 2016). They took reactant concentrations, reaction time, and pH into consideration. At the beginning, Ni^{2+}/Co^{2+} reacted with OH^- to generate nickel/cobalt hydroxide monomer nuclei and then formed the original nanoparticles (step I). The fresh nanoparticles tended to aggregate and produce larger particles because of the existence of high surface energy from nanoscale. Here, a slow but steady reaction rate was sharply controlled by the slow-released OH^- adjusting agent of NH_4Cl and NaOH in the solution, resulting in the separation of nucleation and growth. That was fatal factor for synthesizing high-quality crystal (step II). As continued to aggregate, the generated particles underwent dehydration process due to hydroxylation reaction (step III). Next, they continued to crystallize along c-axis and develop into the petal-like high crystalline LDHs nanosheets (step IV). Finally, because of Ostwald ripening, the self-assembly occurred and resulted in the obtained coordination nanosheets further reacted to a stable Ni/Co-LDHs flower-like microsphere structure (step V). All things considered, it can be seen that this facile but effective strategy can efficiently control the structure of Ni/Co-LDHs by varying the reaction condition, employed to prepare other LDHs nano/micro materials.

At the end of this portion, what we want to introduce is the application of metal oxide frameworks (MOFs) (Yang et al. 2017). As known, MOFs, a category of highly porous materials, have attracted researchers' attention a lot since 1990s (Li et al. 1999). Facing two intrinsic problems of MOFs (insufficient mechanical/chemical stability and low electrochemical conductivity), researchers took measures to enhance the pseudocapacitance of MOFs by doping metal oxides into the system with Chemical Precipitation Method. In the report, as starting materials, MOF-manganese hexacyanoferrate hydrate (MHCF) nanocubes had an obvious change that manganese in the framework reacted to manganese oxides (2016a, b). MHCF nanocubes were synthesized via a simple chemical precipitation method. Besides, MHCF was dissolved in the mixed solvent system with 10 mL C_2H_5OH and 15 mL H_2O under stirring to get a homogeneous solution. At the same time, NH_4F was dissolved in distilled water. Then, the obtained NH_4F solution was added to the MHCF solution. Subsequently, the obtained mixture was stirred for 20 min and then the product was obtained after being washed. As shown in Fig. 5.1e–g, the nanoflower modified the surface of each individual MHCF cubes evenly.

5.2 Sol-Gel Method

In the field of advanced materials, the sol-gel process is an approach to produce solid materials derived from the small molecules. This method is often utilized for the synthesis of oxides, especially the titanium oxides or silicon oxides. This method involves the transformation of monomers into a colloidal solution (sol) that works as the precursor for an integrated network (or gel) of either network polymers or discrete particles.

This method can date from mid-nineteenth century. Ebelman found the glassy appearance of SiO_2 after the hydrolysis of tetraethyl orthosilicate. Meanwhile, the water in SiO_2 gel replaced by organic solvent was found by Graham, attracting chemists' attention. After a long period of exploration, the colloid chemistry was gradually formed. Sol-gel has been an effective method for preparing ultrafine particles.

Hydrothermal carbonization (HTC) of carbohydrate is one of the promising candidates for the fabrication of carbon-based materials, since it can provide a facile, inexpensive, and eco-friendly route (Enterría and Figueiredo 2016). Nevertheless, the preparation of highly porous carbon materials through a direct HTC process is quite difficult. Herein, the solubilizing technique of micelles was proposed by Wang and co-workers to direct the HTC of fructose, utilizing an amphiphilic block copolymer, that is, poly-(4-vinylpyridine)-block-poly-(ethylene glycol) (P4VP-PEG) shown in Fig. 5.2a, as a structure-directing agent (Wang et al. 2014). Based on this strategy, porous hierarchical carbon materials with tunable properties were successfully manufactured. The P4VP-PEG micelles could solubilize fructose and meanwhile restrict the appearance of primary carbon domains upon the sol-gel processes. Moreover, the size of micelle could be easily controlled through adjusting the synthetic conditions. Thereby, the particle size of the resultant carbon materials was effectively tuned at the range of 20–100 nm based on the direction of the initial micelle size.

Besides, uniform and porous yolk-shelled carbon spheres (YS-CSs) with a hierarchical structure have been successfully prepared via a novel gradient sol-gel strategy with surfactant-directing co-assembly by utilizing cationic surfactant cetyltrimethylammonium bromide (CTAB) as a template, resorcinol-formaldehyde (RF) as a carbon source along with tetraethoxysilane (TEOS) as an assistant pore-forming reagent (by Wang et al. 2015). On the basis of the abovementioned observations, the synthesis of yolk-shell-structured carbon spheres undergoing a gradient sol-gel process with surfactant-directing co-assembly has been proposed (Fig. 5.2b). Firstly, in an alkaline solution with ethanol-water mixed solvent, RF precursor forms negatively charged emulsion droplets through the hydrogen bonding of ammonia, water, alcohol, resorcinol, and formaldehyde, silicate oligomers hydrolyzed from TEOS are also negatively charged, while CTAB with positive charged (CTA^+) can bind to the surface of the formed RF emulsion droplets and silicate oligomers via electrostatic interactions. Under the catalysis of ammonia molecules, the cross-linkage of RF droplets and silicate oligomers occurs but with different rates. In our case, the hydrolysis polymerization of RF is much faster than that of TEOS at the sol-gel

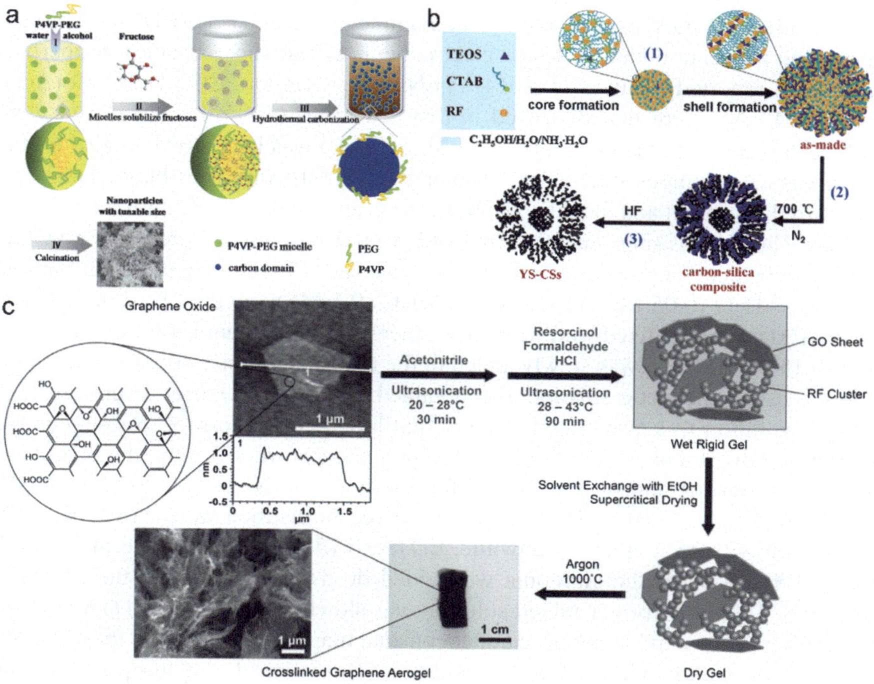

Fig. 5.2 **a** Mechanism for the fabrication of hierarchically porous aggregated carbon materials. Reprinted from Ref. Wang et al. (2014), copyright 2014, with permission from The Royal Society of Chemistry. **b** The schematic illustration of the three-step synthetic processes of the yolk-shell porous carbon spheres: (1) gradient sol-gel process, (2) carbonization process, (3) silica removal by HF. Reprinted from Ref. Wang et al. (2015), copyright 2015, with permission from Elsevier. **c** Diagram of the GO-RF aerogel preparation process. Reprinted from Ref. Lim et al. (2015a, b), copyright 2015, with permission from Elsevier

process beginning. Thus only few silicate oligomer stakes part in the co-assembly of CTAB and RF through the electrostatic interaction, leading to the core formation resulted from the condensation and growth of the assembled hybrid CTAB/RF aggregates. With the sol-gel process prolonged, the concentration of RF emulsion was gradually decreased due to the RF consumed for the core growth, resulting in a slower polymerization rate which well matches the hydrolysis polymerization rate of TEOS. Thus the gradient hydrolysis and condensation of silicates begin. The silicate oligomers from the hydrolysis of TEOS together with RF emulsion droplets can interact with CTAB to co-assemble at the surface of the cores, thereby creating the hybrid silica/CTAB/RF shells and forming core-shell structured products. Obviously, the component differences between the core and shell strongly rely on the gradient sol-gel process of RF and TEOS under alkaline conditions.

Similarly, spinel $NiCo_2O_4$ was successfully synthesized by an easy sol-gel approach with the participation of three crucial chelating agents including citric acid

(CA), oxalic acid (OA) as well as ethylenediamine tetraacetic acid (EDTA) explored during the fabrication and the tested electrochemical and supercapacitor properties (Zhu et al. reported) (Zhu et al. 2014) (1) $Ni(Ac)_2 \cdot 4H_2O$ (0.05 M) and $Co(Ac)_2 \cdot 4H_2O$ (0.1 M) were first dissolved in distilled water (100 mL) and mixed well with each other. Citric acid solution (0.3 M, 100 mL) was later slowly added to the mixture at room temperature under constant magnetic stirring. The obtained mixture was stirred overnight and then the solution was evaporated at 80 °C until the gel was formed. The gel was subsequently dried and ground to obtain the precursor powder. Finally, the precursor powder was further annealed at 375 °C for 2 h in air. (2) Firstly, $Ni(Ac)_2 \cdot 4H_2O$ (0.05 M) and $Co(Ac)_2 \cdot 4H_2O$ (0.1 M) were dissolved in distilled water (100 mL) and mixed well with each other. Then, an ethanol solution of oxalic acid (0.15 M, 100 mL) was slowly added to the mixed solution at room temperature under constant magnetic stirring. The obtained mixture was stirred overnight and then the solution was evaporated at 80 °C until the gel was formed. Subsequently, the gel was dried and ground to obtain the precursor powder. Lastly, the precursor powder was further annealed at 320 °C for 2 h in air. (3) $Ni(Ac)_2 \cdot 4H_2O$ (0.05 M) and $Co(Ac)_2 \cdot 4H_2O$ (0.1 M) were first dissolved in distilled water (100 mL) and mixed well with each other. Meanwhile, EDTA (0.15 M) was dissolved in distilled water (100 mL), and the ammonia was added dropwise to maintain the constant pH at 6.5. Then, the former mixed solution was slowly added to the EDTA ammonium solution at room temperature under constant magnetic stirring and the constant pH = 6.5 was maintained via dropwise addition of ammonia during this process. The obtained mixture was stirred overnight and then the solution was evaporated at 80 °C until the gel was formed. The gel was subsequently dried and ground to obtain the precursor powder. Finally, the precursor powder was annealed for 2 h at 400 °C in air. The chelating agent is a particularly important factor that affects the size of particles, the different pore structures and specific surface area of the $NiCo_2O_4$, thereby leading to the distinctions of their electrochemical properties; such observations are of fundamental importance allowing the tailoring of electrochemical properties through careful choice of chelating agent.

NiO nanomaterials with three different morphologies were obtained via a sol-gel method and their morphology-dependent supercapacitive performance was further explored (Kim et al. reported) (Kim et al. 2013). The special three-dimensional (3D) nanoflower-shaped NiO has the highest pore volume, thus manifesting the superior supercapacitor properties. The nanopores in the flower-shaped nanostructures can greatly improve the contact and transport of the electrolyte, allowing for more 3D nanochannels in the NiO network, thus offering longer electron pathways.

When it comes to carbon material, graphene is worth mentioning. Two-dimensional (2D) graphene, with carbon atoms bonded in a hexagonal lattice, has attracted considerable attention within the scientific community for its outstanding properties such as its electrical and thermal conductivity (1738 S m^{-1} and ~5 × 10^3 W m^{-1} K^{-1} respectively), intrinsic carrier mobility (>2 × 10^5 cm^2 V^{-1} s^{-1}), theoretical surface area (~2600 m^2 g^{-1}), mechanical strength (~118 GPa), and elastic modulus (~1 TPa). Lim et al. have presented an approach for the ultrafast fabrication of graphene aerogels on the basis of the polycondensation of resorcinol, formaldehyde, and graphene oxide with HCl as the catalyst and acetonitrile as the solvent (shown in Fig. 5.2c) (Lim et al. 2015a, b). This synthetic approach greatly reduces the gelation time from many hours or even several days (using the traditional base-catalyzed route) to just 1–2 h, making it easier to produce graphene aerogels on a large scale within reasonable time and potentially reducing their cost. Conceivably, the capacitance of the graphene aerogels could be further improved by tailoring surface area and pore volume using CO_2 activation and varying the amount of GO loading; or by experimenting with aqueous electrolytes with smaller ion size, as opposed to the organic electrolyte used in this study.

Doping element is a good choice for enhancing the property of devices. For instance, Tian et al. synthesized restacking-inhibited N-doped graphene (GN) for the application of supercapacitors based on the sol-gel strategy in Fig. 5.3 (Tian et al.

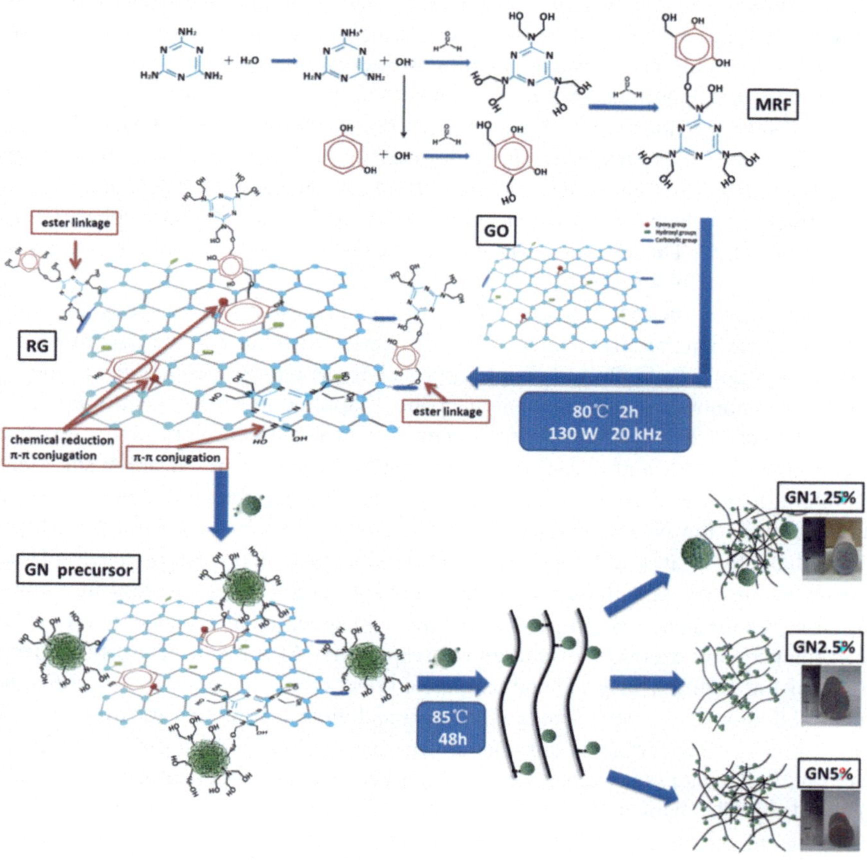

Fig. 5.3 Schematic illustration of in situ polymerization between MRF and GO. Reprinted from Ref. Tian et al. (2015), copyright 2015, with permission from Elsevier

2015). The as-prepared GN2.5% exhibits higher specific capacitance compared to the individual carbon materials and better long-periodic stability after thousands of cycles. Moreover, the conclusions are noteworthy: (1) GN2.5% shows excellent specific capacitance compared to that of MRF, due to the unique electron conduction in graphene; the disorder and defect in graphene sheets are overcome by removing oxygen-containing groups and restoring the conjugated structure; (2) Because N-doped sites could serve as active sites to enhance the affinities with carrier ions in the Faradaic reactions, the ample doped N would improve the pseudocapacitance, meanwhile the oxygen-containing groups could decrease the hydrophobicity of carbon materials and charge transport resistance; (3) Compared to RG and GNs with high GO content, the favorable morphology effect of GN2.5% is conducive to charges stored in the charging process and adsorption capacity of carrier ions; (4) The high capacitance is obtained from carbon material with larger surface area and narrower pores. And the micropore plays the more significant role than mesopore and specific surface in improving capacitance. In short, the approach proposed in this work could open up a general route to prepare the laminated and N-doped electrode material with highly promising application in energy storage.

Similarly, a rich N-doped porous carbon with large specific surface area as well as enhanced specific capacitance for supercapacitors was synthesized from poly(acrylic acid)/methylated melamine-formaldehyde resin through sol-gel process at ambient temperature for 24 h, followed by calcination and carbonization at 350 °C and 500 °C, respectively, for 1 h and KOH activation at 700 °C for 2 h under N_2 atmosphere (from Jiang et al. 2015). The specific surface area of C-700-1.5 with a nitrogen content of 8.8 wt% is up to 2674 $m^2\ g^{-1}$.

What is more, due to the synergy effect, more and more researchers pay attention to combing carbon material with metal oxide in order to achieve better nanomaterial. Dam et al. reported a facile and simple approach to synthesize a composite of mesoporous NiO nanowires and graphene nanosheets applied in supercapacitor applications (Dam et al. 2014). A one-pot sol-gel method in a water/ethylene glycol mixture in combination with a graphene oxide was used to prepare the Ni precursor. Heat treatment in air was carried out to thermally reduce the graphene oxide to graphene and to convert the Ni precursor to NiO. NiO nanowires showed a rough surface, with a diameter of around 60 nm and are homogeneously deposited on the graphene sheets. In comparison with its counterparts, the NiO/graphene nanocomposite shows superior pseudocapacitive properties such as high specific capacitance, good cyclic performance, and excellent discharge rate capability. As a result of the synergistic effect of the addition of graphene as elastic conductive channels, the as-prepared material demonstrates better charge transport and more favorable ionic diffusion. In addition, a novel approach is introduced using sol-gel dip-coating method to fabricate ZnO/rGO/ZnO sandwich as flexible high-performance supercapacitor electrode materials by Ghorbani et al. (2017). This ZnO/rGO/ZnO paper shows amazing flexibility based on freestanding GO paper as a new substrate for deposition of ZnO thin films. Obtained ZnO thin film on GO paper leads to the dense layer with unaffected layered structure of GO paper. The sandwich papers show the most proper morphology and performance as well as homogeneous ZnO thin film formation on rGO paper when sol-gel process parameters were set to 1-PrOH as a solvent, TeA as a stabilizer, sol concentration of 0.2 M, deposition speed of 30 mm min^{-1}, and 10 deposited layers. We aspire that this method of sandwiching of metal oxide and GO paper into flexible electrodes will provide a new approach to the design of flexible electrode for supercapacitor applications. Besides, shown in Fig. 5.4a, Wu et al. controlled synthesized V_2O_5/MWCNT core/shell hybrid aerogels with different MWCNT contents through a facile mixed growth and self-assembly methodology in a sol-gel process (Wu et al. 2015). During the hydrolysis process, core/shell structure V_2O_5 coated MWCNTs and V_2O_5 nanowires were obtained simultaneously through in situ growth and preferred orientation growth, respectively. Then these two kinds of 1D nanowires are self-assembled into 3D V_2O_5/MWCNT hybrid aerogels.

As a result of the expensiveness of the precursors, process, and non-Eco friendliness, the design and development of an economic as well as highly active non-precious electrocatalyst for ethanol electro-oxidation is challenging In Fig. 5.4b,

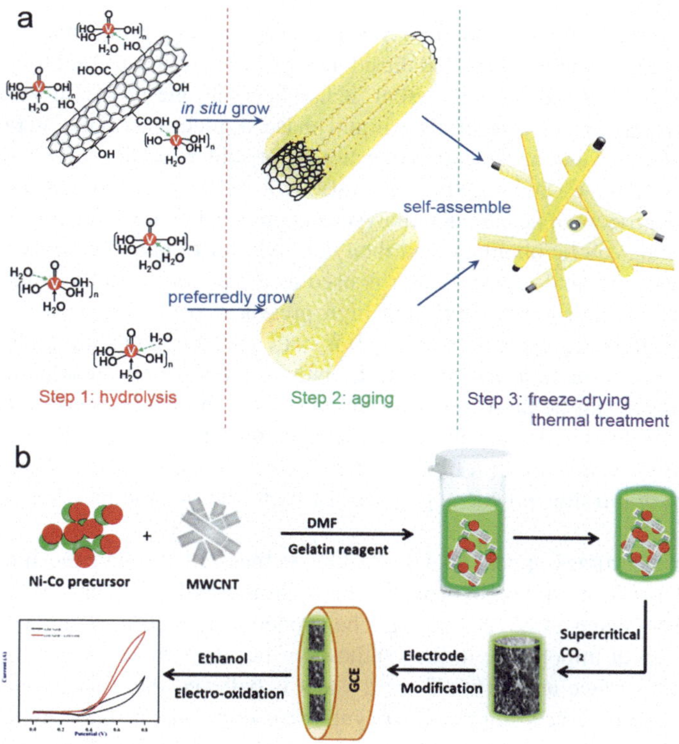

Fig. 5.4 **a** Illustration of the mixed growth and self-assembly methodology for controlled synthesis of 3D V_2O_5/MWCNT core/shell hybrid aerogels. Reprinted from Ref. Wu et al. (2015), copyright 2015, with permission from The Royal Society of Chemistry. **b** Preparation of $NiCo_2O_4$-MWCNT nanocomposite aerogel and electro-oxidation of ethanol. Reprinted from Ref. Jayaseelan et al. (2016), copyright 2016, with permission from Elsevier

a novel preparation of mesoporous $NiCo_2O_4$-MWCNT nanocomposite aerogels based on sol-gel technique is proposed (Jayaseelan et al. 2016). Multiwall carbon nanotube (MWCNT)-supported $NiCo_2O_4$ nanocomposite aerogels used as an efficient catalyst for the ethanol electro-oxidation was reported. The MWCNTs exhibit an interconnected fibrous network with uniform dispersion of $NiCo_2O_4$ nanoparticles. The effects of MWCNT concentration on the ethanol electro-oxidation of MWCNT/$NiCo_2O_4$ aerogels are studied. Jayaseelan et al. found that using a proper loading of MWCNTs allowed it to reach higher current densities for the oxidation of ethanol in an alkaline media. The highly porous and fibrous MWCNT/$NiCo_2O_4$ aerogels are the promising electro-catalysts for the oxidation of a direct ethanol fuel cell.

5.3 Hydrothermal Method

Since 1982, Hydrothermal Method for the preparation of ultrafine powder by hydrothermal reaction has attracted the attention of both domestic and foreign countries (Sasikala et al. 2017). At present, it has become one of the most important methods in preparing nanomaterials in liquid phase. A general term for chemical reactions in fluids, such as water/solution or water vapor, under conditions of high temperature (100 ~ 1000 °C) and high pressure (10 ~ 100 MPa), is Hydrothermal Method. Through the process of accelerated dialysis and physical control, the improved inorganic materials can be obtained, and then the high purity nanomaterials are got by filtration, washing, and drying. Water is used as a reaction medium in the reaction environment of high temperature and high pressure. Water serves as a chemical composition and participate in reaction. It is both a solvent and a mineralizer. It can also be used as a medium for pressure transfer, so that usually insoluble or insoluble substances can be dissolved, and the reaction can also be recrystallized. The general principle of hydrothermal synth plays an important part esis is to ensure that the reactive materials are in a high reactive state (Jain et al. 2016). The minimum particle size of ultrafine powders, prepared by hydrothermal method, has reached up to nanoscale.

Superfine particles appear in UH-Fe_3O_4 synthesized through the intermediate of the Fe(II)-ETA complex, hydrolysis and hydrothermal treatment by Wang et al. in Fig. 5.5A (Wang et al. 2013a, b). The moderate reduction of ETA and ultrasound is of great importance in the synthesis of Fe_3O_4 particles which have a very high specific surface area (165.05 $m^2\ g^{-1}$). It is believed that this study is able to open the way for a new approach to synthesize other metal oxide nanomaterials for supercapacitor applications with excellent electrochemical performances. Meanwhile, nitrogen-doped graphene has been a hot research topic. Further utilizing the excellent properties of graphene macroscopic assemblies are crucial. Herein, Chen et al. first reported an unique and convenient hydrothermal process for controllable synthesis and structural adjustment of the nitrogen-doped graphene hydrogel (GN-GH). By using organic amine and graphene oxide as precursors, readily scaled-up for

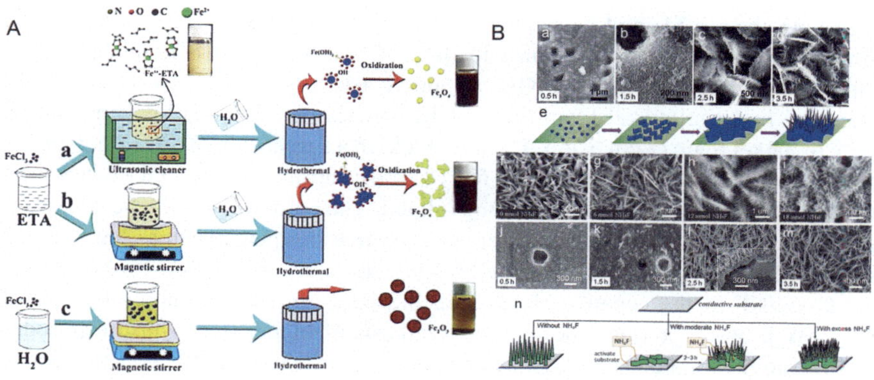

Fig. 5.5 **A** Preparation of routes toward **a** UH-Fe_3O_4, **b** H-Fe_3O_4, and **c** H-Fe_2O_3. Reprinted from Ref. Wang et al. (2013a, b), copyright 2013, with permission from The Royal Society of Chemistry. **B** SEM images of the precursor obtained with the assistance of 12 mmol NH4F at various reaction stages by setting the reaction time to **a** 0.5 h, **b** 1.5 h, **c** 2.5 h, **d** 3.5 h; **e** scheme of the products at various reaction stages; **f–i** SEM images of the products obtained at different concentration of NH4F. **j–m** Morphologies of the precursor at different reaction times without adding NH4F; **n** proposed mechanism for the effect of NH4F on morphology construction. Reprinted from Ref. Chen et al. (2013a, b), copyright 2015, with permission from The Royal Society of Chemistry

mass production of nitrogen-doped graphene hydrogel can be obtained (Chen et al. 2013a, b). The organic amine is not only as nitrogen sources to obtain the nitrogen-doped graphene but also as an important modification to control the assembly of graphene sheets in the 3D structures. Inner structure of the GN-GHs and the content of nitrogen in the graphene are easily adjusted by organic amine. As you can see, the supercapacitor performance of the typical product could be remarkably enhanced in this way. In addition, large-scale production of three-dimensional (3D) hierarchical porous nickel cobaltate nanowire cluster arrays, derived from nanosheet arrays with robust adhesion on Ni foam, were successfully prepared by a simple hydrothermal method by Chen et al. in Fig. 5.5B (Chen et al. 2013a, b). On the basis of the morphology evolution upon reaction time, a possible formation process is reported. The role of NH_4F in formation of the structure has also been investigated on the basis of different NH_4F amounts. This unique structure significantly enhances the electroactive surface areas of the $NiCo_2O_4$ arrays, which brings out better interfacial/chemical distributions at the nanoscale, fast ion, electron transfer as well as good strain accommodation.

Furthermore, Wang et al. developed a supercapacitor electrode in composition with 3D self-supported Co_3O_4@$CoMoO_4$ core-shell architectures which are directly grown on nickel foam (Wang et al. 2016). Co_3O_4 nanocones were grown vertically on the nickel foam as the core and $CoMoO_4$ nanosheets were further engineered to immobilized on the surface of the nanocones as the shell. The unique architecture takes advantage of a large interfacial area, numerous channels for rapid diffusion of electrolyte ions, fast electron transport, and the high electrochemical activity from both the Co_3O_4 and $CoMoO_4$. Zhai et al. reported an effective and simple strategy to prepare large areal mass loading of MnO_2 on porous graphene gel/Ni foam (denoted as MnO_2/G-gel/NF) for supercapacitors (SCs) (Zhai et al. 2013). Graphene hydrogel/nickel foam (G-gel/NF) was simply obtained by immersing a piece of nickel foam (2 cm × cm, 1 × mm thick) in a suspension of GO (3 mg mL^{-1}) and subsequently heating at 180 °C for 12 h. During the process, GO nanosheets were reduced into G-gel and coated on Ni foam. The color of Ni foam became dark after hydrothermal reaction. Moreover, an asymmetric supercapacitor (ASC) on the basis of MnO_2/G-gel/NF (MnO_2 mass: 6.1 mg cm^{-2}) used as the positive electrode and G-gel/NF as the negative electrode achieved a remarkable energy density of 0.72 mW h cm^{-3}. Rational design and synthesis of binder-free hybrid electrodes with hierarchical core-shell structures has been regarded as an effective strategy to improve the electrochemical performance of the supercapacitors by Xu et al. (Wu et al. 2017). In this work, Co_3O_4@$NiCo_2O_4$ core-shell structures with high yield are successfully fabricated on flexible carbon cloth using a facile hydrothermal method for ultra-long Co_3O_4 nanowires and a chemical bath for $NiCo_2O_4$ nanoflakes.

5.4 Solvothermal Method

Instead of water as a medium for organic solvent, the method, similar to the hydrothermal synthesis principle for preparing nanometer material, is called Solvothermal Method. Replacing water with non-aqueous solvent not only expands the application scope of hydrothermal technology, but also realizes the reactions which cannot be achieved under normal conditions, including preparing materials with metastable structure.

As Pang et al. emphasized the significance of morphology, Wang et al. developed a simple morphology-controlled synthesis for hierarchical α-$Ni(OH)_2$ microspheres (Wang et al. 2018; Zhang et al. 2016a, b). Three kinds of α-$Ni(OH)_2$ microspheres with different structures were prepared by one-step surfactant-assisted solvothermal method in Fig. 5.6A, using water, ethanol, and their mixture as solvent, respectively. First, $Ni(NO_3)_2 \cdot 6H_2O$ and sodium dodecyl sulfate (SDS) were dissolved in 50 mL deionized water, 50 mL ethanol and their mixed solution (25 mL deionized water and 25 ml ethanol), respectively, followed by addition of 3.0 g urea with continuous stirring. Then the solutions were transferred into 100 mL Teflon-lined stainless autoclaves and held at 110 °C for 15 h. After cooled down to room temperature naturally, the precipitates were collected by centrifugation and washed several times with deionized water and absolute ethanol. Finally, the products were dried to constant weight at 60 °C. The samples obtained from three different solvents were denoted as $Ni(OH)_2$-a, $Ni(OH)_2$-b, and $Ni(OH)_2$-c correspondingly. The effects of solvent on the structure, morphology, and capacitive performance of α-$Ni(OH)_2$ were investigated. Among the as-prepared samples, $Ni(OH)_2$-c microspheres, obtained from the mixed solvent (ethanol solution), are composed of hierarchical flower-like nanosheets and possess the roughest surface and the largest specific surface area (101.3 m^2 g^{-1}).

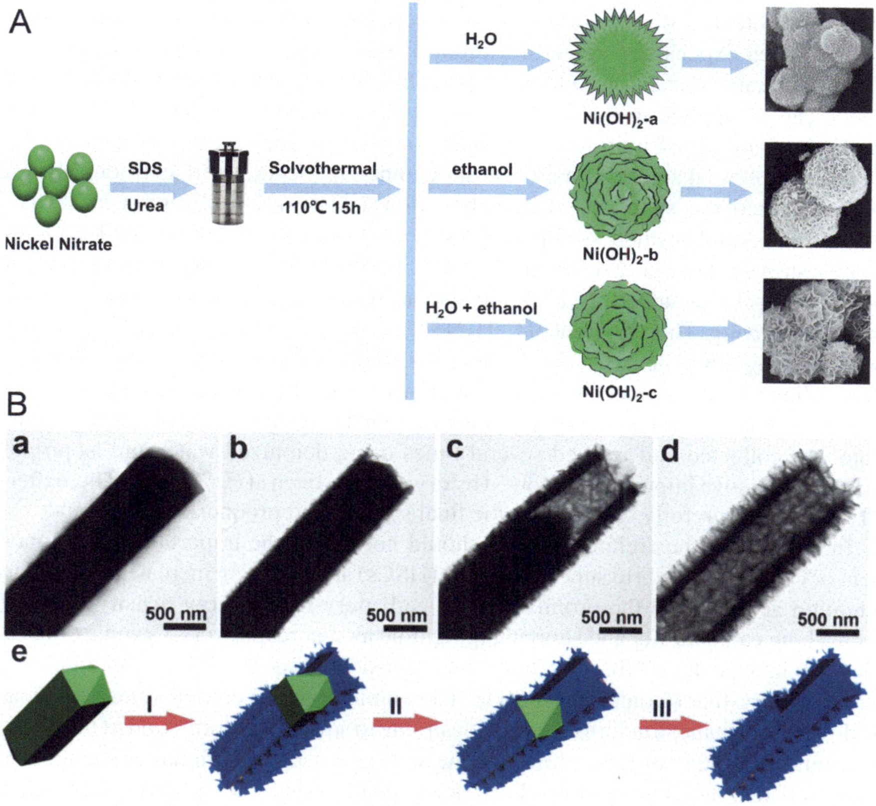

Fig. 5.6 **A** Schematic of hierarchical α-$Ni(OH)_2$ microspheres synthesis. Reprinted from Ref. Wang et al. (2018), copyright 2018, with permission from Elsevier. **B** Representative TEM images of the samples obtained after reaction for **a** 4 h, **b** 6 h, **c** 10 h and **d** 12 h. **e** The corresponding schematic illustration of the formation process for hierarchical NiCo-LDH tetragonal microtubes. Reprinted from Ref. Ma et al. (2016), copyright 2016, with permission from The Royal Society of Chemistry

In addition, ultra-small, crystalline, as well as dispersible NiO nanoparticles were successfully synthesized by Fominykh for the first time. What's more, they are promising candidates as catalysts for electrochemical water oxidation (Fominykh et al. 2014). Very small nickel oxide nanocrystals can be made, through a solvothermal reaction in tert-butano, the sizes of which are tunable from 2.5 to 5 nm and have a narrow particle size distribution. Even after drying, the crystals are perfectly dispersible in ethanol, which shows stable transparent colloidal dispersions. The structure of the nanocrystals is in accord with phase-pure stoichiometric nickel(II) oxide with a partially oxidized surface exhibiting Ni(III) states. The 3.3 nm nanoparticles demonstrate a remarkably high turn-over frequency of 0.29 s^{-1}

at an overpotential of $\gamma = 300$ mV for electrochemical water oxidation, outperforming even expensive rare earth iridium oxide catalysts. The unique features of these NiO nanocrystals provide great potential for the novel composite materials which can be applied in the field of (photo) electrochemical water splitting. Such as the preparation of uniform hole-conducting layers for organic solar cells, the dispersed colloidal solutions may also show other applications. In addition, Mishra et al. reported the synthesis of carnation flower-like SnS_2 (CF-SnS_2) *by* a one-step solvothermal method as supercapacitor electrodes for potential application in energy storage devices (Mishra et al. 2017). In a typical procedure, 1.54 mmol of $SnCl_4 \cdot 5H_2O$ was added to an 80 mL isopropyl alcohol solution under vigorous magnetic stirring. Then, 6 mmol of C_2H_5NS was added to the solution. After vigorous stirring using a magnetic bar for 15 min, the solution was transferred to a 100 mL Teflon-lined stainless steel autoclave, which was sealed tightly and held in a convection oven at 180 °C for 24 h. After cooling to room temperature, the obtained precipitate was collected and washed several times using deionized water and isopropyl alcohol to remove impurities, followed by drying in an oven at 60 °C in air. Thereafter, CF-SnS_2 was carefully collected as the final synthesized product.

In addition to the architecture, we should not ignore the importance of component or composite. Hybrid supercapacitors (HSCs) are attracting increasing scientific attention as they hold the promise to realize battery-level energy density together with a long calendar life and short charging time as supercapacitors. Complex mixed metal oxides such as $NiCo_2O_4$ have been considered as promising candidates to be the battery-like (faradaic) electrode. The employment of complex non-spherical hollow micro-/nanostructures is highly desirable to address the limitation of the redox reaction at the near surface. However, the synthesis faces additional challenges. A novel self-templated method had been developed by Ma et al. to fabricate ultrathin mesoporous $NiCo_2O_4$ nanosheet-assembled hierarchical tetragonal microtubes via thermal transformation of a nickel–cobalt layered double hydroxide (NiCo-LDH) precursor (Ma et al. 2016). Of note, the NiCo-LDH microtubes were prepared by a one-pot solvent-heat method, involving the formation of metal acetate hydroxide solid prisms and a subsequent hollowing process (demonstrated in Fig. 5.6B). Due to their complex structures with ultrathin subunits, these hierarchical $NiCo_2O_4$ microtubes deliver exquisite electrochemical performance, including high specific capacitance as well as a long lifespan as a battery-type electrode for HSCs. The research finding will provide a possibility to eliminate the performance gap between batteries and supercapacitors.

Similarly, Huang et al. devised a route synthesize NiMn LDH@rGO composite with 3D flower-like architecture by a facile one-step solvothermal way (Huang et al. 2018). By manipulating the composition of the precursors, shell structures with different loadings of NiMn LDH on graphene nanosheet were obtained. The whole process was shown in Fig. 5.7a. Electrochemical characterization showed that LDH/rGO-4 with flower-like hierarchical architectures exhibited superior specific capacitance and high rate performance owing to the large accessible specific surface area and high capacitance. The assembled ASC constructed from LDH/rGO-4 and

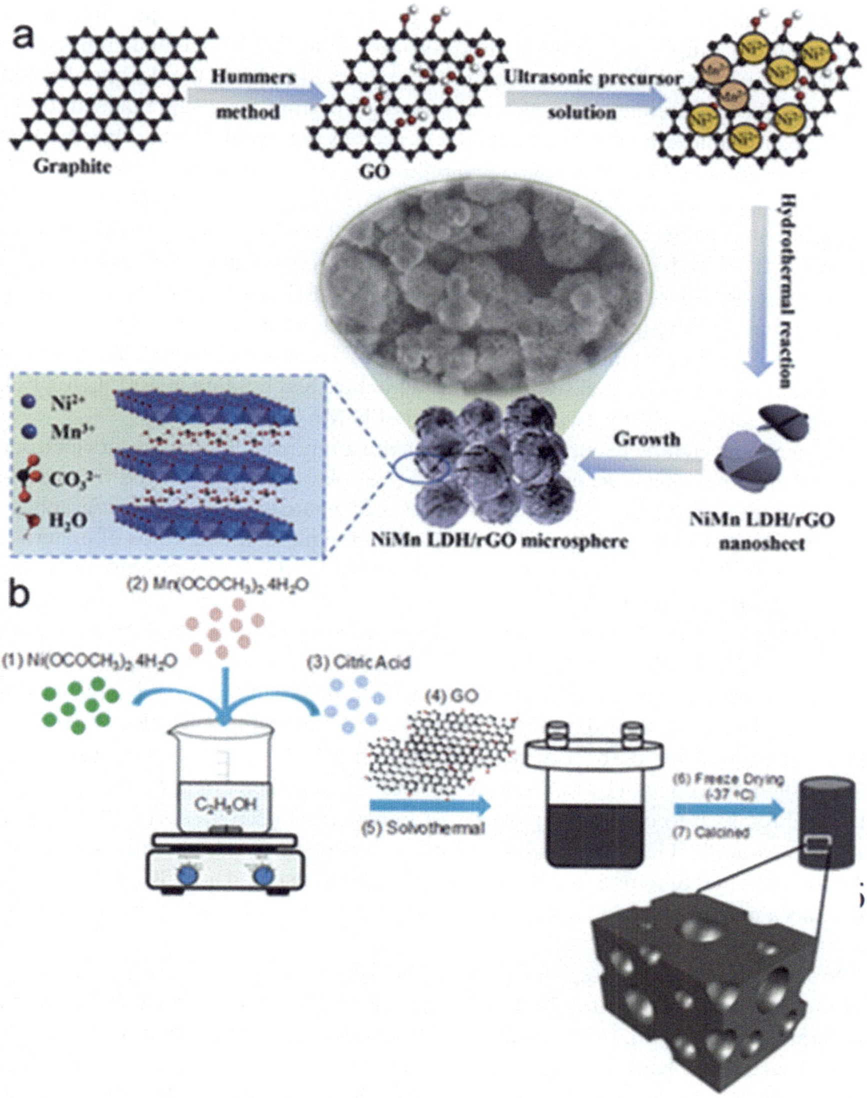

Fig. 5.7 **a** Schematic illustration of the formation of self-assembled 3D flower-like architectured NiMn LDH/rGO microspheres composite by a facile one-step solvothermal synthesis process. Reprinted from Ref. Huang et al. (2018), copyright 2018, with permission from Elsevier. **b** Schematic diagram of the preparation steps of $NiMn_2O_4$/rGOH. Reprinted from Ref. Ngo et al. (2017), copyright 2017, with permission from The Royal Society of Chemistry

rGO electrodes revealed an impressively high energy and power output. It is anticipated that the simple and facile method for fabricating 3D flower-like architecture of NiMn LDH/rGO microspheres composite may shed light on the design of novel hierarchical structure electrode materials for high-performance supercapacitors. In Fig. 5.7b, a simple solvothermal method was used by Ngo et al. to hybridize Nickel-manganese spinel oxide ($NiMn_2O_4$) and reduced graphene oxide hydrogel (rGOH), and a highly porous 3D structure was fabricated (Ngo et al. 2017). $NiMn_2O_4$/rGOH was fabricated using a facile one-step solvothermal method. Typically, 0.250 g of $Ni(OCOCH_3)_2 \cdot 4H_2O$, 0.490 g of $(CH_3COO)_2Mn \cdot 4H_2O$, and 1.267 g of $C_6H_8O_7 \cdot H_2O$ were dissolved in 10 mL of ethanol (C_2H_5OH) with vigorous stirring for 30 min. The resulting mixture was then added to 5 mL of an aqueous GO slurry with a 5 mg mL^{-1} solid content that was synthesized by Hummer's method. Transferring the mixture to a Teflon-lined stainless steel autoclave and holding at 160 °C for 12 h to fabricate a porous 3D networked structure of $NiMn_2O_4$/rGOH.

An activated pyrene decorated graphene nanocomposite, applied as an electrode material in supercapacitor, was synthesized via solvothermal followed by heating method (reported by Li et al. 2018). A series of nanostructured pyrene/graphene composites (PGCs) were prepared by controlling the mass of two kinds of materials in the composite process. The reduced graphene oxide (rGO) was synthesized using pervious literature report (Wu et al. 2015). Pyrene decorated graphene composites were synthesized by solvothermal process and followed by activated process. As a typical procedure, 25 mg of pyrene and 75 mg of GO were dropped into 75 mL of ethanol and then sonication treatment for 30 min. When the mixed solution was uniformly dispersed, the mixture was poured into a Teflon-lined stainless steel autoclave (50 mL in volume) for solvothermal reaction at 160 °C for 16 h. The suspension was washed with ethanol by repeated centrifugation, and then dried in a vacuum oven at 70 °C for 12 h to obtain the powders of pyrene/rGO ($PGO_{1/3}$). The $PGO_{1/3}$ was then heated at 95 °C under argon flow for 2 h to activate pyrene-based composites. The final product of pyrene/graphene composites was denoted as $PGC_{1/3}$. This study provides a new method and a way to learn from it for the study of graphene/organic electrode materials.

Via a solvothermal route, mesoporous $NiCo_2S_4$ nanoparticles were synthesized as remarkable supercapacitor electrode materials by Zhu et al. (Li et al. 2018). In the preparation of $NiCo_2S_4$, 0.195 mmol of $Ni(AC)_2 \cdot 4H_2O$ and 0.39 mmol of $Co(AC)_2 \cdot 4H_2O$ were dissolved in 40 mL of ethylene glycol and stirred for 30 min. Then, 1.17 mmol of thiourea was introduced into the mixed solution under stirring. After stirring for another 30 min, the solution was transferred to a Teflon-lined stainless steel autoclave and heated in an oven at 180 °C for 24 h. The black precipitate was collected by centrifugation after being cooled to room temperature, then washed with distilled water and ethanol several times, followed by drying at 60 °C for 24 h to obtain the products.

5.5 Thermal Decomposition Method

As a chemical decomposition method, thermal decomposition (also called thermolysis) is caused by heat. The temperature of substance decomposition is the temperature of chemically decomposing (Dutta et al. 2014). The reactions are usually endothermic because heat is needed to break chemical bonds in the compound during decomposition. If decomposition is exothermic enough, a positive feedback loop can be created, resulting in thermal runaway and possible explosion. Thermal decomposition method is divided into solid-phase thermal decomposition method, gas-phase thermal decomposition method, and self-propagation high temperature-synthesis method (SHS method). The first one is more commonly used in the field of nanomaterial synthesis so we mainly introduce it in this chapter.

Based on the thermal decomposition and phase transition at elevated temperature, the $Cu_4(SO_4)(OH)_6$ precursor synthesized via a hydrothermal method could be converted into 3D nanoporous CuO by Yang et al. (2018) CuSO4 $\cdot$ $5H_2O$ (0.15 M) and urea (0.052 M) were dissolved in 50 mL deionized water under magnetic stirring to prepare a clear solution. Then, blue transparent solution was transferred into 100 mL Teflon-lined stainless steel autoclave, and subjected to hydrothermal treatment at 90 °C in an electric blast drying oven for 3 h. The precipitates were collected by centrifugation after cooling down to room temperature, then washed with deionized water and elthyl alcohol for several times and dried at 60 °C for 12 h. The precursor was calcined in air at 750 °C for 2 h.

The development of 3D graphene frameworks is usually limited by complex preparation procedure and low specific surface area. Via a facile suitable method including quick thermal decomposition from sodium chloroacetate, Zhu et al. successfully synthesized 3D graphene frameworks (GFs) with desired large specific surface area (up to 1018 $m^2\ g^{-1}$), which is much larger than those of sodium acetate (Zhu et al. 2016). The ruthenium trichloride ($RuCl_3 \cdot nH_2O$) and zirconium chloride ($ZrCl_4$) were dissolved in ethanol with the proportion of Ru:Zr = 4:6 to prepare a mixture of precursor solution. The precursor solution was deposited on one side of the Ti-substrates (the Ti-substrates were pretreated by sand-blasting, etching in 10wt% boiling oxalic acid for 2 h, then washed with distilled water and finally dried at 80–100 °C). The samples were annealed at 280, 290, 300, 325, 350, 400, and 450 °C for 10 min, respectively, after evaporation of the solvent. The operation was repeated until the desired RuO_2 loading (1.0 mg cm^{-2}) was obtained. A final annealing at the corresponding temperature for 1 h was applied to complete the treatment. The chlorine element in the sodium chloroacetate has a strong induction capacity of in situ activation, which can regulate the formation of graphene in one step during pyrolysis. Besides, electrode coatings of 40%RuO_2–60%ZrO_2 binary oxide were formed on Ti substrate by thermal decomposition method with the annealing temperature varying from 280 to 450 °C (from Ma et al. 2017). The XRD and TEM analyses showed that 290 °C was the critical crystallization temperature of RuO_2.

In addition, Wang et al. had successfully utilized a nanocellulose-assisted low temperature (lower than 500 °C) thermal treatment method to synthesize reduced

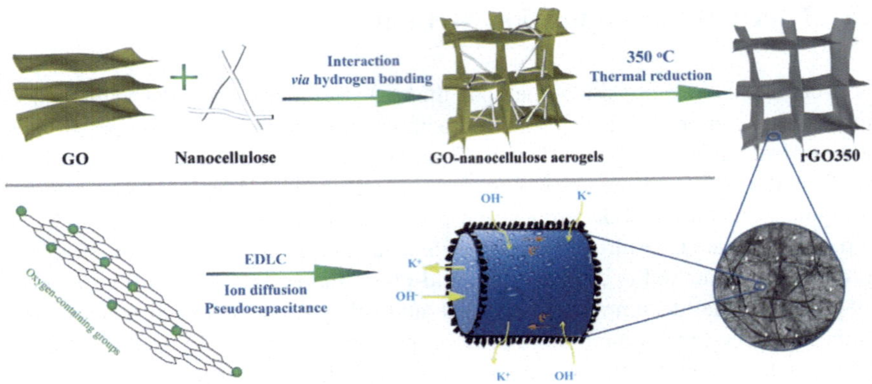

Fig. 5.8 Schematic illustration of the synthesis steps and the ionic and electronic transport mechanisms of the rGO capacitor electrode within the KOH electrolyte. Reprinted from Ref. Wang et al. (2017a, b), copyright 2017, with permission from Elsevier

graphene oxide (rGO) aerogels. The nanocelluloses could promote graphene oxide (GO) solution gelating, which was beneficial to prepare GO aerogel with low concentration dispersion (2.85 mg mL^{-1}). After the thermal decomposition, the residual nanofibers served as spacer, not only preventing the restacking of graphene sheets, but also integrating with rGO sheets to give a special carbon-based aerogel with many defects (holes) in Fig. 5.8 (Wang et al. 2017a, b). When the temperature was beyond 350 °C, thermal decomposition of nanocellulose seems to begin. Therefore its presence in rGO sheets was in the form of amorphous carbon nanofibers. The rGO aerogels, which were synthesized at 350 °C, provide an optimal balance in high content of CO-type functional groups, wide interlayer spacing, and large defects content.

Porous Co_3O_4 materials were synthesized through a solid-state conversion method from freshly prepared Co-MOF crystal by Meng et al. (2013). The unique Co-MOF crystal mentioned here was prepared by the specific chemical coordination of the auxiliary ligand 4,40-bipyridine (bpy) and the carboxylic ligand azobenzene-3,5,40-tricarboxylic acid (H_3ABTC). 2D bilayer structural intermediates were constructed, and then formed a 3D polycatenation supramolecular array architecture by means of hydrogen bonding interactions and p-p stacking. Finally, porous Co_3O_4 particles were obtained by facile thermolysis of the Co-MOF crystals, including two-step calcination process (removing of carbon residue).

5.6 Microemulsion Method

Microemulsion Method is a method of forming microemulsion with mixed metal salts and certain precipitant, and controlling the nucleation and growth of colloidal particles in a smaller microarea (called microreactor), and then obtaining ultrafine particles by heat treatment. Microemulsion is usually a transparent and isotropic thermodynamic stability system composed of surfactants, cosurfactant (usually alcohols), and oil (usually hydrocarbons). Microemulsion is surrounded by tiny monolayer of surfactant and cosurfactant. The size of it is between several to several dozens of nanometers. These tiny pools separate each other and form microreactor.

The microemulsion particles are constantly moving like Brown motion. When different particles collide with each other, the surface active agent and the hydrocarbon chain of cosurfactant can permeate each other. At the same time, the material in the "pool" can go through the interface into another particle. For example, the conductance and percolation phenomenon of microemulsion formed by anionic surfactants is due to the continuous conduction chain formed by the cations in the pool passing through the microemulsion interface and the transition between particles. The properties of this material exchange in microemulsion make it possible to make chemical reactions in the "pool". Obviously, it is another effective technique for the preparation of nanomaterials.

Usually, two kinds of reactants are dissolved in two microemulsions, respectively, which are made up of exactly the same, and then mixed under certain conditions. The two kinds of reactants encounter and react with each other through the exchange of materials, and the growth of the reaction products will be limited when the interfacial strength of the microemulsion is larger. If the size of microemulsion particles is controlled in a few nanometers, the reaction products are dispersed in different microemulsion "pools" in the form of nanoparticles. The study shows that the nanoparticles can be stable in the "pool". The nanoparticles are separated from the microemulsion by adding the mixture of water and acetone to the microemulsion through the overspeed centrifugation or the addition of the mixture of water and acetone to the completion of the reaction. Then the organic solvent is used to remove the oil and surfactants attached to the surface of nanoparticles. Finally, after drying at a certain temperature, the required nanomaterials can be obtained.

One of the main obstacles to develop graphene-based supercapacitors with high energy density is to maintain large ion-accessible surface area and high electrode density (Xu et al. 2015). The system of ionic liquid (IL)-surfactant microemulsion was developed by She et al., which was found to promote IL-filled micelles adsorpt spontaneously onto graphene oxide (GO) (She et al. 2017). This adsorption not only played an important role in distributing the IL over all available surface area, but also act a pivotal part in providing an aqueous formulation which can be slurry cast onto current collectors. As a result, a dense nanocomposite film of GO/IL/surfactant was left. The IL could act as a dual role of electrolyte and spacer, via reducing the GO and removing the surfactant (result from a low-temperature (360 °C) heat treatment).

Through a galvanostatic method, Zhang et al. electrochemically polymerized polythiophene (PTh) onto the multiwalled carbon nanotube (MWCNT) modified carbon paper in an oil-in-ionic liquid microemulsion (O/IL) (Zhang et al. 2014). An O/IL microemulsion was prepared by dispersing an oil phase with reactants into a continuous IL phase. It contains 1 mL *n*-hexane as the oil phase, where 0.3 mL thiophene monomer was dissolved, 2.7 g TritonX-100 as the non-ionic surfactant and 2 mL ionic liquids of (BMIM)PF6 as the continuous phase. After a moderate stirring, a transparent and uniform O/IL microemulsion was formed. Polythiophene was deposited on the multiwalled carbon nanotube in the prepared O/IL microemulsion by a galvanostatic procedure in a one-compartment cell with a two-electrode configuration at the constant current of 3 mA. The MWCNT modified carbon paper electrode of dimension (1 $\times$ 1 cm) was used as the working electrode, and a graphite rod with a diameter of 6 mm was used as counter electrode. After polymerization, PTh/MWCNT composite film was obtained after being washed with ethanol and deionized water, and then dried in air. The neat polythiophene was deposited on carbon paper in the same way to act as contrast tests for further characterization. As-prepared PTh/MWCNT composites had an interlaced framework morphology, at the same time, MWCNTs had been equably coated by PTh with thickness of 2–3 nm.

A novel, valid method has been researched for the fabrication of composites using manganese oxide (MnO_2) grown in situ on 3D graphene through reverse microemulsion (water-in-oil) method (Wei et al. 2016). A uniform coating of nanoscale MnO_2 layers could be surveyed on internal surface of 3D graphene, which might be advantageous for rapid ionic and electronic transport. Those electrochemical properties of MnO_2/3D graphene composites are able to be optimized by controlling composite structures and mass loading of MnO_2. Similarly, MnO_2/3D reduced graphene oxide (RGO) composites had been reached through reverse microemulsion (water/oil) method in Fig. 5.9 (Wei et al. 2015). Initially, the oil system was prepared by mixing 50 mL of cyclohexane (oil), 57.1 mL of isopropyl alcohol (cosurfactant), and 16.7 mL of OP-10 (surfactant). The intermingled solution had been kept stirring through one magnetic bar until it became transparent. 0.01 g of 3D RGO was joined in that solution, and then divided into two equal aliquots. Then, aqueous solutions of $KMnO_4$ (0.1 M) and $MnSO_4 \cdot H_2O$ (0.15 M) were added into the two oil system aliquots, respectively. Homogeneous reverse microemulsions (w/o) were prepared by stirring for 5 min. The reverse microemulsion of $KMnO_4$ was then dropped into the system containing $MnSO_4 \cdot H_2O$ under stirring and the reverse microemulsion reaction was carried out at a constant temperature of 28 °C for 14 h. The product (dark brown precipitate) had been separated through centrifugation and washed with ethanol and deionized water, which is in order to remove organic compounds and surfactant. The MnO_2/3D RGO composite was freeze dried for 48 h. MnO_2 nanoparticles (3–20 nm in diameter) with diverse morphologies had been prepared and dispersed in the same way on macropore surfaces of 3D RGO, providing channels for rapid ionic and electronic transport.

Furthermore, hybrid supercapacitors (battery-supercapacitor hybrid devices, HSCs) transfer high energy in seconds (excellent rate capability), along with

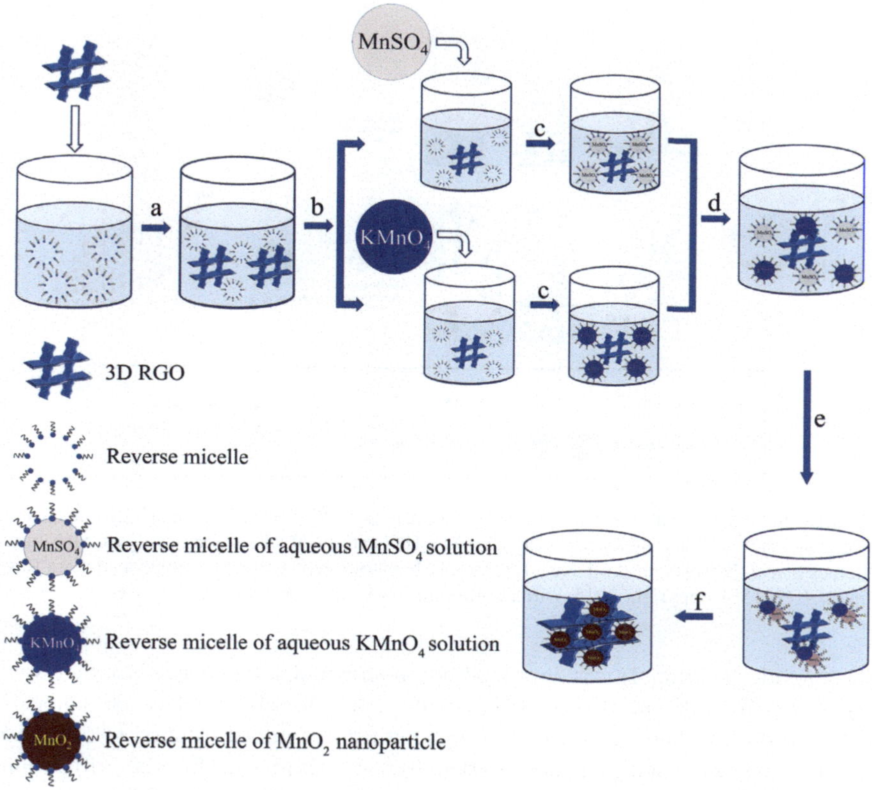

Fig. 5.9 Schematics to illustrate the procedure to fabricate the MnO_2/3D RGO composite via a reverse microemulsion reaction. **a** Mixing process of the 3D RGO and oil system; **b** solution divided into two equal aliquots; **c** emulsification; **d** mixing process of the reverse microemulsions of the $KMnO_4$ and $MnSO_4 \cdot H_2O$; **e** interfacial reaction; **f** reverse microemulsion reaction under stirring for 14 h. Reprinted from Ref. Wei et al. (2015), copyright 2015, with permission from Elsevier

steady catalytic performance. One of those pivotal constraints in exploiting high-performance HSCs is disequilibrium in terms of power performances between sluggish Faradaic lithium-intercalation anode and rapid non-Faradaic capacitive cathode. In order to work out similar challenges, Lim et al. synthesized Nb_2O_5@carbon core_shell nanocyrstals (Nb_2O_5@C NCs) as uprated anode materials with controlled crystalline phases [orthorhombic (T) and pseudohexagonal (TT)] through a convenient one-pot synthesis method, which is based on water-in-oil microemulsion system (Lim et al. 2015a, b). In terms of synthesis of *T*-Nb_2O_5@C NCs, microemulsion had been described in the following. The oil phase which was made up of 11.5 g Igepal CO-520 and 225 mL cyclohexane was intermixed with 1.25 mL 75 mM HCl (or HNO_3) aqueous solution and 3 mL ethanol. Afterwards, 0.375 mL of Niobium

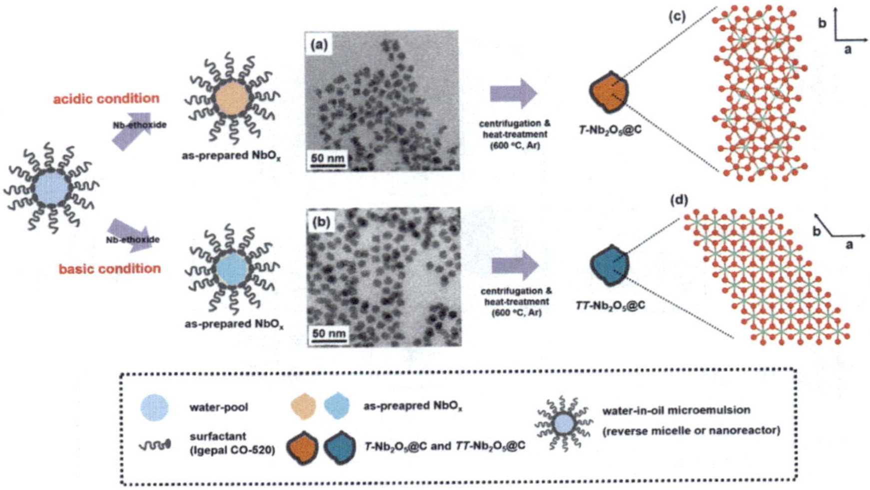

Fig. 5.10 Schematic diagram of the synthesis procedures of Nb_2O_5@C NCs and TEM images of as-prepared NbO_x NCs synthesized using 75 mM **a** HCl and **b** NaOH aqueous catalysts. Structural schemes of **c** T-Nb_2O_5 and **d** TT-Nb_2O_5 (green b, Nb atom; red b, O atom). Reprinted from Ref. Lim et al. (2015b), copyright 2015, with permission from American Chemical Society

(V) ethoxide (1.5 mmol) was appended into microemulsion under appropriate stirring at the environment of room temperature. After 20 min, synthesized nanoparticles had been isolated through centrifugation using a mixed solution of 1:1 (v/v) ether/n-hexane several times. Soon afterwards, materials reached from centrifugation step were kept drying at environment of 100 °C overnight. In the end, heat-treating resulting materials in Ar atmosphere at the environment of 600 °C for 2 h. In order to acquire carbon shell-free T-Nb_2O_5, carbon shell of T-Nb_2O_5@C was divided through calcination in O_2 at the environment of 450 °C for 2 h. The TT-Nb_2O_5@C and TT-Nb_2O_5 had been prepared under identical conditions as mentioned above. Nevertheless, a 75 mM NaOH (or KOH) aqueous solution had been used to take place of 75 mM HCl (or HNO_3) aqueous solution (Fig. 5.10).

5.7 Chemical Vapor Deposition Method

Chemical vapor deposition (CVD) uses volatile metal halides, hydride or organic metal compounds as raw materials, to produce the required compounds by chemical reaction, condenses rapidly under the protection of gas and prepare many kinds of nanomaterials. This method is the most attractive method to synthesize nanomaterials of high melting point inorganic compounds, also known as chemical vapor reaction.

As science and technology advance, CVD technology has been continuously innovating, including high-frequency CVD, plasma CVD, laser CVD, and so on. Most

CVD involves two or more than two kinds of gaseous reactants on the surface of the matrix, and the products are deposited on the substrate. Since 1980s, CVD technology has been gradually applied to the preparation of powdery, lump, and fibrous materials.

The preparation of nanomaterials by CVD has many advantages, such as homogeneous particles, high purity, small size, good diversity, high chemical reaction activity, controllable process, and continuous process.

For example, realization of a highly flexible, lightweight, and high-performance flexible supercapacitor was achieved using three-dimensional graphene on flexible graphite-paper. A simple and fast self-assembly approach was utilized for the uniform deposition of CVD-grown high-quality 3D graphene powders on flexible graphite-paper substrate by Ramadoss et al. (2017). Besides, Wang et al. described one convenient method to fabricate three-dimensional (3D) few-layer graphene/multiwalled carbon nanotube (MWNT) hybrid nanostructures on industrial grade metal foam foils (nickel foam) through a one-step ambient pressure chemical vapor deposition (APCVD) process (Wang et al. 2013a, b). Here are the details. 3D few-layer graphene/MWNT foams were grown through an ambient pressure chemical vapor deposition (APCVD) method though a mixture of acetylene and hydrogen on 1.0 mm thick nickel foam, which is typically used as current collector in the battery industry. Briefly, nickel foam is pretreated with diluted acetic acid and deionized (D.I.) water to ensure the surface is completely clean and free from oxidation. Next, the nickel foam is annealed at 800 °C under ambient pressure with the flow of H_2 and Ar for 45 min in order to release the residue stress in the foam, enlarge the average grain size, and also flatten the surface. After annealing, a mild reactive ion etching (RIE) O_2-plasma is used in the annealed nickel foam for 2 min and 2 nm Fe catalyst layer is deposited on the surface of plasma-treated nickel foam by e-beam evaporation. The as-prepared nickel foam is loaded into aquartz-tube furnace chamber, heated to 750 °C under ambient pressure in an Ar/H_2 (200:200 sccm) atmosphere, and annealed for 5 min. Acetylene is added to stimulate growth of graphene and CNTs simultaneously on nickel foam frame. After growth, the chamber is cooled to room temperature at an average cooling rate of 50 °C min^{-1}.

Meanwhile, high-porosity MnO_2 materials with mesoporous structure were prepared by convenient redox reaction, and polypyrrole (PPy) nano-films had been grown on synthesized mesoporous MnO_2 by chemical vapor deposition, which is in order to form a 3D nanocomposite structure (Wang et al. 2017a, b). Firstly, the as-prepared mesoporous MnO_2 powders were immersed into an ethyl alcohol solution containing 5 wt% ferrous chloride at room temperature for 20 min under tenth atmospheric pressure and then directly dried at 80 °C in atmospheric environment; as a result, ferrous ions were transformed into ferric ions. Secondly, the sample was spread out in a culture dish, which was heated to 80 °C in a vacuum chamber. After the pressure of the vacuum chamber reached one-tenth of the atmospheric pressure, the vapor of the pyrrole monomer was continuously introduced to the vacuum chamber, which was maintained at 0.15 atm. Finally, the deposition of PPy thin films lasted for 20 min to finish the synthesis of PMMO. In addition, Rengaraj et al. presented a non-enzymatic cholesterol sensor based on a nickel oxide (NiO) and high-quality graphene composites initially (Rengaraj et al. 2015). Graphene was grown on 25-mm-thick copper foil by a CVD technique using a mixture of methane and hydrogen. The copper foil was first annealed in hydrogen at ~1000 °C for 20 min. Graphene was then grown over methane and hydrogen at the same temperature for 10 min. A thin poly(methylmethacrylate) (PMMA) film was subsequently spin-coated on a graphene/copper substrate at 2000 rpm. After drying overnight at room temperature, the graphene on backside of Cu foil was divided through plasma etching for 2 min. Cu was then dissolved in ammonium persulfate aqueous solution, after which it was washed with DI water. The floating graphene/PMMA was scooped onto the SiO_2/Si substrate (1 × 1 cm), dried under vacuum overnight and baked at the environment of 150 °C for 10 min. Eventually, PMMA was dissolved in acetone. The one-step electrochemical synthesis of the NiO/graphene nanocomposite was carried out by scanning the potential of the graphene/GCE between −1.2 and 0 V *versus* SCE at a scan rate of 50 mV s^{-1} for 30 cycles in the environment of 0.1 M acetate buffer solution (ABS, pH 4.0) containing 10 mM $Ni(NO_3)_2$ as an electrolyte. After deposition, the electrode was flushed with distilled water, then kept annealed at the environment of 270 °C for 3 h in air.

Furthermore, Xia et al. attempted to meet requirements of universal design for high-performance supercapacitor electrodes through combining with strategies of lightweight substrate, porous nanostructure design, and conductivity modification. The whole process is shown in Fig. 5.11. To begin with, they fabricated highly porous electrodeposited 3D Ni films as the template for chemical vapor deposition (CVD)-grown 3D porous graphite foams. Besides, the 3D porous Ni films are acted as template for following the growth of CVD-graphite foams. After CVD and etching

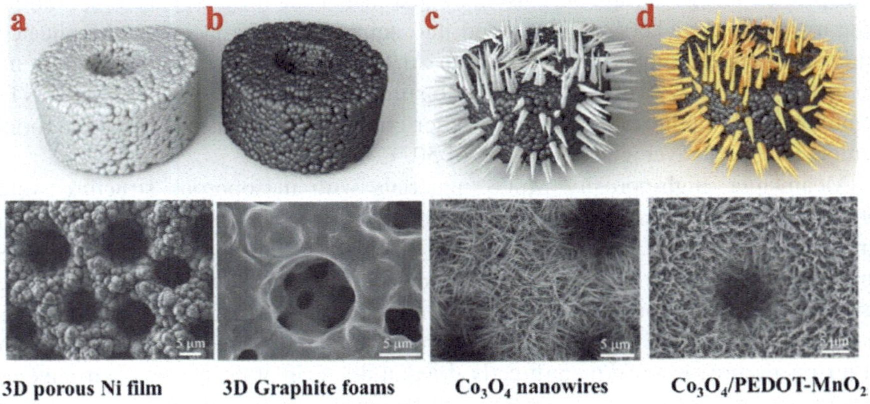

Fig. 5.11 Schematics of the fabrication process of thin 3D porous graphite foams (GF) and their integrated composites with Co_3O_4/PEDOT-MnO_2 core/shell nanowire arrays. **a** Porous Ni films as the substrate. **b** 3D GF by CVD growth from the Ni film substrate followed by etching. **c** Co_3O_4 nanowires by hydrothermal growth on the GF. **d** PEDOT-MnO_2 composite shell by co-electrodeposition. The bottom row is the corresponding SEM images of the structures. Reprinted from Ref. Xia et al. (2014), copyright 2014, with permission from American Chemical Society

Ni template, it presents that whole 3D porous structure of graphite foams is well preserved and still keeps pore size of 5–10 μm. They assembled Co_3O_4/PEDOT-MnO_2 core/shell nanowire arrays on the ultrathin 3D porous graphite foams by the combination of hydrothermal synthesis and anodic electrodeposition methods to form an integrated electrode for supercapacitor applications.

References

Chen Y, Qu B, Hu L, Xu Z, Li Q, Wang T (2013a) High-performance supercapacitor and lithium-ion battery based on 3D hierarchical NH_4F-induced nickel cobaltate nanosheet–nanowire cluster arrays as self-supported electrodes. Nanoscale 5:9812

Chen P, Yang JJ, Li SS, Wang Z, Xiao TY, Qian YH, Yu SH (2013b) Hydrothermal synthesis of macroscopic nitrogen-doped graphene hydrogels for ultrafast supercapacitor. Nano Energy 2:249–256

Dam DT, Wang X, Lee JM (2014) Graphene/NiO nanowires: controllable one-pot synthesis and enhanced pseudocapacitive behavior. ACS Appl Mater Interfaces 6:8246–8256

Du W, Wang Z, Zhu Z, Hu S, Zhu X, Shi Y, Pang H, Qian X (2014) Facile synthesis and superior electrochemical performances of $CoNi_2S_4$/graphene nanocomposite suitable for supercapacitor electrodes. J Mater Chem A 2:9613–9619

Dutta S, Bhaumik A, Wu KC-W (2014) Hierarchically porous carbon derived from polymers and biomass: effect of interconnected pores on energy applications. Energy Environ Sci 7:3574–3592

Enterría M, Figueiredo JL (2016) Nanostructured mesoporous carbons: tuning texture and surface chemistry. Carbon 108:79–102

Fominykh K, Feckl JM, Sicklinger J, Döblinger M, Böcklein S, Ziegler J et al (2014) Ultra-small dispersible crystalline nickel oxide nanoparticles as high-performance catalysts for electrochemical water splitting. Adv Func Mater 24:3123–3129

Ghorbani M, Golobostanfard MR, Abdizadeh H (2017) Flexible freestanding sandwich type ZnO/rGO/ZnO electrode for wearable supercapacitor. Appl Surf Sci 419:277–285

Huang L, Liu B, Hou H, Wu L, Zhu X, Hu J, Yang J (2018) Facile preparation of flower-like NiMn layered double hydroxide/reduced graphene oxide microsphere composite for high-performance asymmetric supercapacitors. J Alloy Compd 730:71–80

Jain A, Balasubramanian R, Srinivasan MP (2016) Hydrothermal conversion of biomass waste to activated carbon with high porosity: a review. Chem Eng J 283:789–805

Jayaseelan SS, Ko TH, Radhakrishnan S, Yang CM, Kim HY, Kim BS (2016) Novel MWCNT interconnected $NiCo_2O_4$ aerogels prepared by a supercritical CO_2 drying method for ethanol electrooxidation in alkaline media. Int J Hydrogen Energy 41:13504–13512

Jiang J, Bao L, Qiang Y, Xiong Y, Chen J, Guan S, Chen J (2015) Sol-gel process-derived rich nitrogen-doped porous carbon through KOH activation for supercapacitors. Electrochim Acta 158:229–236

Kim SI, Lee JS, Ahn HJ, Song HK, Jang JH (2013) Facile route to an efficient nio supercapacitor with a three-dimensional nanonetwork morphology. ACS Appl Mater Interfaces 5:1596–1603

Li H, Eddaoudi M, O'Keeffe M, Yaghi OM (1999) Design and synthesis of an exceptionally stable and highly porous metal-organic framework. Nature 402:276–279

Li T, Li GH, Li LH, Liu L, Xu Y, Ding HY, Zhang T (2016) Large-scale self-assembly of 3D flower-like hierarchical Ni/Co-LDHs microspheres for high-performance flexible asymmetric supercapacitors. ACS Appl Mater Interfaces 8:2562–2572

Li Z, Zhang W, Li Y, Wang H, Qin Z (2018) Activated pyrene decorated graphene with enhanced performance for electrochemical energy storage. Chem Eng J 334:845–854

Lim MB, Hu M, Manandhar S, Sakshaug A, Strong A, Riley L, Pauzauskie PJ (2015a) Ultrafast sol-gel synthesis of graphene aerogel materials. Carbon 95:616–624

Lim E, Jo C, Kim H, Kim MH, Mun Y, Chun J et al (2015b) Facile synthesis of Nb_2O_5@Carbon core-shell nanocrystals with controlled crystalline structure for high-power anodes in hybrid supercapacitors. ACS Nano 9:7497–7505

Ma F-X, Yu L, Xu C-Y, Lou XW (David) (2016) Self-supported formation of hierarchical $NiCo_2O_4$ tetragonal microtubes with enhanced electrochemical properties. Energy Environ Sci 9:862–866

Ma J, Wu Y, Zuo J, Jiang C, Khan DF, Zhang H, Zhu J (2017) Effects of temperature on the capacitive performance of Ti/40%RuO_2-60%ZrO_2 electrodes prepared by thermal decomposition method. J Electroanal Chem 789:133–139

Meng F, Fang Z, Li Z, Xu W, Wang M, Liu Y et al (2013) Porous Co_3O_4 materials prepared by solid-state thermolysis of a novel Co-MOF crystal and their superior energy storage performances for supercapacitors. J Mater Chem A 1:7235

Mishra RK, Baek GW, Kim K, Kwon HI, Jin SH (2017) One-step solvothermal synthesis of carnation flower-like SnS_2 as superior electrodes for supercapacitor applications. Appl Surf Sci 425:923–931

Ngo Y-LT, Sui L, Ahn W, Chung JS, Hur SH (2017) $NiMn_2O_4$ spinel binary nanostructure decorated on three-dimensional reduced graphene oxide hydrogel for bifunctional materials in non-enzymatic glucose sensor. Nanoscale 9:19318–19327

Pang H, Yan Z, Wang W, Chen J, Zhang J, Zheng H (2012) Facile fabrication of $NH_4CoPO_4 \cdot H_2O$ nano/microstructures and their primarily application as electrochemical supercapacitor. Nanoscale 4:5946

Ramadoss A, Yoon KY, Kwak MJ, Kim SI, Ryu ST, Jang JH (2017) Fully flexible, lightweight, high performance all-solid-state supercapacitor based on 3-Dimensional-graphene/graphite-paper. J Power Sources 337:159–165

Rengaraj A, Haldorai Y, Kwak CH, Ahn S, Jeon K-J, Park SH, Han Y-K, Huh YS (2015) Electrodeposition of flower-like nickel oxide on CVD-grown graphene to develop an electrochemical non-enzymatic biosensor. J Mater Chem B 3:6301–6309

Sasikala SP, Poulin P, Aymonier C (2017) Advances in subcritical hydro-/solvothermal processing of graphene materials. Adv Mater. https://doi.org/10.1002/adma.201605473

She Z, Ghosh D, Pope MA (2017) Decorating graphene oxide with ionic liquid nanodroplets: an approach leading to energy-dense, high-voltage supercapacitors. ACS Nano 11:10077–10087

Tian G, Liu L, Meng Q, Cao B (2015) Facile synthesis of laminated graphene for advanced supercapacitor electrode material via simultaneous reduction and N-doping. J Power Sources 274:851–861

Wang W, Guo S, Penchev M, Ruiz I, Bozhilov KN, Yan D, Ozkan M, Ozkan CS (2013a) Three dimensional few layer graphene and carbon nanotube foam architectures for high fidelity supercapacitors. Nano Energy 2:294–303

Wang L, Ji H, Wang S, Kong L, Jiang X, Yang G (2013b) Preparation of Fe_3O_4 with high specific surface area and improved capacitance as a supercapacitor. Nanoscale 5:3793

Wang S, Liu R, Han C, Wang J, Li M, Yao J, Li H, Wang Y (2014) A novel strategy to synthesize hierarchical, porous carbohydrate-derived carbon with tunable properties. Nanoscale 6:13510–13517

Wang J, Feng S, Song Y, Li W, Gao W, Elzatahry AA, Aldhayan D, Xia Y, Zhao D (2015) Synthesis of hierarchically porous carbon spheres with yolk-shell structure for high performance supercapacitors. Catal Today 243:199–208

Wang J, Zhang X, Wei Q, Lv H, Tian Y, Tong Z et al (2016) 3D self-supported nanopine forest-like Co_3O_4@$CoMoO_4$ core-shell architectures for high-energy solid state supercapacitors. Nano Energy 19:222–233

Wang J, Ran R, Sunarso J, Yin C, Zou H, Feng Y, Li X, Zheng X, Yao J (2017a) Nanocellulose-assisted low-temperature synthesis and supercapacitor performance of reduced graphene oxide aerogels. J Power Sources 347:259–269

Wang N, Zhao P, Liang K, Yao M, Yang Y, Hu W (2017b) CVD-grown polypyrrole nanofilms on highly mesoporous structure MnO_2 for high performance asymmetric supercapacitors. Chem Eng J 307:105–112

Wang D, Guan B, Li Y, Li D, Xu Z, Hu Y, Wang Y, Zhang H (2018) Morphology-controlled synthesis of hierarchical mesoporous α-$Ni(OH)_2$ microspheres for high-performance asymmetric supercapacitors. J Alloy Compd 737:238–247

Wei B, Wang L, Miao Q, Yuan Y, Dong P, Vajtai R, Fei W (2015) Fabrication of manganese oxide/three-dimensional reduced graphene oxide composites as the supercapacitors by a reverse microemulsion method. Carbon 85:249–260

Wei B, Wang L, Wang Y, Yuan Y, Miao Q, Yang Z, Fei W (2016) In situ growth of manganese oxide on 3D graphene by a reverse microemulsion method for supercapacitors. J Power Sources 307:129–137

Wu Y, Gao G, Yang H, Bi W, Liang X, Zhang Y, Zhang G, Wu G (2015) Controlled synthesis of V_2O_5/MWCNT core/shell hybrid aerogels through a mixed growth and self-assembly methodology for supercapacitors with high capacitance and ultralong cycle life. J Mater Chem A 3:15692–15699

Wu X, Han Z, Zheng X, Yao S, Yang X, Zhai T (2017) Core-shell structured Co_3O_4@$NiCo_2O_4$ electrodes grown on flexible carbon fibers with superior electrochemical properties. Nano Energy 31:410–417

Xia X, Chao D, Fan Z, Guan C, Cao X, Zhang H, Fan HJ (2014) A new type of porous graphite foams and their integrated composites with oxide/polymer core/shell nanowires for supercapacitors: structural design, fabrication, and full supercapacitor demonstrations. Nano Lett 14(3):1651–1658

Xu Y, Shi G, Duan X (2015) Self-assembled three-dimensional graphene macrostructures: synthesis and applications in supercapacitors. Acc Chem Res 48:1666–1675

Yang Q, Xu Q, Jiang H-L (2017) Metal–organic frameworks meet metal nanoparticles: synergistic effect for enhanced catalysis. Chem Soc Rev 46:4774–4808

Yang F, Zhang X, Yang Y, Hao S, Cui L (2018) Characteristics and supercapacitive performance of nanoporous bamboo leaf-like CuO. Chem Phys Lett 691:366–372

Zhai T, Wang F, Yu M, Xie S, Liang C, Li C et al (2013) 3D MnO_2–graphene composites with large areal capacitance for high-performance asymmetric supercapacitors. Nanoscale 5:6790

Zhang H, Hu Z, Li M, Hu L, Jiao S (2014) A high-performance supercapacitor based on a polythiophene/multiwalled carbon nanotube composite by electropolymerization in an ionic liquid microemulsion. J Mater Chem A 2:17024–17030

Zhang YZ, Cheng T, Wang Y, Lai WY, Pang H, Huang W (2016a) Flexible supercapacitors: a simple approach to boost capacitance: flexible supercapacitors based on Manganese Oxides@MOFs via chemically induced in situ self-transformation. Adv Mater 28:5241

Zhang J, Gu P, Xu J, Xue H, Pang H (2016b) High performance of electrochemical lithium storage batteries: ZnO-based nanomaterials for lithium-ion and lithium–sulfur batteries. Nanoscale 8:18578–18595

Zhang J, Xu J, Wang Y, Xue H, Pang H (2017) Our contributions in nanochemistry for antibiosis, electrocatalyst and energy storage materials. Chem Rec 18:1–15

Zhao J, Wu J, Li B, Du W, Huang Q, Zheng M, Xue H, Pang H (2016) Facile synthesis of polypyrrole nanowires for high-performance supercapacitor electrode materials. Progr Nat Sci Mater Inter 26:237–242

Zhu Y, Ji X, Wu Z, Song W, Hou H, Wu Z, He X, Chen Q, Banks CE (2014) Spinel $NiCo_2O_4$ for use as a high-performance supercapacitor electrode material: understanding of its electrochemical properties. J Power Sources 267:888–900

Zhu Y, Cui H, Meng X, Zheng J, Yang P, Li L, Wang Z, Jia S, Zhu Z (2016) Chlorine-induced in situ regulation to synthesize graphene frameworks with large specific area for excellent supercapacitor performance. ACS Appl Mater Interfaces 8:6481–6487

Chapter 6
Nanomaterials for Batteries

Owing to the energy crisis, batteries have captured numerous attentions due to their large energy density with stable electrochemical properties, and they have been successfully applied in power electric vehicles, hybrid electric vehicles, and millions of electronic devices. Batteries, regardless of their chemistry-aqueous, non-aqueous, Li, or Na based, store energy within the electrode structure through charge transfer reactions. Therefore, speeding up charge transfer reactions is the key to improve the performance of batteries device. Downsizing the materials' particles could shorten the ion diffusion distance and lead to an improved rate performance. So, nanomaterials have become a hot research area for electrode materials. In this chapter, we provide an overall summary in evaluation of nanostructured materials for batteries, including lead-acid batteries, lithium-ion batteries, sodium-ion batteries, metal-air battery, and lithium-sulfur battery.

6.1 Lead-Acid Batteries

6.1.1 Lead-Acid Battery Classification, Structure, and Working Principle

Lead-acid batteries are often called lead accumulator. It is a kind of electrode mainly made of lead and its oxide; the electrolyte is sulfuric acid. When Lead-acid battery is at discharge state, the main component of the positive electrode is lead peroxide, the main component of the negative electrode is lead; at charging state, the main component of the positive electrode and negative electrode are lead sulfate.

Lead-acid battery was invented by French Plante in 1859 and has been developed for nearly 150 years. Lead-acid battery has made great progress in theoretical research and product variety, product electrical performance, etc. Lead-acid batteries play an indispensable role not only in transportation, communications, electric power,

H. Pang et al., *Synthesis of Functional Nanomaterials for Electrochemical Energy Storage*,
https://doi.org/10.1007/978-981-13-7372-5_6

and military but also in all fields of navigation and aviation. Lead-acid batteries have the largest amount of civilian use in industry, because of its stable quality and reasonable price. Lead-acid batteries are being improved for using as a power source for electric vehicles and power storage. On the other hand, lead-acid batteries made great progress in miniaturized, high performance and were required ease of use and simple maintenance.

6.1.1.1 Classification of Lead-Acid Batteries

Lead accumulator can be divided the following according to purposes: ① The accumulator for motor vehicle start-up, for example, starting of cars, tractors, and internal combustion engine. ② Fixed-type accumulator was being communications equipment power supply, switch off power plants and substations and it was being computer and other non-power standby power supply. ③ Traction battery for vehicle drive power, such as the train station transport battery car, industrial and mining electric locomotive power supply; ④ Motorcycle with batteries. ⑤ Shipping with batteries. ⑥ Aviation with batteries. ⑦ Tanks with batteries. ⑧ Railway passenger car with batteries. ⑨ Beacon with batteries. ⑩ Miner's lamp with batteries.

According to the plate structure classification: ① The anointing type, positive and negative plates are all made of lead alloy grille and coated with lead paste. ② Tube type, the conductive skeleton of the positive plate is covered with woven fiber pipe, and the active substance is put into the tube; the negative pole is made of ordinary anointing pad. ③ Form, the positive pole is made of pure lead, its active substance is the thin layer of lead itself; the negative electrode USES the anointing pad. ④ Semi-form, with pure lead cast into tight small square grid, then coated with lead paste. ⑤ The negative plate is anointed.

According to electrolyte and charging maintenance classification: ① Dry discharge battery, the plate is in a dry discharge state, and the electrolyte is injected and the initial charge can be used. ② Dry charged battery, the plate is in a dry charging state, and the injection of electrolyte can be used for a short time. ③ With liquid storage battery, it can be used. ④ No maintenance accumulator, normal use process without maintenance plus water. ⑤ Low maintenance battery, under normal operating conditions, only a long time to add water once. ⑥ The wet charge accumulator, after charging the electricity, pour out most of the electrolyte and can be used during the storage period. The structure of lead accumulator is mainly composed of positive plate group, negative plate group, electrolyte, and container. The positive plate and the negative plate are composed of a grid and an active material. In addition to supporting the active material, the plate grid also conducts electricity. The grid generally uses lead antimony alloy, sometimes with pure lead or lead calcium alloy.

6.1.1.2 Structure of Lead-Acid Batteries

Lead accumulator in charge state, is extremely spongy lead, is extremely lead dioxide; both positive and negative are lead sulfate. At present, the accepted flow reaction is the theory of disulfation, and the equation is shown in Eqs. (6.1) and (6.2). Its correctness is verified from the following three aspects: ① Chemical analysis was used to confirm the composition of positive active substances as PbO_2, and the composition of negative active substances was lead. ② When using 2F power, the change of H_2SO_4 concentration was measured, equivalent to 2 mol H_2SO_4, and 2 mol H_2O, which was consistent with the battery reaction. ③ The thermodynamic data calculates the electromotive force of the battery, which is consistent with the measured value.

Battery symbol and discharge reaction of lead-acid battery:

$$\begin{aligned}&\mathrm{Pb, PbSO_4|H_2SO_4|PbSO_4, PbO_2(or\ Pb|H_2SO_4|PbO_2)}\\&\mathrm{PbO_2 + 3H^+ + HSO_4^- + 2e = PbSO_4 + 2H_2O}\end{aligned} \tag{6.1}$$

$$\mathrm{Pb + HSO_4^- = PbSO_4 + H^+ + 2e} \tag{6.2}$$

$$\begin{aligned}&\mathrm{Pb + PbO_2 + 2H^+ + 2HSO_4^- = 2PbSO_4 + 2H_2O}\\&\left(\mathrm{Pb + PbO_2 + 4H^+ + 2SO_4^{2-} = 2PbSO_4 + 2H_2O}\right)\end{aligned} \tag{6.3}$$

The causes of self-discharge of lead-acid batteries can be analyzed from thermodynamics. From the potential of lead to pH, $\mathrm{Pb + HSO_4^- = PbSO_4 + H^+ + 2e}$, $\mathrm{2H^+ + 2e = H_2}$. This leads to a negative self-discharge in the conjugate reaction. $\mathrm{2H_2O = 4H^+ + O_2 + 4e}$, $\mathrm{PbO_2 + 3H^+ + HSO_4^- + 2e = PbSO_4 + 2H_2O}$. This causes a positive self-discharge in the conjugate reaction (Liu et al. 2015).

6.1.2 The Performance of Lead-Acid Battery

Open-circuit voltage: the electromotive force of the battery is 2.045 V, so its rated voltage is 2.0 V. The relationship between battery open-circuit voltage and electrolyte density can be calculated in the next type.

$$\begin{aligned}&\text{Open-circuit voltage} = 1.850 + 0.917(\rho_l - \rho_w) \text{ or}\\&\text{Open-circuit voltage} = \rho_l + 0.84\end{aligned} \tag{6.4}$$

ρ_l in liquid is the density of the electrolyte, ρ_w is the density of water.

Capacity: The capacity of lead-acid battery is a function of temperature and discharge current. The hourly rate and temperature of the discharge are clearly defined in the standard. The starting type lead-acid battery usually uses 20-h rate capacity, the fixed type is commonly used in 10-h rate capacity, and the battery for power traction uses 5-h rate. The relation between the capacity of the battery and the temperature is

$$C = CT/[1 + K(T - T_s)] \tag{6.5}$$

The efficiency and lifetime of the battery capacity: efficiency (input/output capacity) * 100%, also known as ah efficiency, this is more commonly used. The electric energy efficiency is (output power/input power) * 100%, also known as the watt hour efficiency.

After the battery is repeatedly charged and discharged multiple times, due to loss and shrinkage of active material, micropores of electrode plate are reduced, the capacity is reduced, and the battery life is gradually shortened. In general, the capacity of the battery can be reduced to 70–80% of the rated value and cannot be used again. The battery life is related to the quality of the manufacturing, and is also affected by the use and maintenance methods. A battery with the same rated capacity, such as a large current discharge, is lower than the capacity of a small current discharge at the later stage. The cycle life of the lead-acid battery is 200–400 times, and the duration of the battery is 3–10 years.

Self-discharge: no matter whether the lead battery works or not, there is a discharge phenomenon in the battery, which consumes the electric energy in white. In addition to the above, the cause of self-discharge is also due to the presence of impurities in the battery.

6.1.3 Sealed Lead-Acid Battery

After more than ten years of rapid development, China has become the world's largest lead-acid battery (LAB) industry, accounting for more than 30% of global LAB output, and the market demand has also increased dramatically. Lead-acid batteries occupy more than a 60% market share of the secondary power supply. China's large market demand comes from a substantial increase in the number of electric bikes (e-bikes). Besides, lead-acid batteries dominate the rechargeable batteries market, especially as the starting power sources in the electric bicycles. Starting power sources, which are the key component of electric bicycles, typically last less than three years, requiring frequent replacement. Currently, electric bicycle batteries including lead-acid, nickel-metal hydride, and lithium batteries which are mainly composed of heavy metal ions, such as lead, copper, cobalt, zinc, manganese, nickel, and electrolytes. These substances have strong negative effects on soil and water, endanger ecological safety, and present a persistent public nuisance upon entering the environment. Battery recycling is an essential part of the e-bicycle industry, and is of great significance to sustainable development and resources and to environmental protection.

In 2017, the number of e-bikes in China reached 29.96 million and the social adoption rate of e-bikes was above 10%. At present, the main concern for consumers of e-bikes is the battery performance. The power source for electric bicycles is a

VRLA battery, consisting of three to four 12 V modules with a total voltage of 36 or 48 V, and the capacity is 10–20 A h. It is a valve-regulated lead-acid (VRLA) battery made up of two parallel-connected strings of 75–12.5 V modules connected in series. Electromechanical relays are used to interconnect lower voltage modules into a single battery when operational. The relays also double as an operational safety that isolates modules in the event of improper battery operation. Control of these relays in fault isolation conditions is non-trivial.

Valve-controlled sealed lead-acid batteries are new types of lead-acid batteries. They use the "lean liquid" design. The glass fiber is used as a separator. The sulfuric acid electrolyte is completely absorbed by the glass fiber membrane. There is no free acid in the interior. The electrolyte in lead-acid batteries is gel-like, and there are no free electrolytes inside the battery. Electrolytes that cannot move freely greatly reduce the risk of lead-acid batteries during use.

Since the end of the 1970s, a fully closed lead-acid battery has been developed internationally, which is divided into air tight and full density type. The battery has the advantages of maintenance free, no pollution, and low price. There are three ways to make the battery air tight.

(1) gas phase catalysis: with the palladium catalyzed Ambrose catalyst installed in the battery cover, the electrode on precipitation of hydrogen and oxygen and recombine into water, and returned to the inside of the storage battery, thereby reducing water loss, reach maintenance free.
(2) auxiliary electrode type: a pair of auxiliary electrodes that absorb hydrogen and oxygen in the battery or an auxiliary electrode containing only one hydrogen in the battery. When the hydrogen generated by the battery is adsorbed on the hydrogen auxiliary electrode, it forms a hydrogen electrode. It forms a self-discharge small battery with PbO_2, which reacts $2H_2 + PbO_2 + H_2SO_4 = PbSO_4 + 2H_2O$, and the water returns to the battery.
(3) the cathode absorption mode: the oxygen generated by the positive electrode when it is charged, diffused through the diaphragm to the negative electrode, reacts with the lead of active substances, forms PbO_2, and then reacts with H_2SO_4 to generate $PbSO_4$ and H_2O.

The cathodic absorption battery can use the appropriate diaphragm to make the battery limit liquid or lean liquid, or use colloidal electrolyte (SiO_2 silica powder and a certain amount of H_2SO_4 to form silica gel). It can make electrolyte fixed, and no gas escapes, to achieve the requirement of full sealing. But considering the self-discharge of the battery and the possibility of hydrogen precipitation in the later period of charging, the battery is equipped with a safety valve. When the air pressure in the battery increases to a certain value, the gas is discharged, so it is also called a valve-controlled closed lead battery.

Sealed maintenance free lead-acid batteries have all the advantages of open lead-acid batteries. The so-called maintenance free is the relatively open battery that needs regular water addition. The whole storage battery is totally enclosed (the redox reaction of the battery is carried out inside the closed enclosure), so there is no "harmful gas" overflow in the battery. No need to add water and other daily operation and maintenance. It can be installed in the host room, which is suitable for the unmanned hand guard machine room (Liu et al. 2015).

The number of valves regulated lead-acid batteries in automobiles, telecommunications, emergency services, and other applications has been increasing, causing great concern. In order to meet the performance requirements of countless applications, a deep understanding of the attribute design of valve-regulated lead-acid batteries is essential. The separator in the form of an absorbent glass mat is a key component of valve-regulated lead-acid batteries. Absorptive glass mat is a versatile material that not only separates the electrodes, but also retains sulfuric acid through wetting and wicking properties. These characteristics are mainly achieved by using glass fiber as a constituent material and by the structure of an absorbent glass mat. The contact angle between glass fiber and acid is zero, which has the added advantage of durability to the acid environment. In addition, the structure–property relationship of the absorbent glass mat depends on the porosity, uniformity, fiber size, and fiber orientation. The synergistic effect of the structural parameters of the fiber and the absorbent glass mat affects the saturation of the acid (the proportion of pore volume filled with acid). This is because the fiber diameter affects the size of the pores and the distribution of the electrolyte depends on the porous structure of the absorbent glass mat. Fill the smaller holes and then the larger holes because the contact area with glass fibers is the largest and the contact area with the gas phase is the smallest. However, changes in saturation have a significant impact on both discharge performance and recharge characteristics. For example, batteries cannot be effectively charged below a critical saturation level. Qualitatively, it is proposed that when the structure of the absorbent glass mat is anisotropic and has a minimum degree of tortuosity and should have a low bulk density, a high performance of the battery can be obtained. However, a detailed analytical model for the wicking properties of absorbent glass fiber mats is yet to be developed, which combines actual fiber and structural parameters. Although, Kamenev and others. The height and capillary rate of the electrolyte in the absorbent glass mat have been predicted, but the actual structural parameters, namely fiber orientation and bulk density, were not considered in the study.

6.1.4 Safety and Elimination Mechanism of Sealed Lead-Acid Batteries

Valve-controlled sealed lead-acid battery, if the gas compound performance is not good or the efficiency of the sealing reaction is low, it will lead to increased battery electrolyte loss, a sharp decrease in battery capacity, and a drastic reduction in battery cycle life, electrolyte drying up, and battery life termination. There are many harmful impurities in the practical application that reduce the hydrogen overpotential of the negative plate. Once these harmful impurities reach the negative plate, they will reduce the hydrogen overpotential to release hydrogen, and some additional reactions will occur on the positive plate. These reactions include plate corrosion, the formation of residual lead oxide or lead sulfate on the positive plate, and organic substances such as dissolved lignin are oxidized. In addition, an increase in internal pressure can cause the battery to rupture, resulting in safety and environmental impacts.

Elimination measures.

(1) Select a grid alloy containing lead–bismuth alloy with limited niobium content, improve the structure of positive electrode plate and separator, increase the density of positive electrode active material, and add phenolic resin that can capture (absorb) antimony in micro glass fiber separator;
(2) Increase hydrogen overpotential and prevent hydrogen evolution from overpotential;
(3) The use of high strength, high toughness, and flame-resistant materials with good resistance to acid and corrosion to handle the battery case, thereby increasing the toughness of the battery case and greatly reducing the risk of battery short circuits;
(4) The use of ultrafine glass fibers as a separator increases the specific surface area of the separator, enhances the electrolyte's adsorption capacity, and reduces electrolyte flow;
(5) Double layer combination terminals and plastic seals prevent electrolyte leakage from the gap;
(6) Put a protective cover on the battery terminal to prevent the battery from short-circuiting.
(7) Hydrogen is used to precipitate a negative grid alloy with a high overpotential. When hydrogen gas has not been precipitated on the negative electrode plate during overcharging, oxygen precipitated on the positive electrode plate undergoes chemical recombination on the negative electrode plate, and is combined with undercomposed hydrogen to form water. In the electrolyte, the loss of electrolyte is minimized (Yang et al. 2018a; Chen et al. 2017a).

6.1.5 Research Progress of Sealed Lead-Acid Batteries

Valve-regulated lead-acid batteries have encountered many problems in the early stage of use. The most serious ones are short battery life, liquid leakage, thermal runaway, short-circuiting of the battery, corrosion of negative busbars, and other battery failures, which has caused the importance of the lead-acid battery industry. Researchers through a lot of battery failure mode of research, from the battery structure, manufacturing processes, battery parts materials for improvement, battery short circuit, leakage, thermal runaway, negative bus corrosion, and other issues have been very good to improve and solve. At present, the research of VRLA batteries mainly focuses on the grid structure, materials, positive electrode active material additives, and separator materials that affect the performance of battery energy and service life (Liu et al. 2015; Yang et al. 2018a).

The research of VRLA battery aims to improve specific capacity and specific power of the battery, improve the cycle life, and improve the rapid charging capability. There are currently six research directions, namely bipolar ear wound battery technology, horizontal VRLA battery technology, bipolar ceramic VRLA battery technology, Pb-CVRLA battery technology, super battery technology, graphite-foam lead-acid battery technology (Chen et al. 2017a; Kumar et al. 2017; Sawai et al. 2007; Chang et al. 2009).

The bipolar ear wound type battery technology uses a very thin lead foil as a polar plate. The positive electrode plate, the separator, and the negative electrode plate are alternately stacked and tightly wound together to form a single-cylindrical battery. This battery has been successfully used as a power battery in hybrid electric vehicles, showing a very good market prospect. The horizontal VRLA battery technology has abandoned the grid and replaced it with a lead cloth that is woven from glass fibers coated with Pb–Sn alloy on the surface. One end is coated with a positive lead paste and the other end is coated with a negative lead paste. The negative connection reduces internal resistance of the battery. The battery board is placed horizontally, which avoids concentration polarization phenomenon and is conducive to heat dissipation and enhanced oxygen compounding efficiency. The bipolar ceramic VRLA battery technology uses ceramics as a separator, adopts a bipolar structure, and has the advantages of long cycle life, small internal resistance, high rate charge and discharge, low cost, and easy recycling. The Pb-CVRLA battery technology is a negative electrode that incorporates activated carbon. Pb–C battery has the advantage of long cycle life, the high specific energy and power, no sulfating of the negative electrode, and high rate charge and discharge. At present, this technology has been applied to wind energy batteries, solar energy storage batteries, and communication backup power batteries. Super battery technology is a combination of super capacitors and lead-acid batteries, so that capacitors in the battery charge and discharge process of buffering, enhance battery power and increase life. This technology can quickly provide or absorb electricity when the vehicle accelerates or brakes. Graphite-foam lead-acid battery technology replaces lead with a foamed graphite material, and the active material is retained, reducing lead content by 2/3 compared to ordinary lead-acid batteries. This battery technology is comparable in performance to advanced nickel-metal hydride batteries and lithium-ion batteries, and also has the advantages of small size, light weight, and a much lower price than nickel-metal hydride batteries and lithium-ion batteries (Soria et al. 2004).

6.2 Lithium Batteries and Lithium-Ion Batteries

6.2.1 Lithium Batteries

Lithium is the ideal negative active substance for high-energy batteries, because it has the most negative standard electrode potential and quite low electrochemical equivalent (Yanxia and Lijuan 2017). The development of lithium battery began in the 1960s, and it has become a very important chemical power supply, application in the aerospace, defense, civil, science and technology, such as cardiac pacemaker, electronic watches, calculator, tape recorder, aircraft, missiles ignition system, torpedoes, etc. (Lingzhi 2009).

Lithium is very reactive. When lithium is used as an electrode, water cannot be used as a solvent (Hong 2014). We can use organic solvent or non-aqueous inorganic solvent electrolyte to make lithium non-water batteries, use molten salt to make molten salt batteries, and use solid electrolyte to make Lithium solid electrolyte batteries.

The commonly used organic solvents are acetonitrile, dimethylformamide, propylene carbonate, and butadiene. $LiClO_4$, $LiAlCl_4$, $LiBF_4$, LiBr, and $LiAsF_6$ can be used as support electrolytes. Non-aqueous inorganic solvents include $SOCl_2$ (thionyl chloride) SO_2Cl_2 (thionyl chloride) $POCl_3$ (phosphoryl chloride), which can also be used as positive active substances. Compared with the traditional battery, the lithium battery has the advantages of high charger/discharger voltage, relatively high power and stable discharge voltage, long Li-storage life, and wide working temperature range, but there are also some deficiencies, one of the primary problems is security. Some lithium non-aqueous solutions batteries may explode when the battery is under heavy load conditions. In addition, the low conductivity of the organic electrolyte solution, the low energy density, and power of the batteries are the problems that need to be solved.

6.2.1.1 Common Types of Lithium Batteries

Lithium organic electrolyte battery: the common electrolyte of Li/MnO_2 battery is $LiClO_4^-$ polypropylene (PC)-glycol dimethyl ether. The open-circuit voltage and working voltage are 3.5 and 2.9 V. The specific energy can reach up to 250 Wh kg^{-1} and 500 Wh L^{-1}. The electrolyte of Li/SO_2 battery is PC-acetonitrile solution containing lithium bromide. The discharge voltage is stable, the specific energy of the cell is up to 520 Wh L^{-1}, the specific power is high, the low-temperature performance is good, and the storage life is long, but the security is poor. The electrolyte of $Li/(CF_x)_n$ battery is γ-butylene solution containing $LiBF_4$, whose open-circuit voltage is 3.1 V, and it has a relatively high actual specific energy. PFC $(CF_x)_n$ is chemically stable and thermally stable, but the cost is high.

Lithium inorganic electrolyte battery: using the inorganic solvent $SOCl_2$. SO_2Cl_2 . $POCl_2$ both as the positive active substance and electrolyte. The performances of $Li/SOCl_2$ and $Li/POCl_3$ batteries are better than the Li/SO_2, which possess the best comprehensive performance in the organic electrolyte battery. The discharge curve is very flat, and its specific power is quite high (As shown in the following Table 6.1).

Room temperature lithium battery: the organic electrolyte lithium battery is widely studied. Its cathode materials are transition metal sulfides, such as CuS, FeS, MnS, Ag_2S, TiS_2, VS_2, MoS_2, VSe_2, $NbSe_2$, $TiSe_2$. The transition metal two sulfide is a layered structure, and the electrode reaction is an embedded reaction. At the time of discharge, Li^+ enters the interlayer, embedded in the lattice of the positive material. For example, the open-circuit voltage of Li/TiS_2 (using 1 mol L^{-1} $LiAsF_6$–2MeTHF as an electrolyte) battery is 2.47 V and its theoretical specific energy is 481Wh kg^{-1}. The average working voltage of the AA type Li/TiS_2 at a discharge current of 200 mA is 2.2 V, and the cycle life is up to 200 times under 80% depth charge–discharge process.

Molten salt lithium battery: This is a promising high-energy battery. Its electrolyte is LiCl–KCl eutectic mixture. The conductivity of 450C is 1.57 S cm^{-1}, which is higher than that of organic electrolyte (about 2–3 orders of magnitude). The negative materials are Li, Li–Al, Li–B, Li–S,i and so on. Alloying can reduce the corrosion of lithium. Li–Al is the most stable; Li–B, Li–Si can increase the Li-storage capacity. The negative materials are transition metal sulfides, such as FeS, FeS_2, and TiS_2. The performances of the Li/FeS_2 and LI/FeS batteries are summarized in Table 6.2.

6.2.2 Lithium-Ion Batteries

The similarities between lithium-ion battery and lithium battery are as follows: two kinds of batteries, all use a metal oxide or sulfide that can make lithium ion intercalate and de-intercalate as the positive electrode and use an organic solvent inorganic salt system as electrolyte (Choi et al. 2018; Lee et al. 2018; Liu and Cui 2018; Liu et al. 2018; Miao et al. 2018; Wang et al. 2018a,b; Yang et al. 2018b,c). The difference is that in lithium-ion batteries, the carbon material which can intercalate and deintercalation the lithium ion is used instead of the pure lithium as the negative electrode. The anode of the lithium battery uses metal lithium. The lithium metal will be deposited on the lithium anode and produces the lithium dendrite during the charging process. The lithium dendrite may penetrate the diaphragm and cause a short circuit inside the battery to explode. To overcome the shortage of lithium battery and enhance its safety, lithium-ion battery came into being. In the late 1980s and early 90s, carbon materials with graphite structure were used as lithium insertion negative electrodes to avoid the safety problem caused using lithium metal (Feng et al. 2013). At the same time, lithium is in the form of Li^+ ions in this battery system. In the charge/discharge cycle of the battery, the Li^+ is continuously intercalated and deintercalated between the positive and negative poles. Therefore, the battery is called the "lithium-ion battery". In 1980, Texas State University professor Goodenough et al. proposed lithium cobalt oxide ($LiCoO_2$) can be considered as cathode

Table 6.1 The performances comparison of lithium batteries and other batteries

Battery	Specific energy/Wh kg^{-1}	Specific power/W kg−1	Open-circuit voltage/V	Work temperature/°C	Storage life/year (20 °C)
Li/SO_2	330	110	2.9	−40 ~ +70	5–10
Li/$SOCl_2$	550	550	3.7	−60 ~ +75	5–10
Zn/MnO_2	66	55	1.5	−10 ~ +55	1
Zn/MnO_2 (Alkalinity)	77	66	1.5	−30 ~ +70	2
Zn/HgO	99	11	1.35	−30 ~ +70	>2

Table 6.2 The performance of LiAl/FeS, Li_4Si/FeS_2 batteries

Battery	LiAl/FeS	Li_4Si/FeS_2
Cell reaction	$2LiAl + FeS = Li_2S + Fe + 2Al$	$Li_4Si + FeS_2 = 2Li_2S + Si + Fe$
Voltage/V	1.33	1.8
Theoretical specific energy/Wh kg^{-1}	458	944
Specific energy/Wh kg^{-1}	90	180
Specific power/W kg^{-1}	100	100
Life expectancy/h	5000	15,000

materials for lithium-ion batteries, subsequently, the Sony Corp of Japan in 1991 launched the first $LixC_6$/organic electrolyte/$Li_{1-x}CoO_2$ system of lithium-ion battery and successfully commercialized it. So far, lithium-ion batteries are not only used in all kinds of electronic consumption products, but also in electric vehicles and large military products.

6.2.2.1 The Working Principle of Lithium-Ion Batteries

Figure 6.1 shows the working principle of lithium-ion battery consisting of commercialized layered $LiCoO_2$ material as cathode material, graphite as anode material, and $LiPF_6$ as electrolyte. Schematic (Dunn et al. 2011). As shown in the picture, when charging, the Li^+ ion is removed from the $LiCoO_2$ material and passes through the diaphragm through the electrolyte. On the negative electrode material, an electron is reduced to metal lithium and embedded into graphite negative electrode material, and that makes anode in the state of lithium rich. At the same time, the Co^{3+} in the positive pole is oxidized to Co^{4+}, and the equal number of electrons can be compensated from the external circuit to the negative pole. When discharging, lithium ions are removed from the graphite negative electrode and go back to the positive electrode through the electrolyte. Meanwhile, the Co^{4+} in the cathode is reduced to Co^{3+}, the cathode is in the lithium rich state, and the current is output from the positive pole to the negative electrode, thus converting the chemical energy into electric energy. Obviously, the lithium-ion battery is a concentration cell. The larger voltage difference in positive/negative materials causes the higher output voltage. Therefore, when selecting the positive material, the potential should be as high as possible; when the negative material is selected, the potential should be as low as possible. Taking the above commercialized battery as an example, the electrode reaction in the charging and discharging process is as follows:

$$\text{Positive reaction: } LiCoO_2 \rightarrow Li_{1-x}CoO_2 + xLi^+ + xe^-$$
$$\text{Anode reaction: } 6C + xLi^+ + xe^- \rightarrow Li_xC_6$$

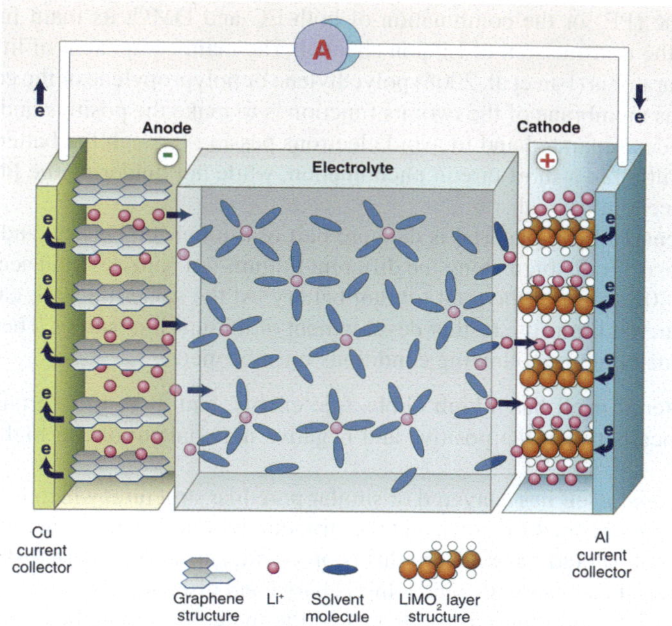

Fig. 6.1 The working principle of lithium-ion battery, reprinted from Ref. (Dunn et al. 2011), copyright 2011, with permission from Science

$$\text{Total battery reaction: } LiCoO_2 + 6C \rightarrow Li_{1-x}CoO_2 + Li_xC_6$$

6.2.2.2 Structural Composition of Lithium-Ion Batteries

Lithium-ion battery is usually composed of the following parts (A cylindrical lithium-ion battery is used as an example): Positive pole, negative pole, diaphragm, air-bleed hole, protective valve, shim, etc. Among them, the main four major components are positive, negative, electrolyte, and diaphragm. The commercialized production of cathode materials has three main categories. They are classified according to their structure. They are layered lithium cobalt oxide (Jeong et al. 1998), spinel structure lithium manganate material (Kumagai et al. 1997), and olivine structure lithium iron phosphate material (Zhang et al. 2011). The common negative materials mainly include the traditional graphite material (Shim and Striebel 2007), amorphous silicon and nano silicon material (Wang et al. 2013) and transition metal oxide material (Zhang et al. 2012). The electrolyte consists of electrolyte and organic solvent, wherein the electrolyte generally can be dissolved in organic solvent in lithium (such as $LiPF_6$, $LiClO_4$, etc.) and the solvent composition is mainly ethylene carbonate (EC), propylene carbonate (PC), two methyl carbonate (DMC) and polyethylene (PE) or the combination of both EC and DMC; its main function is to provide the transmission of Li^+ ion channel. The main component of lithium-ion battery separator is (Fan et al. 2008) polyethylene or polypropylene or the compound microporous membrane of the two; its function is to make the positive and negative two electrodes separated and to avoid electrons passing through the battery's inner circuit, resulting in a short-circuit phenomenon, while not hindering the lithium ion in which freedom through.

At present, cathode material is the core part of lithium-ion battery, and it is also the main basis for distinguishing the different lithium-ion batteries, and accounts for more than 40% costs of the total lithium battery. At the same time, the capacity of cathode material limits the further development of lithium-ion battery. Therefore, as a positive material, the following conditions must be met:

1. the material itself has a high Gibbs free energy, and there is a certain voltage difference between the positive and negative materials to obtain higher output voltage.
2. the material itself has a layered or similar pore like structure, which is beneficial to the de-embedded Li^+ ions, and the structure will not change when it is inlaid.
3. the Li^+ ion should have a large diffusion coefficient in the internal structure of the material and can withstand a high charge and discharge current.
4. the chemical and physical properties of the material are all stable, so that the battery has good reversibility.
5. the material and the electrolyte coexist steadily without any reaction.
6. the material has good thermal stability, nontoxic, environmentally friendly, and easy to prepare.

6.2.2.3 Cathode Material

At present, the capacity of the commercialized lithium-ion battery anode material is over 300 mAh g^{-1} (Li et al. 2010), and the specific capacity of the alloy anode material has even exceeded 1000 mAh g^{-1} (Lee et al. 2012) at the experimental stage. However, the actual specific capacity of the corresponding commercial positive material is less than 200 mAh g^{-1}. Therefore, the key to further enhance the Li-storage capacity of the whole lithium-ion battery is to study the higher Li-storage capacity of the cathode material.

Layered Lithium Cobalt ($LiCoO_2$) Cathode Material

$LiCoO_2$ is the earliest commercialized layered oxide cathode material, which has the advantages of simple production technology, high working voltage, stable charging and discharging performance, and so on. The research of $LiCoO_2$ started in 1980. J.B. Goodenough and others put forward that $LiCoO_2$ can be considered as the cathode material for lithium-ion battery and was commercialized (Ozawa 1994) by Japanese Sony Corp in 1991.

$LiCoO_2$ has a layered structure (the structure of alpha-$NaFeO_2$), which belongs to the R-3 m space group (Perkins et al. 2010), Li^+, Co^{3+} and O^{2-} occupy 3a, 3b, and 6C positions in the spatial structure, respectively. Co^{3+} ions and Li^+ ions are all in the eight faces of O^{2-} ions. In the direction of C-axis, there is a layer of space structure with Co^{3+} ion layer and a lithium-ion layer alternately arranged. During charging/discharging process, the lithium-ion can be reversibly removed/embedded from its layer. The theoretical specific capacity of $LiCoO_2$ cathode material is about 274 mAh g^{-1}, but in practical application, the specific capacity (only 140 mAh g^{-1}) is about half of the theoretical value. This is because when the charging voltage reaches 4.3 V, there will be some side reactions, resulting in an irreversible transformation of the structure, and the increase of the battery impedance. Only about one half of lithium ions can be removed from the structure (Aurbach et al. 2003). Therefore, the charging voltage of $LiCoO_2$ is generally less than 4.4 V. Besides, cobalt resources are scarce, expensive, and unfriendly to environment, which makes $LiCoO_2$ have many limitations in the application of Li-ion batteries.

Layered Lithium Nickel ($LiNiO_2$) Cathode Material

Compared with $LiCoO_2$, $LiNiO_2$ has higher actual capacity, and has more advantages in price and resources. It has been considered as extremely promising cathode material for $LiCoO_2$. The $LiNiO_2$ structure, like $LiCoO_2$, belongs to the layered structure of the alpha-$NaFeO_2$, and the R-3 m space group. The theoretical specific capacity of $LiNiO_2$ is 275 mAh g^{-1}, meanwhile, the actual specific capacity is between 190 mAh g^{-1} and 210 mAh g^{-1}, and the self-discharge rate is lower (Liu et al. 2001). However, $LiNiO_2$ also has shortcomings (Kanno et al. 1994), for example, the ionic radius of Ni^{2+} is very close to the ionic radius of Li^+. In the process of material synthesis, the Ni^{2+} ions in the transition metal layer are very easy to migrate to the Li^+ layer and have cation mixing with Li^+. So far, no pure, structurally stable, strictly stoichiometric $LiNiO_2$ materials have been synthesized. Many $LiNiO_2$ are synthesized in the form of nickel rich compounds ($Li_{1-y}Ni_{1+y}O_2$). At the same time, this is also the reason why $LiNiO_2$ is still not commercialized. In addition, the thermal stability of $Li_{1-x}NiO_2$ is poor. Under the same conditions (such as electrolyte composition and termination voltage), the thermal decomposition temperature of $Li_{1-x}NiO_2$ is about 200 °C, and the heat release is more than that of $Li_{1-x}CoO_2$. This is because, at the later stage of charging, the Ni^{4+} at high price is unstable. It is not only easy to oxidize and decompose electrolytes, but also releases heat and gas from the collector, and it is unstable and easy to heat up and generate O_2. When the heat and gas are gathered to a certain extent, there may be an explosion.

The common synthetic methods of $LiNiO_2$ are solid phase method and liquid phase method (Song and Lee 2002). The solid phase method usually combines lithium compounds (such as LiOH, $LiNO_3$) with nickel compounds (such as $Ni(OH)_2$, $Ni(NO_3)_2$) and then roasted at high temperature in oxidizing atmosphere, then cooled and ground to obtain layered $LiNiO_2$. Because nickel is difficult to oxidize to + 3 values, it must be carried out at higher temperature. However, too high temperature is easy to generate lithium deficient $LiNiO_2$, so it is difficult to batch produce ideal $LiNiO_2$ layered structure. Usually, in the process of synthesis, we should reduce the synthetic temperature as far as possible and use oxygen atmosphere or lithium excess to stabilize Ni^{3+}, reduce lithium volatilization and inhibit lithium deficiency. Besides, the electrochemical properties of $LiNiO_2$ can be enhanced by doping other elements such as Mg (Muto et al. 2012), Al (Chen et al. 2004), Co (Cai et al. 2001), Ti and so on.

Spinel Structure ($LiMn_2O_4$) Cathode Material

$LiMn_2O_4$ material was first reported by Thackeray research group (Thackeray et al. 1983). The price of manganese is cheaper than cobalt and nickel, and manganese has the advantages of innocuity, less pollution, and easy recycling. Therefore, the $LiMn_2O_4$ cathode material of the spinel structure has aroused wide attention and research. The $LiMn_2O_4$ positive material has a tetragonal symmetry Fd-3 m structure. In one cell, there are 8 lithium atoms, 16 manganese atoms, and 32 oxygen atoms, of which Mn^{3+} and Mn^{4+} account for half each. The lithium ion is in the 8A position of the tetrahedron, and the manganese ion is in the 16d position of the eight-surface body, and the oxygen ion is in the 32e position of the eight-surface body. The tetrahedral 8a, 48F and eight-hedral 16d coplane form a three-dimensional ion channel for interworking, which is convenient for lithium ion to release and embed (Jow 2002) freely in the channel. In addition, there is enough Mn^{3+} spinel $LiMn_2O_4$ cathode material space structure, even in the delithiated state, and still can maintain the stability of the cubic close-packed oxygen distribution, to ensure the smooth slippage and embedded Li^+, so the materials have high capacity and voltage platform. The theoretical specific capacity of $LiMn_2O_4$ is 148 mAh g^{-1}, In the meantime, the actual specific capacity can reach 120 mAh g^{-1}.

$LiMn_2O_4$ is usually prepared by a high temperature solid state reaction technique. Calcination lithium hydroxide and manganese oxide mixture can obtain spinel $LiMn_2O_4$ under 700 °C. However, the synthetic method has some shortcomings, such as heterogeneous phase, large synthetic particle, wide range of particle size distribution, and long calcination time. The biggest drawback of spinel $LiMn_2O_4$ is the capacity attenuation, especially the high temperature capacity decay. The reason is derived from the change of spinel structure, the main reason can be summarized as follows.

1. manganese will dissolve in the electrolyte at high temperature
2. the Jahn–Teller effect and the formation of the passivation layer
3. high oxidation of manganese
4. the electrolyte is decomposed at high potential, destroying the spinel structure.

The modification of spinel $LiMn_2O_4$ is usually through ion doping, surface coating, and other methods. Doping is one of the most widely reported methods at present. It mainly inhibits the Jahn–Teller distortion effect by doping some transition metal elements like Mn ions, such as Fe (Bang et al. 2003), Cr (Taniguchi and Research 2005), Al (Taniguchi and Research 2005), Mg (Kakuda et al. 2007), and so on. The surface coating is mainly used to protect the electrode active substances, inhibit the side reaction from the active substances to the electrolyte, reduce the probability of manganese dissolved in the electrolyte during the reaction process, thereby improving the overall cycle performance of the material. The commonly used coating materials include ZnO (Tu et al. 2007), SiO_2 (Arumugam and Kalaignan 2008), Al_2O_3 (Lee et al. 2004), TiO_2 (Yu et al. 2006), and so on.

Olivine Structure ($LiFePO_4$) Positive Material

$LiFePO_4$ is used as a new cathode material for the past few years. Since it has stable structure and high reversible capacity, it is mainly used in high-rate lithium-ion batteries. People used to call it lithium iron phosphate. In 1997 (Padhi et al. 1997), John. B. Goodenough et al. Of Texas State University reported that $LiFePO_4$ had the function of reversible out/in Li^+, but the material didn't cause much attention because of its low electronic conductivity and poor charge–discharge performance. Since 2002, ion doping modification of $LiFePO_4$ materials has greatly improved its electrical conductivity and high current charge and discharge performance, which has caused widespread research and rapid development.

$LiFePO_4$ is an olivine structure with orthonormal symmetry. The space point group is Pbnm, the cell parameter is a = 0.6008 nm, b = 1.0334 nm, c = 0.4693 nm, and the cell volume is 0.2914 nm^3 (Sources 2011). In the olivine structure of $LiFePO_4$, O atoms are distributed in the form of six party dense heap. P atoms and O atoms form PO_4 tetrahedron through covalent bonds, Li and Fe, respectively, form LiO_6 and FeO_6 eight-hedron with the ionic bonds of O atoms, respectively. The FeO_6 octahedron is connected to the bc plane at a specific angle. A PO_4 tetrahedron, two LiO_6 octahedron and a FeO_6 octahedron are connected to form a three-dimensional structure. When Li^+ is removed, the $Li_xFePO_4/Li_{1-x}FePO_4$ phase interface is produced. With the continuous withdrawal of Li^+, the interfacial area decreases. When reaching a critical surface area, Li^+ is limited by the interface. Therefore, the Li-storage capacities of $LiFePO_4$ are affected by diffusion rate of Li^+, especially under the influence of high current.

$LiFePO_4$ is favored by researchers, mainly because of the following advantages:

1. Excellent safety, high temperature property, and thermal stability are the safest cathode material for lithium-ion battery at present.
2. High reversible specific capacity, its theoretical specific capacity is 170 mAh g^{-1}.
3. It has no pollution to the environment and does not contain any heavy metal elements that are harmful to the human body.
4. It has excellent overcharge resistance and no memory effect.
5. The resources are abundant, and the cost is low.

At the same time, $LiFePO_4$ also has some shortcomings. The electronic conductivity and ionic diffusion rate of materials are low, so the material performance requires higher particle size, and the density of material is relatively low. To overcome these shortcomings, the modification of $LiFePO_4$ materials is mainly focused on three aspects: improving the electronic conductivity, ion diffusion rate, and compaction density of materials. The main methods are doping and coating, such as common doping elements Mg (Wang et al. 2004), Nb, Ti, Co (Wang et al. 2005), Zn (Liu et al. 2006), etc., and common coating C (Cui et al. 2014), TiO_2 (Chang et al. 2008) and conductive polymerization (PPy, PANI).

Layered $LiNiCoMnO_2$Cathode Material

Among the cathode materials, $LiMn_{1/3}Ni_{1/3}Co_{1/3}O_2$, which are characterized by high voltage and high capacity, have become the research focus in recent years. The obvious synergistic effect has been shown in this kind of positive material after introducing Ni, Co, and Mn. The introduction of Co can effectively suppress the cation mixing phenomenon of Li^+ and Ni^{2+}, stabilize the structure of materials, and improve the conductivity of materials. However, too high Co concentration leads to the reduction of Li capacity. The introduction of Ni as an electronic active material can effectively enhance the electrochemical capacity and the current density of materials. The introduction of Mn can effectively reduce the cost of the materials and enhance the safety of the materials. $LiNi_{1/3}Co_{1/3}Mn_{1/3}O_2$ and $LiCoO_2$ have the similar structure of alpha-$NaFeO_2$, R-3 m space group, trigonal system, O^{2-} accounted for a cubic close-packed structure of the lattice, in position 6C. Li^+ and transition metal ions occupy octahedral voids of the close-packed structure and are arranged alternately on the (111) surface of cubic packing structure, which is located at the position of 3a and 3b, respectively. The chemical bonds formed between transition metal ions and O^{2-} are stronger and are combined with Li^+ in the way of electrostatic interaction, so that Li^+ can be reversibly embedded and deactivated, thus forming a two-dimensional lithium-ion diffusion channel.

There are two models for the spatial structure of the transition metal layer of the $LiNi_{1/3}Co_{1/3}Mn_{1/3}O_2$ material. Figure A shows the spatial model based on Wood's notation theory. Transition metal Ni, Co, Mn are three elements orderly and regularly arranged in the transition metal layer plane, forming triangular lattices. Figure B shows another model, "piled-up model", in which CoO_2, NiO_2, and MnO_2 have a regular accumulation in the transition metal layer. In the layered $LiNi_xCo_yMn_{1-x-y}O_2$ cathode materials, the valence of Co is + 3, which is consistent with the electronic structure of Co in the $LiCoO_2$ material. However, Ni and Mn valence, respectively, are + 2, + 4 value, which show that the electronic structures were different from that of $LiNiO_2$ and $LiMnO_2$. Take the $LiNi_{1/3}Co_{1/3}Mn_{1/3}O_2$ material charge and discharge process as an example, the redox reaction of Ni^{2+}/Ni^{3+} electron pair is mainly in the range of $0 \leq x \leq 1/3$; in the range of $1/3 \leq x \leq 2/3$, mainly Ni^{3+}/Ni^{4+} electrons occur Redox reaction; Co^{3+}/Co^{4+} electron pair in the range of $2/3 \leq x \leq 1$ redox reaction. The valence of manganese throughout the charge/discharge process does not change. It is generally believed that Mn^{4+} does not participate in the redox process during the entire charge/discharge process and plays a role of stabilizing the material structure.

Vanadium Phosphate

Vanadium phosphate is used as a potential cathode material, researchers have also done a lot of researches on many vanadium phosphates, including those with stable structure, such as, $Li_3V_2(PO_4)_3$, $VOPO_4$, and $LiVPO_4$. There are two main forms of $Li_3V_2(PO_4)_3$, one is a monoclinic with thermodynamic stability and the other is a rhombohedral system formed by the exchange of ions in a stable Na^+ and Nasicon structure. At about 3.80 V, two lithium ions in the rhombohedral system de-insert, but only about 1.3 can be re-embedded in the reverse process. Monoclinic $Li_3V_2(PO_4)_3$ is currently a hot spot in the research of cathode materials, in which all three lithium ions are well deintercalated and can be reversely inserted at a higher rate. So, we focus on the monoclinic $Li_3V_2(PO_4)_3$, however, the electrochemical activity of this material is complicated. A series of scanning curves will appear during charging.

Monoclinic $Li_3V_2(PO_4)_3$ belongs to the monoclinic space group ($m - Li_3V_2(PO_4)_3$), $P2_1/n$ space group with unit cell parameters of a = 0.86199 nm, b = 0.86095 nm, and c = 1.20559 nm. Li and V occupy the octahedral sites, P occupy the tetrahedral sites, and each of the VO_6 octahedrons has 6 PO_4 tetrahedrons surrounding it, while each PO_4 tetrahedron has 4 VO_6 octahedron. The polyhedron, PO_4 tetrahedron, and VO_6 octahedron make up the spatial three-dimensional network structure by sharing the vertex oxygen atoms. The vanadium atom has two positions V (1) and V (2) in the crystal structure of lithium vanadium phosphate, and the lithium atom has three different positions in lithium vanadium phosphate.

The electrochemical characteristics of lithium vanadium phosphate are closely related to the charge–discharge potential. Figure 6.2 shows the charge–discharge curve of vanadium lithium phosphate in different potential range and the relationship between the number of lithium ion inlay and the valence state of the element. From

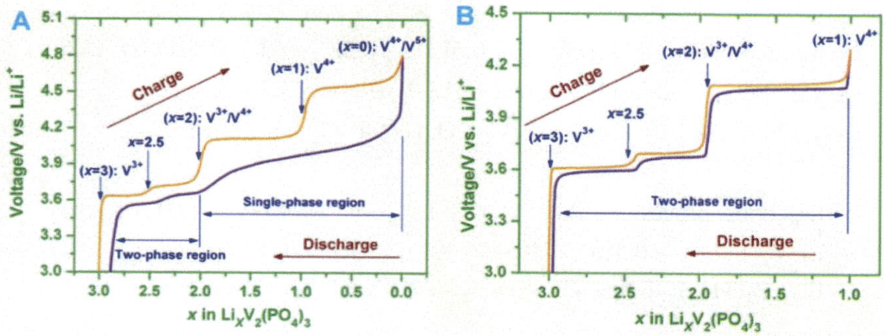

Fig. 6.2 The charge–discharge curve of vanadium lithium phosphate in different potential range, reprinted from Ref. (Wang et al. 2004), copyright 2004, with permission from American Chemical Society

the graph, we can see that the potential between 3.0 and 4.8 V, vanadium lithium phosphate has four charging platforms, and the plateau potential is 3.6, 3.7, 4.1, and 4.6 V, respectively. The corresponding phase transition reactions are $Li_3V_2(PO_4)_3 \rightarrow Li_{2.5}V_2(PO_4)_3 \rightarrow Li_2V_2(PO_4)_3 \rightarrow LiV_2(PO_4)_3 \rightarrow V_2(PO_4)_3$.

Among them, the first lithium ion removal is divided into two steps, corresponding to the 3.6 and 3.7 V potential platforms. When three lithium ions are removed from the lithium phosphate, the volume of the lithium phosphate is reduced by ~7.8% and $V_2(PO_4)_3$ is generated. In $V_2(PO_4)_3$, V (1) and V (2) have the same bond energy, bond distance and oxidation state (+4.5) resulting in the lithium ion disorderly embedded $V_2(PO_4)_3$ during the discharge process, which leads to a similar solid dissolve discharge effect, making the discharge curve appear S type. When two lithium ions are embedded, i.e., $Li_2V_2(PO_4)_3$ is generated, the ordered state of lithium vanadium phosphate is restored, and the discharge behavior changes to a phase change transition corresponding to the discharge platform of 3.6 V and 3.7 V in the discharge curve. Specific chemical reaction is as follows:

$$\text{Charging process: } Li_3V_2(PO_4)_3 - 0.5Li^+ - 0.5e^- \rightarrow Li_{2.5}V_2(PO_4)_3\ (3.6\ V)$$
$$Li_{2.5}V_2(PO_4)_3 - 0.5Li^+ - 0.5e^- \rightarrow Li_2V_2(PO_4)_3\ (3.7\ V)$$
$$Li_2V_2(PO_4)_3 - Li^+ - e^- \rightarrow LiV_2(PO_4)_3\ (4.1\ V)$$
$$LiV_2(PO_4)_3 - Li^+ - e^- \rightarrow V_2(PO_4)_3\ (4.6\ V)$$

$$\text{Discharge process: } V_2(PO_4)_3 + 2Li^+ + 2e^- \rightarrow Li_2V_2(PO_4)_3\ (3.9 - 4.5\ V)$$
$$Li_2V_2(PO_4)_3 + 0.5Li^+ + 0.5e^- \rightarrow Li_{2.5}V_2(PO_4)_3\ (3.7\ V)$$
$$Li_{2.5}V_2(PO_4)_3 + 0.5Li^+ + 0.5e^- \rightarrow Li_3V_2(PO_4)_3\ (3.6\ V)$$

When the charge–discharge potential range is in the range of 3.0–4.3 V (Zhai et al. 2010), no solid solution phenomenon occurs in the lithium vanadium phosphate, and the charge–discharge curve shows three pairs of charge–discharge platforms at 3.6, 3.7 and 4.1 V. The corresponding charge and discharge reactions are

$$\text{Charging process: } Li_3V_2(PO_4)_3 - 0.5Li^+ - 0.5e^- \rightarrow Li_{2.5}V_2(PO_4)_3\ (3.6\ V)$$
$$Li_{2.5}V_2(PO_4)_3 - 0.5Li^+ - 0.5e^- \rightarrow Li_2V_2(PO_4)_3\ (3.7\ V)$$
$$Li_2V_2(PO_4)_3 - Li^+ - e^- \rightarrow LiV_2(PO_4)_3\ (4.1\ V)$$

$$\text{Discharge process: } LiV_2(PO_4)_3 + Li^+ + e^- \rightarrow Li_2V_2(PO_4)_3\ (4.1\ V)$$
$$Li_2V_2(PO_4)_3 + 0.5Li^+ + 0.5e^- \rightarrow Li_{2.5}V_2(PO_4)_3\ (3.7\ V)$$
$$Li_{2.5}V_2(PO_4)_3 + 0.5Li^+ + 0.5e^- \rightarrow Li_3V_2(PO_4)_3\ (3.6\ V)$$

Although lithium vanadium phosphate in the range of 3.0–4.3 V only reversibly deintercalates two ions, it has no $V_2(PO_4)_3$ phase formation with poor kinetic properties, so its cycle stability and electrochemical performance are good.

Layered LiV_3O_8 Cathode Material

In 1957, Wadsley first reported the layered $Li_{1+x}V_3O_8$ and proposed its application potential as cathode materials for lithium-ion batteries (Wadsley 2010). Until the 1980s, Besenhard et al. confirmed that $Li_{1+x}V_3O_8$ has excellent lithium insertion capability and found that $Li_{1+x}V_3O_8$ as a lithium-ion battery cathode material has the advantages of high Li-storage capacity and to a large extent overcomes Disulfide clusters that cannot withstand deep discharges and moisture absorption in the air. Since then, a group of people represented by Rome and Sofia have conducted in-depth studies on the electrochemical properties of LiV_3O_8. Later, Picciotto and Zhang used modern instrumental analysis techniques to systematically analyze the crystal structure of $Li_{1+x}V_3O_8 (x = 0 \sim 0.2)$ respectively.

LiV_3O_8 is a typical layered compound belonging to the monoclinic space group $P2_1/m$ (Picciotto et al. 1993). The crystal structure of LiV_3O_8 is a layered structure composed of octahedron and deformed trigonal bipyramid, and the $(V_3O_8)^-$ layer of the B–C plane is arranged in an orderly arrangement along a-axis. In the LiV_3O_8 crystal, there are two types of structural units, the deformable VO_5 trigonal bipyramids and the VO_6 octahedra, and the two are connected to each other through a common-angle oxygen atom to form a V–O layer. There are different octahedral and tetrahedral voids between the V–O layers for lithium ion occupation. Interestingly, $Li_{1+x}V_3O_8$ crystal materials are fixedly connected by some of the interlayer lithium ions, in contrast to other layered compounds interconnected by weak Van der Waals forces. Lithium ions (Li_1) pre-existing in the material occupy octahedral sites. Due to the high energy barrier at this location, lithium ions cannot easily escape. This part of "dead lithium" (Li_1) takes on the role of structural support by a strong ionic bond $(V_3O_8)^-$ layer. Excess lithium (corresponding to x in $Li_{1+x}V_3O_8$ occupies the interlaminar tetrahedral gap, and this portion of lithium is a movable "live lithium" (Li_2). Studies have shown that lithium ions at the octahedral sites do not block the migration and diffusion of lithium ions at the tetrahedral sites. Therefore, this part of the lithium can be freely embedded/de-embedded.

Since the pre-existing lithium of the LiV_3O_8 basically occupies the interlaminar octahedral position and belongs to the immobile "dead lithium", when using LiV_3O_8 as the positive electrode, it is necessary to use a negative electrode material capable of providing a lithium source. Taking Li/LiV_3O_8 system as an example, the electrode reaction during charging and discharging is as follows:

$$\text{Positive reaction: } x\text{Li} \rightleftarrows x\text{Li}^+ + xe$$

$$\text{Anodic reaction: } \text{LiV}_3\text{O}_8 + x\text{Li}^+ + xe \rightleftarrows \text{Li}_{1+x}\text{V}_3\text{O}_8$$

At 2.63 V, theoretically 1 mol of LiV_3O_8 can reversibly deintercalate more than 3 mol of lithium, corresponding to a specific capacity of up to 300 m Ah g^{-1}.

Jin et al. (1998), through systematic experimental research, found that temperature and current density had a great influence on the discharge capacity of $Li_{1+x}V_3O_8$ during charging and discharging. When the current density is small, and temperature is relatively high, the specific capacity is the largest. The lithium insertion process of

LiV_3O_8 material has obtained a generally accepted result, through numerous studies that divided into the following three stages: (1) When the amount of lithium ion intercalation is less than 1.5, the material is in the original LiV_3O_8 single-phase region, and the diffusion of lithium ions in it is very fast, and is basically not affected by temperature; (2) When $1.5 < x < 3.2$, LiV_3O_8 coexists with $Li_4V_3O_8$. The formation of new phase $Li_4V_3O_8$ slows down the diffusion rate of lithium ions, and the Li^+ diffusion after entering $Li_4V_3O_8$ phase is greatly influenced by temperature; (3) When $x > 3.2$, the material shows a single-phase region of $Li_4V_3O_8$.

6.2.2.4 Anode Material

The anode materials of lithium-ion batteries are mainly used as a main body for storing lithium, and it realizes the insertion and extraction of lithium ions during charge–discharge (Wang et al. 2017b). Currently, the research of anode material plays a decisive role in the emergence of lithium-ion batteries. Just because of the emergence of carbon materials that the safety of metal lithium electrodes is solved, which directly leads to the application of lithium-ion battery. The anode materials are mainly various carbon materials, such as natural graphite, modified graphite, graphitized mesocarbon microbeads, and soft charcoal (such as coke) and some hard carbon, and so on. Other non-carbon anode materials include nitrides, silicon-based materials, tin-based materials, titanium-based materials, alloy materials, and the like. Nanoscale materials have also attracted attention in the research of negative electrode materials. Thin film formation of negative electrode materials is a requirement for high-performance negative electrodes and the development of microelectronics industry in recent years for chemical power sources, especially lithium secondary batteries.

Carbon Anode Material

Graphite

Graphite material has good conductivity, high crystallinity, excellent layered structure, suitable for insertion and extraction of lithium, and forms Li-GIC, a lithium-graphite intercalation compound that has specific charge/discharge capacity of over 300 mAh g^{-1} (Xu et al. 2017). The efficiency is above 90%. The deintercalation reaction of lithium in graphite occurs between 0 and 0.25 V (vs. Li^+/Li) and has a good charge–discharge potential platform, which can be matched with the lithium source cathode materials $LiCoO_2$, $LiNiO_2$, $LiMn_2O_4$, and the like.

Graphite includes artificial graphite and natural graphite. Artificial graphite is prepared by graphitizing high-graphitizable carbon (such as pitch coke) in N_2 atmosphere at 1900–2800 °C. Common artificial graphites include mesocarbon microbeads (MCMB) and graphite fibers. Natural graphite includes amorphous graphite and flake graphite. Amorphous graphite is of low purity and has a crystal plane spacing (d002) of 0.336 nm. It is mainly an ordered structure of 2H crystal planes, that is, arranged in the order of ABAB…, with the reversible specific capacity of only 260 mAh/g and an irreversible specific capacity of more than 100 mAh/g. The flake graphite interplanar spacing (d002) is 0.335 nm, which is mainly 2H + 3R crystal surface ranking structure, that is, the graphite layers are arranged in two orders of ABAB.... and ABCABC... Flake graphite with more than 99% carbon has reversible charge/discharge capacity of 300–350 mAh/g. Since the graphite pitch (d002 = 0.34 nm) is smaller than the interlaminar spacing (d002 = 0.37 nm) of the lithium-interlayer compound Li-GIC, the graphite layer spacing during charge/discharge process changes and the graphite layer is easily peeled off and powdered. Also accompanied by the phenomenon of lithium and organic solvents co-embedded in the graphite layer, will affect the battery cycle performance. Therefore, people have studied other graphite materials, such as modified graphite and graphitized carbon fiber.

Soft Carbon

Soft carbon, which is easily graphitized carbon and the amorphous carbon can be graphitized at 2500 °C or higher. The soft carbon has low crystallinity (i.e., graphitization degree), small crystal grain size, large interplanar spacing (d002), and good compatibility with the electrolyte, but the first charge and discharge have higher irreversible capacity and lower output voltage There is no apparent charge and discharge plateau potential. Common soft carbons include petroleum coke, needle coke, carbon fiber, and carbon microspheres.

Hard Carbon

Hard carbon refers to hardly graphitized carbon and is a pyrolytic carbon of a high molecular polymer. This kind of carbon is also hard to graphitize at a high temperature of 2500 °C or higher. Common hard carbons are resinous carbon (such as phenolic resin, epoxy resin, polycaprolactone PFA-C, etc.), organic polymer pyrolytic carbon (PVA, PVC, PVDF, PAN, etc.), carbon black (acetylene black). Among them, Polycarbonate resin carbon PFA-C, Japan Sony Corporation has been used as the lithium-ion batteries anode material. The capacity of PFA-C is up to 400 mAh/g, and the PFA-C crystal plane spacing (d002) is suitable. This facilitates the insertion of lithium without causing significant expansion of the structure and has a good charge–discharge cycle performance. Another type of hard carbon material is polyacene (PAS), an amorphous semiconductor material obtained by pyrolysis of phenolic resin at a temperature below 800 °C. Its capacity is about 800 mAh/g and the interplanar spacing is 0.37–0.40 nm, which is favorable for lithium intercalation and deintercalation resulting in good cycle performance.

Tin-Based Materials

Although most of the anode material is a carbon-based material; the low charge/discharge capacity and initial charge–discharge efficiency, and insufficient embedding of organic solvents, other non-carbon materials with high specific capacity have been developed. One of them is tin-based materials. Japan was the first country to study tin-based anode materials. Companies such as Sanyo Electric, Matsushita Electric, and Coats have conducted research in succession. Tin-based negative electrode materials include tin oxides, tin-based composite oxides, tin salts, tin alloys, and the like.

Tin Oxide

The simple oxides of tin include tin oxide, stannous oxide and its mixture. Compared with the theoretical capacity of 372 mAh/g of carbon material, the specific capacity of tin oxide is much higher than that of 500 mAh/g, but the first irreversible capacity is larger. There are two views on the mechanism of the Sn's oxide storage: one is an alloy type and the other is an ionic type. The deinserting process of Li in the ionic mechanism is considered as follows:

$$x\text{Li} + \text{SnO}_2(\text{SnO}) \rightleftharpoons \text{Li}x\text{SnO}_2(\text{Li}x\text{SnO})$$

That is, lithium reacts with (sub)tin oxide in a one-step reversible reaction to form lithium (I) stannate. The mechanism of alloying lithium storage is that the reaction of Li and tin oxide or stannous oxide is carried out in two steps in charge and discharge process:

$$\text{Li} + \text{SnO}_2(\text{SnO}) \rightarrow \text{Li}_2\text{O} + \text{Sn}$$
$$x\text{Li} + \text{SnLi} \rightleftharpoons \text{Li}_x\text{Sn}(0 < x < 4.4)$$

The first step is lithium substitute for Sn in tin oxide or stannous oxide, forming metal Sn and Li_2O. This step is irreversible. Next, the metal Sn reacts with metal Li to form LiSn alloy.

Almost all the experimental phenomena support the mechanism of alloy-type storage. In ionic mechanism, the reaction only regenerates a (stannate) lithium phase without Li_2O generation, and the first charge and discharge efficiency are higher. The alloy-type mechanism has an irreversible Li_2O generation in the first step, so the first charge/discharge efficiency is very low. XRD analysis observed the separated metals Sn and Li_2O, and no uniform $Li_xSnO_2(Li_xSnO)$ phase was observed. Electron paramagnetic resonance spectroscopy and XPS analysis also showed that Li exists as an atom in the oxide of Sn. By XRD, Raman, and high-resolution electron microscopy analysis of Sn oxides represented by SnO, it was proved that the mechanism of Sn-oxide deintercalation is an alloy-type mechanism. The mechanism of alloy-type deintercalation is that the formation of Li_2O causes the initial irreversible capacity, as well as the decomposition or condensation of Sn oxides and organic electrolytes. The formation of alloys of Sn and Li also causes the reversible capacity. Before the substitution reaction and the alloying reaction proceed, the organic electrolyte decomposes to form an amorphous passivation film. The thickness of the passivation film is several nanometers and the composition are Li_2CO_3 and alkyl $Li(ROCO_2Li)$. In the substitution reaction, fine Sn particles are formed in the nanometer size, highly dispersed in lithium oxide. In the alloying reaction, the produced Li_xSn also has a nanometer size. The reason why Sn oxides have a very high capacity as a negative electrode material is that there are nanosized particles of Li in the reaction product.

The main problem of tin oxide anode materials is that for the first time the irreversible capacity is large. The result is mainly due to the generation of Li_2O during the first charge and discharge and the formation of SEI film; another problem is the volume change of the material itself (the density of SnO_2, Sn, Li is 6.99, 7.29, 2.56 g/cm^3, causes the volume of material before and after the reaction to change greatly) to cause the electrode to "powder" or "reunite". Consequently, the specific capacity of the material decreases and the cycle performance decreases. To reduce the "volume effect" of tin oxide electrode materials, the following measures are usually taken:

1. Preparation of tin oxides with special morphology (such as thin films, nanoparticles or amorphous form), so that the volume expansion rate is minimized.
2. Select a suitable battery operating voltage window to reduce the occurrence of the side reactions.
3. Doping the electrode, for example, the incorporation of Mo, P, and B elements, to prevent the formation of tin clusters in charge and discharge reactions.

Tin-Based Composite Oxide

The study of tin-based composite oxide (TCO) began at Japan's Coats. The researchers found that amorphous tin-based composite oxides have better cycle life and higher reversible specific capacity which result in great attention. Subsequently, there are many studies on this area in succession. Tin-based composite oxides can solve the problem of large volume change of Sn oxide anode materials, high irreversible capacity for first charge and discharge, and unsatisfactory cycling performance to some extent by adding some metal or non-metal oxides to Sn oxides. Oxides of elements such as B, Al, Si, Ge, P, Ti, Mn, Fe, etc. are then obtained by heat treatment.

Tin-based composite oxides have an amorphous structure, and other oxides added to make the mixture form an amorphous glass body, and so it can be the general formula $SnM_xO_y(x = 1)$, in which M represents a group of metal or non-metal elements. Structurally, the tin-based compound oxide consists of an active-site Sn–O bond and a surrounding random network structure. The random network consists of added metal or non-metal oxides, which separate the active centers from each other. Therefore, Li can be effectively stored, and the capacity is related to the active center. The reversible specific capacity of tin-based composite oxides can reach 600 mAh/g, and the volume specific capacity is greater than 2200 mAh/cm^3, which is around two times the charge/discharge capacity of the highest carbon negative electrode material (amorphous carbon and graphitized carbon, respectively, less than 1200 and 500 mAh/cm^3).

There are two types of lithium storage mechanism for tin-based compound oxides: one is ion type and the other is alloy type. The ionic mechanism suggests that the Li is embedded with the TCO electrode, and Li exists in the ion form in the product. Taking $SnB_{0.5}P_{0.5}O_3$ as an example, the mechanism can be expressed as

$$x\text{Li} + \text{SnB}_{0.5}\text{P}_{0.5}\text{O}_3 \rightleftharpoons \text{Li}x\text{SnB}_{0.5}\text{P}_{0.5}\text{O}_3$$

The alloying mechanism of TCO is like the mechanism of tin oxide alloy. It is also the mechanism of the two-step reaction. First, TOC and Li react to form Li_2O, other oxides and tin, then tin react with Li to form lithium tin alloy. Taking Sn_2BPO_6 as an example, its mechanism can be expressed as:

$$4Li + Sn_2BO_6 \rightarrow 2Li_2O + 2Sn + 1/2B_2O_3 + 1/2P_2O_5$$

$$8.8Li + 2Li_2O + 2Sn + 1/2B_2O_3 + 1/2P_2O_5 \rightleftarrows 2Li_{4.4}Sn + 2Li_2O + 1/2B_2O_3 + 1/2P_2O_5$$

The amorphous structure of tin-based composite oxides has little change in volume before and after charging and discharging. The structure is stable, and it is not easy to be destroyed. So, the cycle performance of tin-based composite oxides is relatively good. Moreover, compared with the oxide of crystalline Sn, the structure of the tin-based compound oxide is beneficial to insertion and removal of lithium, and the increase of the diffusion coefficient of lithium.

Silicon-Based Materials

Silicon (Si)-based materials have been widely studied as the anode materials for Li-ion batteries (LIB) due to its excellent discharge/charge capacities. However, the anode materials have a large volume change in the cycle process, causing pulverization of Si, loss of the electric contact, as well as plenty of side reactions. These disadvantages cause the poor long cycling life and slow the extensive commercial applications of Si for LIBs. The lithiation/delithiation reaction and interphase reaction mechanism are gradually studied. First, Si maintains a discharge/charge capacity of about 4200 mAh g^{-1} during full lithiation progress. Second, Si anode material has a considerably low discharge voltage (about 0.4 V vs. Li^+/Li), which causes the formation of a high working potential and the current density in the LIBs. Third, there are some advantages of Si, such as abundant Si element, low cost, good environmentally friendly, excellent chemical stability. Therefore, Si can be considered as a promising candidate material.

A battery of Si–Li phases consist in the courses of thermal alloying of Si and Li such as LiSi material, $Li_{12}Si_7$ material, $Li_{13}Si_4$ material, and $Li_{22}Si_5$ material. For the crystalline phase, it has more stable dynamics than the amorphous phase owes to the poor formation energy. However, the crystalline phase is not necessarily beneficial during the lithiation of Si progress. Liu et al. reported an atomic model of the crystalline Si lithiation. The result vividly confirmed the peeling process of Li-ions on Si (111) atomic plane. The formation of Li_xSi can be investigated by the in situ transmission electron microscopy (TEM), the Si nanowires (SiNWs) (the diameter of 130 nm), and the growth direction of (111) plane is used in lithiation process. The amorphous Li_xSi shells are first developed during lithiation progress. The gradual migration of amorphous and crystalline interface is exhibited in Fig. 6.3. Meanwhile, the thickness of amorphous and crystalline interface has only $\approx$1 nm.

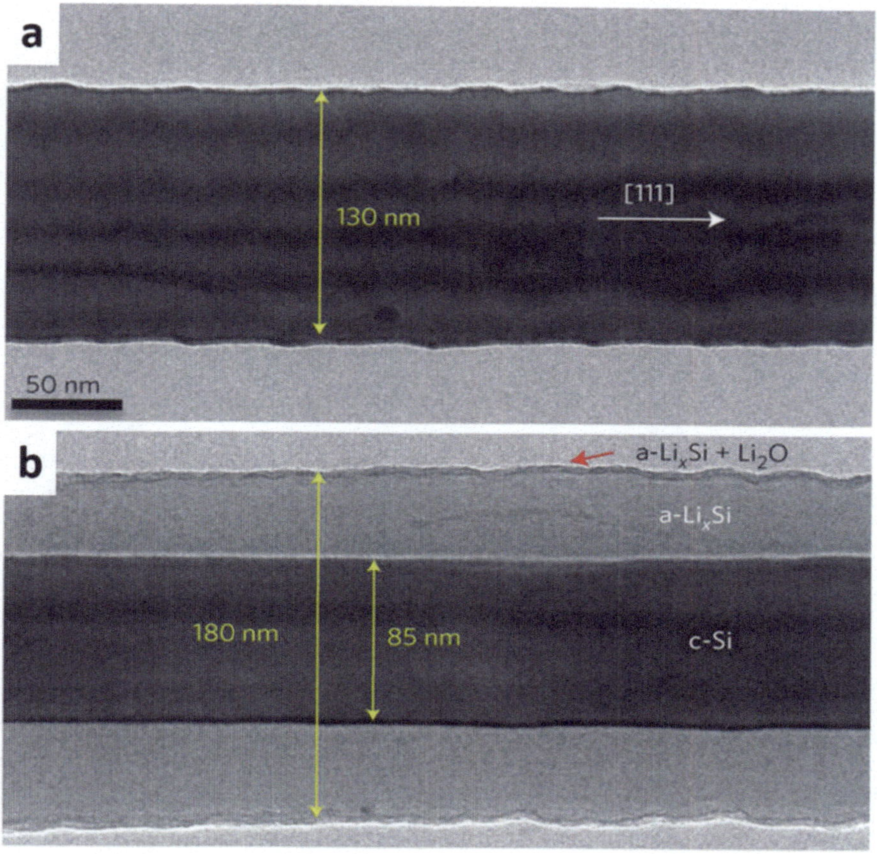

Fig. 6.3 **a** The SiNW with crude sidewall owe to faceting, **b** SiNW of partially lithiated with the a-Li_xSi layer encircling a c-Si core, reprinted from Ref. (Liu et al. 2012), copyright 2012, with permission from Nature

The volumes of wires increased are about 280% after the SiNW full lithiation, and the theoretical capacity is 3579 mAh g^{-1}.

Si and alloy-type anode materials have an inevitable challenge such as the poor cycle stability that owe to the significant volume change. Si has bigger volume change during full lithiation progress when compared to the original value. Therefore, Si electrodes can retain a good morphological feature during cycling progress, which is a very big challenge in the future. Figure 6.4a displays the discharge/charge performance of the bare Si anode (particle size is about 10 μm). The Si delivers an excellent initial discharge/charge capacity when the metallic Li is used as a counter electrode. However, the initial irreversible discharge/charge capacities are very high (about 2650 mAh g^{-1}), this reveals that amounts of Li^+ are irreversibly transferred between Li and Si electrode or the Li^+ is consumed due to the side reactions. These

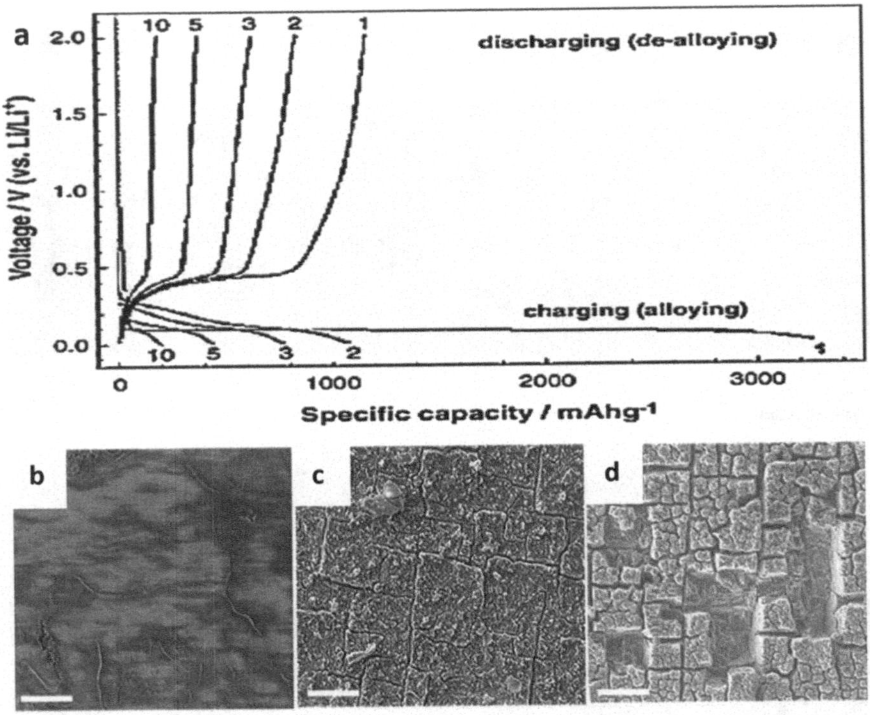

Fig. 6.4 **a** Discharge/charge curves of a Si anode, reprinted from Ref. (Lee et al. 2004), copyright 2004, with permission from The Electrochemical Society. **b–d** The SEM images of Si electrode surface at 3rd, 8th, and 50th cycle, respectively, reprinted from Ref. (Liu et al. 2016), copyright 2016, with permission from Nature

disadvantages cause that the Si electrode cannot be used in commercial field. Even worse, its discharge/charge capacity reduces very fast at 10 cycles. Figure 6.4a shows discharge/charge curves of Si electrode at different cycling cycles. Meanwhile, Fig. 6.4b–d displays many cracks at different cycling. Hence, the active material may not contact with electrode surface.

Many research groups try to investigate the potential natural phenomenon to solve these problems. Hence, some previous works have revealed the failure modes of Si anodes materials. The electrode failure modes due to the serious volume change are mainly divided into the following three points.

(1) Serious volume changes cause strong internal stress on the surface of Si and pulverization of the Si morphology. By the way, several of alloy-type materials have the same phenomenon.
(2) Some active materials lose the electrical contact with the contiguous units, the conductive networks, and the current collector due to serious volume change. Therefore, the active materials exhibit the self-isolation and low electrical conductivity.

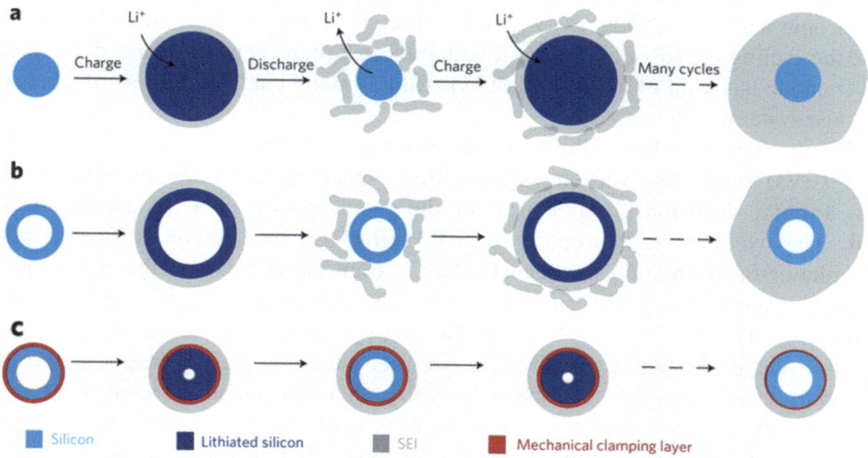

Fig. 6.5 The schematic illustration of lithiation processes of **a** solid nanowire with repetitive SEI layers, **b** Si nanotube with SEI layers of a repetitive grow, **c** Si nanotube with a stable SEI layer due to the mechanical restraining layer, reprinted from Ref. (Cheng et al. 2012), copyright 2012, with permission from The Royal Society of Chemistry

(3) The SEI layer is repetitively formed owing to the serious volume change and pulverization of Si (see Fig. 6.5). As the electrolyte is decomposed on the Si anode surface, it causes the self-passivating SEI layers formation. The SEI layers are mainly composed of the polycarbonate, the Li-based salt, and oxide material during the initial lithiation progress, and it obviously limits the flow of electrons. In the meantime, the formation of the SEI layers can prevent the electrolyte from immediate contact with the Si anode and can avert further decomposition. The fresh SEI layers can be continuously formed because of fracture of the Si anode material. The stable SEI layers avail to increase long cycling life of Si electrodes. However, the excess SEI layers can consume amounts of Li^+ and block the electron conduction.

The fabrication of Si electrode faces a severe challenge due to the serious volume change. In the meantime, the full cell battery displays a poor Li-storage capacity. Some methods can accommodate the serious volume change such as the engineering void space which usually can decrease the power density of Si anode.

Spinel Structure $\boldsymbol{Li_4Ti_5O_{12}}$ ***Material***

Murphy et al. were the first to report the Li^+ intercalation ability of $Li_4Ti_5O_{12}$ (LTO) material in 1983 (Xu et al. 2017) after that, LTO was considered as a very prominent anode material of Li-ion batteries (LIBs) due to its "zero strain" structure during cycling processes, the stable voltage (1.5 V vs. Li^+/Li), the high safety, and long cycling life (Han et al. 2017). However, LTO anode material has poor electronic and ionic conductivity. Hence, many researchers try to adopt series methods such as the carbon coating, the size reduction, and the structural doping to improve the electronic and ionic conductivity of LTO. Therefore, the electrochemical property and the cycling stability of LTO anode material can be enormously enhanced. Despite LTO anode has a lot of advantages, LTO still has many challenges in some energy storage systems.

Firstly, how to enhance the power property is a challenge in the electric vehicles (EVs) field. As everyone knows, the low electrical conductivity and the low lithium diffusion coefficient limit the rate capacity of LTO anode material. The excellent rate property of LTO relies on the lithium diffusion and the phase transformation mechanism. Li^+ is transferred between the tetrahedral 8a sites (spinel $[Li_3]^{8a}[Li_1Ti_5]^{16d}[O_{12}]^{32e}$) and the octahedral 16c sites (rock-salt $[Li_6]^{16c}[Li_1Ti_5]^{16d}[O_{12}]^{32e}$) upon the phase transition, in the meantime, the stable voltage plateau is 1.55 V [10]. The cell structure hardly changes during charge/discharge progress. Compared with the other active material such as olivine $LiFePO_4$, there are a little of papers to study the insertion/deinsertion processes mechanisms of LTO due to the similar lattice constant ($Li_4Ti_5O_{12}$ and $Li_7Ti_5O_{12}$). Therefore, $Li_4Ti_5O_{12}$ and $Li_7Ti_5O_{12}$ materials are difficult confirmed by the conventional characterization technique. However, recently, the in situ characterization methods can provide more evidences about the mechanism of Li^+ insertion/deinsertion for LTO, which also can provide a method to enhance high capacity of LTO.

Secondly, how to control the charge/discharge cycles and the emission of gas (CO_2, H_2, and CO) of LTO electrodes is another challenge of LIBs, because the gas can further cause the safety issues. Therefore, many researchers look for the cause of the gas. The lack of solid electrolyte interface (SEI), the decomposition of electrolyte, and the special electrochemical potential will cause safety issues. However, there are still some methods to control the emission of gas such as online in situ gas analysis approaches, which are available to the large-scale applications of LTO storage batteries in the future.

Li_3VO_4 Material

The anode material plays a vital role in the property of LIBs. In 2013, Li_3VO_4 was considered as a very promising candidate material. Li_3VO_4 materials have some advantages compared with other anode material. Firstly, the Li_3VO_4 lattice consists of the corner-sharing VO_4 and the tetrahedra LiO_4. The Li_3VO_4 structure is very similar to Li_3PO_4. Hence, Li_3VO_4 is used as an ionic-conductor, which facilitates the transfer of Li^+. Secondly, Li_3VO_4 anode materials have a small volume change upon cycling processes. Therefore, Li_3VO_4 anode materials can display excellent cycling stability. Thirdly, Li_3VO_4 materials have the lower Li^+ intercalation voltage (from 0.5 to 1.0 V vs. Li^+/Li) and the higher discharge/charge capacity (about 590 $mAhg^{-1}$) compared to LTO. In addition, since the voltage range of Li_3VO_4 exceeds than other the graphite anode material, the Li_3VO_4 can effectively avoid the growth of lithium dendrites. However, Li_3VO_4 material has low electronic conductivity, which limits its wide application of in LIBs. Recently, some researchers try to adopt series methods to enhance conductivity and rate performance of Li_3VO_4 material, such as the carbon coating or the forming composite or doping with metal ions.

The doping with ions method is a simple way to improve the electrochemical property of the electrode materials. For instance, Mg^{2+} or Mo^{6+} doping can enhance electric conductivity of electrode material. Besides, Ni-doped Li_3VO_4 can display the superior Li-storage capacity and cycling stability. Cu-doped Li_3VO_4 is another method to improve electrical conductivity. For example, Wang et al. adopt the Cu-doped Li_3VO_4 to improve the electrochemical properties of the electrode materials. In the work, Li_3VO_4 doped with 10% Cu content displays the ultrahigh Li-storage capacity (about 335 mAh g^{-1}) at 8 Ag $^{-1}$, which is more than twice as high as the pristine Li_3VO_4.

6.2.3 *Electrolyte*

Organic electrolyte is a key factor restricting the development of lithium-ion battery. The organic electrolyte of Lithium-ion battery mainly has three parts: (1) lithium electrolyte salt, (2) organic solvent, (3) additives. In addition, the organic electrolyte also contains some other impurities such as water, hydrogen fluoride, metal ions, and so on. Currently, $LiPF_6$ is used in commercial Li-ion battery. To date, no single solvent has met the requirements of lithium-ion batteries. Therefore, typical mixed solvents and alkyl carbonate mixed solvents have been applied in commercial Li-ion battery.

6.2.3.1 Lithium Electrolyte Salt

Chemical and electrochemical performance of electrolyte lithium salts lithium-ion battery electrolyte lithium salt, according to different types of anions, can be divided into inorganic anion lithium salt and organic anion lithium salt as the two major categories. The inorganic anionic lithium salts mainly include $LiClO_4$, $LiBF_4$, $LiAsF_6$, and $LiPF_6$, etc. The organic anionic lithium salts mainly include $LiCF_3SO_3$ and $LiN(SO_2CF_3)_2$ and their derivatives and the like. It can also be divided into fluorine-containing lithium salts and non-fluorine-containing lithium salts simply based on whether the anions contain fluorine or not. The anion structure is a vital factor affecting the property of lithium salt, and the smaller lattice energy is the first condition for the lithium salt to obtain certain solubility in organic solvents. Therefore, the lithium anion must first have a larger anion radius and the second is easy to dissociate with the lithium ion to enhance the electronic conductivity of an electrolyte. In addition, the lithium salt anion should also have good electrochemical stability, thermal stability, and decomposition products to form a particularly stable SEI membrane on the negative electrode surface. The fluorinated lithium salt anion has a charge delocalization effect, it inhibits the formation of ion pairs and improves the conductivity of the electrolyte; besides, it also improves the electrochemical stability of the electrolyte system; the decomposition product of salt facilitates the formation of a particularly stable SEI membrane. Therefore, fluoride-containing lithium salts have been the main body of lithium salts in Li-ion battery electrolytes, and it is also an important direction of development.

6.2.3.2 Inorganic Anion Electrolyte Lithium Salt

Many simple lithium salts such as LiF, LiCl, and LiBr have not been used in Li-ion batteries that owe to the low solubility in lithium salts. Although LiI-based electrolytes have moderate conductivity, LiI is difficult to prepare in non-aqueous conditions, and I is easily oxidized. Li_3AlF_6 and Li_2SiF_6 have low solubility in organic solvents (only about 0.1 mol/L), and their electrical conductivity is generally at 10–5 S/cm. $LiSbF_6$ and $LiAlCl_4$ have larger anions and smaller lattice energies. They have good conductivity as lithium salt organic electrolytes, such as the conductivity of 1 mol/L of $LiSbF_6$ + THF and $LiAlCl_4$ + THF electrolyte is 16×10^{-3} S/cm. However, Sb(V) and Al(III) are easily reduced, and the SEI membrane formed is permeable to Li^+ but also permeable to $SbF6^-$, $AlCl4^-$. $LiTaF_6$ and $LiNbF_6$ based organic electrolytes also have suitable electrical conductivity, but they are very expensive and not easy to obtain the high purity, and on lithium electrodes, like $LiSbF_6$ and $LiAlCl_4$, Ta(V) and Nb (V) is easily reduced to metal Ta and Nb. In addition, $LiSbF_6$ and $LiTaF_6$ were also found to initiate polymerization of cyclic ethers in the electrolyte. Therefore, among the numerous lithium salts, only $LiClO_4$, $LiPF_6$, $LiBF_4$, and $LiAsF_6$ may be used in lithium-ion batteries. $LiClO_4$ is the longest researched lithium salt with appropriate electrical conductivity, thermal stability, and oxidation stability. However, it is generally accepted in the international lithium battery industry that it is only suitable for research work systems and cannot be used in practical applications. In batteries, this is because $LiClO_4$ itself is a strong oxidant, and it is feared that under certain uncertain conditions, it may cause safety problems; $LiBF_4$ not only has poor thermal stability, is easily hydrolyzed, but also has relatively low electrical conductivity. Among the known lithium salts, the $LiAsF_6$-based electrolyte has the best cycle efficiency, relatively good thermal stability, and almost the highest conductivity. However, the potential carcinogenic effect of the As(V) reductant has limited its application. Therefore, $LiPF_6$ has been applied in commercial Li-ion battery and Li AsF_6 is mainly used in military lithium batteries. $LiClO_4$ is mostly used in experimental studies because it does not have good hydrolysis and thermal stability.

$LiPF_6$ has the following advantages as a lithium-ion battery electrolyte lithium salt: (1) It can form an appropriate SEI film on the electrode, especially on the carbon negative electrode; (2) It can effectively passivate the positive electrode current collector to prevent it from dissolving; (3) It has a wider power chemically stable window; (4) It has appropriate solubility and high conductivity in various non-aqueous solvents; (5) It has relatively good environmental friendliness. $LiPF_6$ has outstanding oxidation stability. In a single solvent DMC electrolyte system, the oxidation potential of several electrolyte lithium salts changes according to the following rules: $LiPF_6$ > $LiBF_4$ > $LiAsF_6$ > $LiClO_4$; the change of conductivity in EC/DMC electrolyte system is $LiAsF_6$ ≈ $LiPF_6$ > $LiCLO_4$ > $LiBF_4$. Tarascon reported that when $Li_{1+x}Mn_2O_4$ was used as a battery cathode material, the oxidation stability of several lithium salts in the DMC + EC (1:1) solvent system changed according to the following rules: $LiPF_6$ > $LiClO_4$ > $LiBF_4$ > $LiAsF_6$ > $LiN(SO_2CF_3)_2$ > $LiCF_3SO_3$, Conductivity changes according to the following rules: $LiAsF_6$ ≈ $LiPF_6$ > $LiClO_4$ ≈ $LiN(SO_2CF_3)_2$ > $LiBF_4$ > $LiCF_3SO_3$. The decreasing order of thermal stability of several electrolyte lithium salts is $LiCF_3SO_3$ > $LiN(SO_2CF_3)_2$ > $LiAsF_6$ > $LiBF_4$ > $LiPF_6$. In PC or EC-based electrolytes, the ion-association was reduced by $LiCF_3SO_3$ > $LiBF_4$ > $LiClO_4$ > $LiPF_6$ > $LiN(SO_2CF_3)_2$ > $LiAsF_6$.

6.2.3.3 Organic Anion Electrolyte Lithium Salt

The organic anion electrolyte lithium salts mainly include $LiCF_3SO_3$, $LiN(SO_2CF_3)_2$ and $LiC(SO_2CF_3)_3$, and their derivatives. So far, $LiN(SO_2CF_3)_2$ and $LiC(SO_2CF_3)_3$ have the highest conductivity in all anionic lithium salts. $LiCF_3SO_3$ shows good cycle efficiency in some secondary lithium battery systems, but not as good as $LiClO_4$, $LiPF_6$, and $LiAsF_6$, whose conductivity is only about half of $LiPF_6$. At the same time, when $LiCF_3SO_3$ based organic electrolyte is used in Li-ion batteries, there is also the problem of corrosion of aluminum or copper electrode current collectors and compatibility with carbon negative electrodes and layered transition metal oxides. In all three-fluorine alkyl and perfluoroaryl sulfonic salts, $LiCF_3SO_3$ has the highest electrical conductivity and the lowest price. $LiN(SO_2CF_3)_2$ (LiTFSI for short) was first proposed by Armand as a Li salt for Li-ion battery electrolyte, having close electrical conductivity to $LiPF_6$ and having inherent electrochemical stability and thermal stability. It is not easy to hydrolyze, and its thermal decomposition temperature exceeds 360 °C. LiTFSI is considered to be the most attractive electrolyte lithium salt for highly graphitized electrodes such as MCMBs, and it can ensure a stable discharge energy close to the maximum energy even in repeated cycles. In each electrolyte system, the Coulomb efficiency of almost every charge–discharge cycle except the first cycle is close to 100%, which is mainly owing to the impact that LiTFSI can form a low resistance and stable SEI film on MCMBs. However, when LiTFSI-based organic electrolytes are used in Li-ion battery, there is also corrosion of copper or aluminum electrode current collectors. This is mainly due to the fact that TFSI-salts such as Al^{3+}, Cu^{2+}, and Fe^{2+} are highly soluble in many organic solvents, preventing the deposition of salt and the passivation process. There are mainly three ways to improve the corrosion potential of LiTFSI on positive current collectors. (1) Add a perfluorinated inorganic anion salt such as $LiPF_6$ to the electrolyte to form the passivation membrane containing fluorine to prevent TFSI-adsorption; (2) Use a low-viscosity ether solvent to reduce the solubility of TFSI complexes of Al^{3+}, Cu^{2+}, and Fe^{2+} ions; (3) Replacement of LiTFSI with an imine salt with a larger molecular radius, such as the $Li + (SO_2CF_2CF_3)_2EC + THF$ electrolyte system with a potential of above 4.4 V relative to the aluminum current collector (relative to Li/Li$^+$). $LiC(SO_2CF_3)_3$ was first prepared by Dominey. It also has good thermal stability and electrochemical stability. At temperatures above 300 °C, decomposition begins to occur. It also has good conductivity which is only about 10% smaller than LiTFSI in the same electrolyte system. When $LiC(SO_2CF_3)_3$ based organic electrolytes are used in lithium-ion batteries, there is no corrosion problem for aluminum or stainless-steel electrode current collectors. This superior stability is mainly due to $LiC(SO_2CF_3)_3$ that has a large anion, a steric hindrance effect attached to the central carbon atom and a high degree of delocalization of the negative charge.

6.2.3.4 New Type Electrolyte Lithium Salt

Although $LiPF_6$ has good electrical conductivity and oxidation/reduction stability, meanwhile, $LiPF_6$ has the disadvantages of poor thermal stability and easy hydrolysis, which has brought considerable difficulties for the production and use of $LiPF_6$. Therefore, finding cheap electrolyte lithium salt as a substitute for $LiPF_6$ is the development direction of electrolyte lithium salt. The research work in this area mainly includes two aspects: (1) complex lithium borate compound; (2) complex lithium phosphate compound.

The complex lithium borate has good environmental friendliness, so it has received some attention in the research of electrolyte lithium salt. There are more than 10 complex lithium borate compounds that have been studied successively, but the complex lithium borate compounds have low conductivity, which limits their application. Li-BOB (lithium bisoxalatoborate) is a representative of this type of lithium salt, and not only has the Al current collector better oxidation stability in LiBOB-based electrolytes than in $LiPF_6$-based electrolytes, but also in PC-based electrolytes, it exhibits a unique SEI film-forming function. The complexed lithium phosphate compound is basically a new compound formed by replacing fluorine in $LiPF_6$ with a trifluoromethyl group or a perfluoroalkyl group. LiFAP($liPF_3(C_2F_5)_3$) is a typical representative of this type of lithium salt. Compared with $LiPF_6$, LiFAP has certain advantages because it does not exist HF, and FAP—has high stability and low reactivity, which prevents the malignant interaction between electrodes and electrolytes and thus protects the electrode active materials. Although the research and development of new electrolyte lithium salts have achieved a lot of results, $LiPF_6$ will still be the main electrolyte lithium salt in consideration of performance, price, and production process.

6.2.4 Separator

The separator itself is not only a good conductor of electrons, but also has the property of passing through the electrolyte ions. The separator material must have good chemical and electrochemical stability, good mechanical properties, and high infiltration of the electrolyte during repeated charge and discharge. The interface compatibility between the separator material and the electrode, and the retention of the separator on the electrolyte all have important effects on charge and discharge performance and cycling property of the lithium-ion battery. Lithium-ion battery commonly used diaphragm materials that include fiber paper or non-woven fabric, synthetic resin made of porous membrane. The common diaphragms are polypropylene and polyethylene porous membranes, and the basic requirement for the membrane is a high stability in the electrolyte. Because polyethylene, polypropylene microporous membrane has higher porosity, lower resistance, higher tear strength, better resistance to acid and alkali, good elasticity, and retention of aprotic solvents, so the goods of Lithium-ion battery separator material mainly uses polyethylene, polypropylene microporous membrane. Polyethylene and polypropylene separators have defects that have poor affinity for electrolytes. For this, they need to be modified, such as grafting hydrophilic monomers on the surface of polyethylene, polypropylene microporous membranes, or changing the organic solvents of the electrolytes, etc. Some studies have found that cellulose composite membrane materials have good lithium-ion conductivity and good mechanical strength and serve as lithium-ion battery separator materials.

Lithium-ion battery separator preparation methods are mainly melt-spinning and cold-stretching (MSCS), or dry and thermally induced phase separation (TIPS) or wet method, the two major categories of methods. Because the MSCS method does not include any phase separation process, the process is relatively simple and there is no pollution during the production process. At present, most manufacturers in the world use this method for production, such as Ube, Mitsubishi, Tonen, and Celanese of the United States. The TIPS process is more complex than the MSCS process and requires the addition and removal of diluents. Therefore, the production cost is relatively high and may cause secondary pollution. Currently, the world's companies that use this method to produce membranes include Asahi Kasei of Japan, Akzo of the United States, and 3M. Company, etc.

In lithium batteries, the basic function of the separator is to block electron conduction while conducting ions between the positive and negative electrodes. Polypropylene microporous membranes are commonly used in lithium primary batteries and separators common to lithium-ion secondary batteries that are polypropylene and polyethylene microporous membranes which have good chemical and electrochemical stability in secondary batteries.

In summary, the requirements for the separator material for lithium-ion battery are as follows:

1. Thickness
   Lithium-ion batteries commonly use thinner diaphragms (<25 pm); diaphragms used in electric cars and hybrid cars are thicker (about 40 pm). In general, the thicker the diaphragm, the greater its mechanical strength and the less likely it is to puncture the battery during assembly. However, the same type of battery, such as a cylindrical battery, has less active material that can be added to it; With less space occupied by thinner diaphragms, more active material is added,

which can increase the capacity and specific capacity of the battery (due to the increased interface area), and the thin diaphragm also has lower impedance.

2. Permeability
   The separator has almost no effect on electrochemical performance of the batteries. For example, the presence of the separator can increase the resistance of the electrolyte by 6–7 orders of magnitude but has almost no effect on performance of the batteries. The impedance coefficient generated by the electrolyte flowing through the effective pores of the membrane is usually distinguished from the impedance coefficient of the electrolyte, the former being known as the MacMullin coefficient. In commercial batteries, the MacMullin coefficient is generally 10–12.
3. Air permeability
   For a given form of membrane material, its air permeability and resistance are proportional. Lithium-ion battery separators should have good electrical properties and low air permeability.
4. Pore volume rates
   The porosity product rate and permeability are closely related, and the porosity of lithium-ion battery separator is about 40%. For the lithium-ion battery, controlling the porosity of separators is very important. The standard porosity rate is an integral part of the diaphragm standard. High porosity and uniform pore size distribution will not hinder the flow of ions. Inhomogeneous pore size distribution will cause uneven current density and affects the activity of the working electrode. Due to the inconsistent working load of some parts of the electrode and other parts, eventually the damage of the battery core is faster.
5. Wettability
   The separator should have fast, complete wetting characteristics in the battery electrolyte.
6. Absorb and retain electrolyte
   In lithium-ion batteries, the separator mechanically absorbs and retains the electrolyte in the battery without causing swelling, because the absorption of the electrolyte is the need for ion transport.
7. Chemical stability
   The separators can exist stably in the battery, are chemically inert to strong oxidation and strong reduction environments, do not degrade under the above conditions, and do not lose mechanical strength. At temperatures up to 75 °C, the separator should be able to withstand the oxidation of highly oxidizing positive electrodes and corrosion of highly corrosive electrolytes. The stronger the antioxidant capacity, the longer the life of the diaphragm in the battery. Polyolefin-based separators (e.g., polypropylene, polyethylene, etc.) are resistant to most chemicals, have good mechanical properties, and can be used in the medium temperature range. Polyolefin-based separators are the ideal choice for commercial lithium-ion battery separators. In contrast, the polypropylene film has better oxidation resistance when it is in contact with the cathode materials of the lithium-ion batteries. Therefore, in the three-layer separator (PP/PE/PP),

polypropylene (PP) is placed in the outer layer and polyethylene (PE) is placed in the inner layer, which increases the oxidation resistance of the separator.

8. Space stability
   When the diaphragm is removed, the edges must be flat and not curled to prevent the battery assembly from becoming complicated. Diaphragms do not collapse when they are immersed in the electrolyte.
9. Puncture strengths
   The separator used for the winding battery has a high requirement for the puncture strength so that the electrode material does not pass through the separator. If part of the electrode material penetrates the separator, a short circuit will occur, and the battery will be wasted. For the separator, the lithium-ion batteries require higher puncture strength than the lithium primary battery.
10. Mechanical strength
    The sensitivity of the separator to the penetration of the electrode material particles is characterized by the mechanical strength. During the winding process, the electrode generates a large mechanical stress between the anode-separator-negative electrode interface, and some looser particles may forcibly penetrate the separator. Short the battery.
11. Thermal stability
    Water in lithium-ion batteries is detrimental, so the electrodes are usually dried under vacuum drying at 80 °C. Therefore, under these conditions, the separators must not shrink significantly. Each battery manufacturer has its own unique dryness. The process requirements for lithium-ion secondary battery separators are drying at 90 °C for 60 min, the transverse and longitudinal shrinkage of the diaphragm should be less than 5%.
12. Aperture
    For lithium-ion battery separators, because the most critical requirement is to not allow lithium branches to pass through, separators with submicron pore sizes are suitable for lithium-ion batteries.
13. High-temperature stabilities
    In high temperature conditions, the diaphragm is needed to stem the mutual transmission between the electrodes. The high temperature stability of the separator was characterized using thermomechanical analysis (TMA). The so-called TMA is to determine the ratio of diaphragm growth and temperature under certain load conditions.
14. Electrode interfaces
    The separator and the electrode should provide a good interface for electrolyte flow.

In addition to the above requirements, the diaphragm should also overcome the following defects: pinholes, wrinkles, gels, dirt, etc. Before the application of the lithium-ion battery, all the above characteristics of the diaphragm should be optimized.

6.3 The Theory and Research Progress of Sodium-Ion Batteries

6.3.1 Cathode Material

Sodium-ion batteries (SIBs), superior to rich reserve and low cost of sodium, have been of great concern in energy storage systems. However, the primary hindrance—large ionic radius of sodium and its slow electrochemical kinetics—make it difficult to develop suitable and high-performance electrodes for SIBs, which is key to improving the energy density of SIBs and facilitating their commercialization. Recently, a lot of research into cathode materials for SIBs have been made, including transition metal oxides, polyanions, organic materials and polymers, and amorphous materials.

6.3.1.1 The Transition Metal Oxides

Compared with other electrode materials, the transition metal oxide is attractive because of their species materials, diversity, structure stability, and good electrochemical performances. In LIBs, layered oxide $LiMO_2$ (M is transition metal Co, Mn, Fe, Ni, etc.) has received widespread attention and application due to the excellent electrochemical properties. Similarly, in the sodium-ion batteries, a similar transition metal oxide Na_xMO_2 is also a research hotspot (Hamani et al. 2011; Pan et al. 2013; Ma et al. 2011). The radius of sodium ion (1.02 Å) is larger than that of lithium ion (0.76 Å), so the difference between radius enables them to occupy oxygen ions in the crystal structure of the oxide electrode. In the transition metal oxide structure, lithium ions occupy tetrahedral oxygen vacancies and octahedral oxygen vacancies, whereas sodium ions are unable to occupy tetrahedral oxygen vacancies due to their large ionic radius. Sodium transition metal mainly contains three different oxide structures: P2 type, P3 type, O3 type. The crystal structure of three types of materials is shown in Fig. 6.6.

The type O_3 material is a typical a-$NaFeO_2$ structure. This structure of material is widely used as positive electrodes in LIBs such as $LiCoO_2$, Li-rich, ternary, and so on. Similarly, type O3 material also received widespread attention in SBIs. For example, $LiCrO_2$ cathode material in LIBs shows a very poor electrochemical performance, but $NaCrO_2$ in SIBs has a reversible specific capacity of 120 mAh/g (0.5 Na). For $LiCrO_2$, $Cr^{4+}O_4^{2-}$ tetrahedrons are easily formed in the $LiCrO_2$ crystal lattice, thus destroying the original layered crystal structure. For $NaCrO_2$, the larger interlayer distances will prevent the movement of Cr^{4+}, thereby avoiding the formation of tetrahedral coordination. The inert gas is used in the $NaCrO_2$ preparation process, and Cr(III) is reduced hard. The material can be further modified by in situ carbon coating, leaving more space for improvement in electrochemical performance (Kim et al. 2012).

P2 and O3 are layered materials, while P2-type layered oxides tend to have higher specific capacity and better cycling properties. For example, Kamaba's research

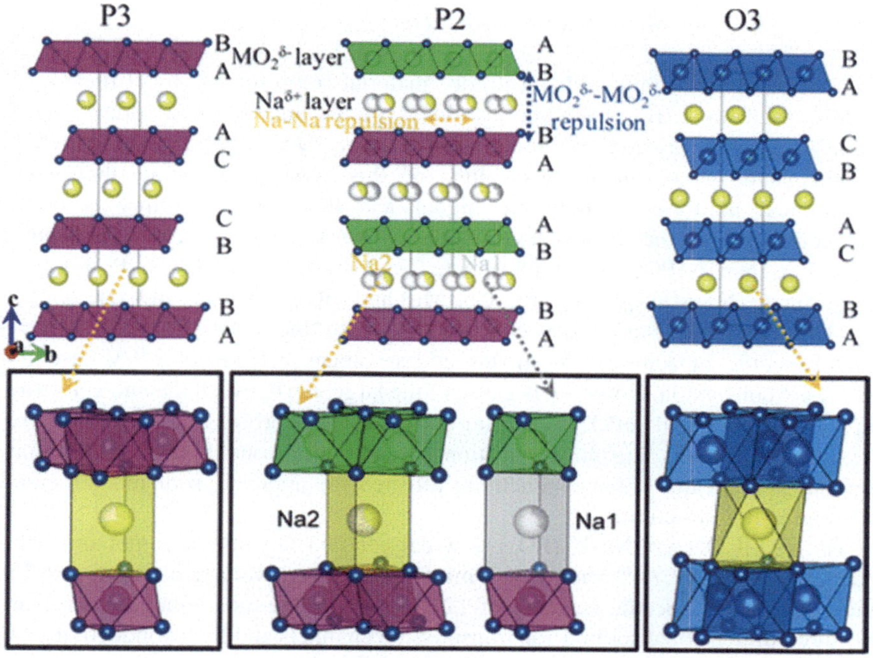

Fig. 6.6 P2 type, P3 type, O3 type crystal structure of three types of material model diagram. Adapted with permission, reprinted from Ref. (Sathiya et al. 2017), copyright 2017, with permission from American Chemical Society

group on $Na_xFe_{0.5}Mn_{0.5}O_2$ has found that the electrochemical performance of P2 is better than that of O3 because more prismatic vacancies are occupied by Na-ions in P2-type, which is conducive to sodium ion prolapse and embedding. In the meantime, it is more difficult for the P2 phase to undergo other phase transitions, accompanied by the rotation of MO_6 octahedron and the breakage of M–O key bits, so P2 phase is more stable than O3 phase. However, the problem is that the initial discharge capacity exceeds the charge capacity, seriously affecting the matching of the whole battery.

Research has shown that P2 materials containing Li can effectively inhibit the phase transition of P2−O2, even if charged to the extreme conditions of 4.4 V. It can effectively maintain the P2 structure, which greatly improves the stability of the material. For example, P2−$NaLi_{0.2}$ $Ni_{0.25}Mn_{0.75}O_8$ has 100 mAh/g reversible specific capacity in the 2–4.2 V electrochemical range with virtually no attenuation after 50 long cycles. P2−$Na_{0.8}[Li_{0.12}Ni_{0.22}Mn_{0.66}]$ has 118 mAh/g reversible specific capacity in the 2–4.2 V electrochemical interval, $Na_{0.66}Li_{0.18}Mn_{0.71}Co_{0.21}O_{2+d}$ has 200 mAh/g reversible specific capacity, while cycling and magnification are also better.

6.3.1.2 The Polyanionic Cathode Materials $A_xM_y[(XO_m)^{n-}]_z$

The polyanionic compound electrode material represented by $LiFePO_4$ is widely used in LIBs. In addition, this type of material has drawn much attention in the research of sodium-ion battery. Due to the good structural stability and strong anion inducibility of this type of material, their electrochemical platforms usually are higher and their cycle stability is better. In recent years, the common research systems mainly include phosphate ($NaFePO_4$, $Na_3V_2(PO_4)_3$), pyrophosphate ($Na_2MP_2O_7$, $Na_4M_3(PO_4)_2P_2O_7$), fluorophosphate (Na_2MPO_4F)(M = Fe, Co, Mn) and sulfate ($Na_2Fe_2(SO_4)_3$).

The olivine-type $NaFePO_4$ is concerned because this material has a high theoretical specific capacity of 154 mAh/g and a voltage platform of 2.9 V based on Fe^{2+}/Fe^{3+} conversion. However, because of the large gap between the unit cell structures of $NaFePO_4$ and $FePO_4$ in charging and discharging process, the material is irreversible when undergoing phase transition. Meanwhile, the one-dimensional sodium-ion channel and the lower conductivity affected the application of this structural material in this structure.

The NASICON type $Na_3V_2(PO_4)_3$ is widely studied polyanionic compound material based on the V^{3+}/V^{4+} variant; this material has a high voltage platform of 3.7 V and a theoretical specific capacity of 117 mAh/g, in the meantime, this structure has three-dimensional sodium-ion transmission channels and high conductivity. On the other hand, a similar $Li_3V_2(PO_4)_3$ has a wealth of research experience that can be directly used for reference, which are conducive to $Na_3V_2(PO_4)_3$ as electrode materials in SIB for research and development. Because of $Na_3V_2(PO_4)_3$, material is prepared in an inert or reducing atmosphere, it can also be modified by means of various carbon compounds and carbon coatings, in order to exhibit excellent electrochemical performance. At the same time, this material can also be modified by cation replacement. In addition, there are reports in the literature that $Na_3V_2(PO_4)_3$ can also be applied as negative electrodes. Based on the V^{2+}/V^{3+} valence change, it has a reversible specific capacity of approximately 100 mAh/g at both 0.3 and 1.6 V platforms.

Pyrophosphate $Na_2MP_2O_7$(M = Fe, Mn, Co) is a large material system with various structural types, high structural stability, and high ionic conductivity. Based on different transition metal cations and different preparation method, this type of material has triclinic, orthorhombic, tetragonal, and other different crystal structures. Between $Na_2FeP_2O_7$ and $Na_2MnP_2O_7$ materials, the triclinic is a more stable structure, but for the $Na_2CoP_2O_7$, the orthorhombic is the most stable structure. Its reversible specific capacity is 80 mAh/g at 3.0 V.

Fluorophosphate Na_2MPO_4F(M = Fe, Co, Mn) is another kind of polyanionic compound system, which also has the characteristics of stable structure and carbon composite modification. In this structural material, sodium ion occupies a pseudo-octahedron with 2-dimensional sodium-ion channels. Among them, the carbon composite Na_2FePO_4F material has a reversible specific capacity of 100 mAh/g at 3.0 V. It has been reported in the literature that the graphene composite $NaVPO_4F$ has a high specific capacity of 120.9 mAh/g with retention of 97.7% after 50 cycles.

There are few reports and studies on sulfate materials, in which $Na_2Fe_2(SO_4)_3$ has a reversible specific capacity of 100 mAh/g and an average operating voltage of 3.8 V.

6.3.1.3 Organic Compounds and Conductive Polymer Cathode Materials.

I. The conductive polymer is a type of polymer with a large π-bond conjugated structure, the delocalized π-bonded electrons can move freely on the polymer chain, become insulative or semiconducting in this state, and undergo oxidative doping (p-doping) or reducing doping (n-doping), can obtain electrical conductivity comparable to metal and is therefore called "conductive" polymer. The first conductive polymer, polyacetylene, was discovered in collaboration with Japanese scientist Hideki Shirakawa and American scientists MacDiarmid and Heeger in 1977, and they received the 2000 Nobel Prize in Chemistry.

The common conductive polymers are oxyacetylene, poly-benzene, polyaniline, polypyrrole, polythiophene, and their derivatives. They are generally prepared by chemical polymerization and electrochemical polymerization methods, except for polyacetylene, the polymerization products of the remaining polymers are all doped. Its specific structure and energy storage capacity are shown in Table 6.3.

Early studies have found that the doping/demising of conductive polymers at room temperature is accompanied by a reversible redox process. This phenomenon has attracted interest in secondary batteries. Subsequent studies show that conductive polymers have a certain energy storage capacity in both organic and aqueous systems, and their capacity is determined by the degree of doping. According to the data in Table 6.3, the doping of other polymers except polyaniline is not higher than 40%, and the corresponding theoretical specific capacity is not higher than 150 mAh/g, while polyaniline with high doping degree is fully charged. The state is unstable.

Table 6.3 The structure and properties of several common conductive polymers

Polymer name	Structure	Maximum doping	Theoretical specific capacity(mAh g^{-1})	Potential (V vs. Na^+/Na)
Oxyacetylene(PAc)	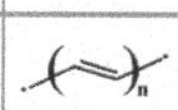	0.07	144	0–2.0(n) 3.5–4.0(p)
Poly-benzene(PPP)	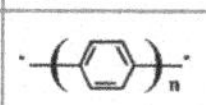	0.4	141	0–1.0(n) 4.0–4.5(p)
Polyaniline(PAn)	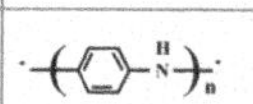	1.0	295	2.5–4.0
Polypyrrole(PPy)	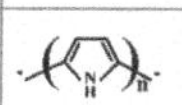	0.33	136	2.5–4.0
Polythiophene(PTh)	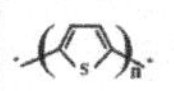	0.25	82	0–2.0(n) 3.0–4.2(p)

II. Radical polymers are polymers that can stably exist and graft a large number of radical groups. This type of polymer is usually composed of two parts: the polymer backbone that determines its processing properties and the radical groups that determine its electrochemical performance. In general, the higher the redox potential used as a side chain radical group, the smaller the molecular weight of the polymer main chain, and the higher the electrode density. The structural and electrochemical characteristics of free radical polymers are as follows:

Radical polymer doping is high but specific capacity is not high. In order to exist a large number of single electrons on the radical polymer, it is necessary to realize electron delocalization and extremely high spatial positions on the radical group, for example, the active center is in a conjugated structure or a ring structure, which is bound to increase the redox single. The molecular weight of the bulk, currently reported electrode material, the PTVE highest specific capacity of 135 mAh/g.

The electrode reaction rate is fast. Unlike other organic reactions, due to the electronic delocalization of the electrochemically active sites of the free radical polymer, only the outer shell electrons participate in the electrode reaction and do not involve the cleavage and generation of chemical bonds, so the charge transfer rate is extremely fast (10^{-1} cm s^{-1}), and shows that the redox potentials of the free radical polymers are similar ($\Delta E \approx 60$ mV), the peak currents are almost equal, the reversibility is high, and the rate performance is good.

Perfect cycle performance. The main chain of the radical polymer mainly includes polystyrene, polymethacrylate, cross-linked polynorbornene, polyvinyl alcohol, polyethylene oxide, and other structures. These main chain structures have a wide electrochemical window, are structurally stable in common organic electrolytes or aqueous systems, and do not involve the cleavage and generation of chemical bonds during the reaction process. Therefore, free radical polymer electrode materials generally have excellent cycle performance.

Special electrode reaction process. The electrode reaction of the free radical electrode firstly transfers the charge between the current collector and the conductive carbon, and then the charge exchange between the free radical site connected to the conductive carbon and the current collector, and then the charge between the free radical site and the conductive carbon moving, the last is the movement of the charge between the radicals on the polymer chain. In general, the main chain structure of the polymer is not conductive, so it is necessary to add a large amount of fibrous conductive agent (VGCF) to build a conductive network to increase the electrochemical performance of free radicals in polymer.

6.3.2 *Anode Material*

6.3.2.1 Carbon-Based Anode Material

I. Graphite carbon-based anode materials

Structured graphite has good electrical conductivity and is suitable for the insertion and extraction of Li^+. It has a wide range of sources and low cost, which is commonly used in LIB systems. Through the electrochemical reduction process, Li^+ is embedded in the Van der Waals gap between the graphitic carbon layers and forms a first-order lithium-graphite intercalation compound (LiC_6). Its reversible capacity is greater than 360 mAh/g (C_{the} = 372 mAh/g) (Chen et al. 2017b). In contrast, the results of graphite as a cathode material in SIBs are not optimistic. Early first-principles calculations showed that Na^+ ions are unable to form intercalated graphite compounds than other alkali metals (DiVincenzo and Mele 1985).

Studies have shown that the insertion of Na^+ forms a higher order compound of NaC_{64}, and the possibility of electrochemical reduction to form low-order sodium-graphite remains to be explored. In addition, the theoretical capacity of the anode material in SIBs is only 35 mAh/g, due to the fact that the distance between the graphite layers of the graphite ink is about 0.335 nm, which is smaller than the minimum interlayer spacing of Na^+ (0.37 nm). In recent years, researchers have found that by increasing the interlamellar spacing of graphite and selecting suitable electrolyte systems (such as ether-based electrolytes), it is possible to increase the sodium storage capacity of graphite and improve its electrochemical performance. Expanded Graphite (EG) is a superior carbon-based anode material in SIB. EG is a graphite-derived material formed by a two-step redox process that retains the long-range ordered layered structure of graphite. The interlaminar distance of 0.43 nm can be obtained by modulating oxidation and reduction processes. These features provide electrochemical insertion of Na^+. The favorable conditions are shown in Fig. 6.7. EG may be a promising carbon-based anode material in SIBs.

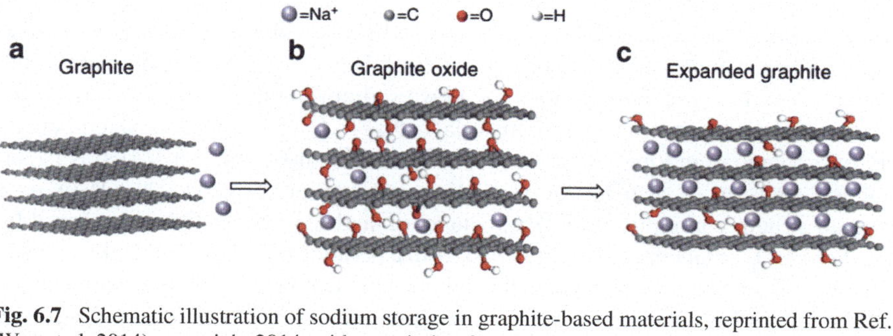

Fig. 6.7 Schematic illustration of sodium storage in graphite-based materials, reprinted from Ref. (Wen et al. 2014), copyright 2014, with permission from Nature

II. Non-graphite carbon-based anode materials

The non-graphitic carbon-based negative electrode material has a large interlayer spacing and disordered structure, which is advantageous for the insertion/absorption of Na^+, and is currently the most studied type of sodium-ion battery carbon-based negative electrode material. According to the degree of difficulty in graphitization and the arrangement of graphite crystallites, non-graphitic carbon-based negative electrode materials are mainly classified into soft carbon (graphitized mesocarbon microbeads, coke, etc.) and hard carbon (carbon black, resin carbon, etc.). Two major categories. Both soft carbon and hard carbon are amorphous carbons and are mainly composed of graphite-like microcrystals having a small width and thickness, but the arrangement is more disordered than graphite and has a relatively large distance between carbon layers.

In general, carbon materials that can be graphitized above 2800 °C are called soft carbons, and the internal arrangement of graphite crystallites is relatively orderly, and the width and thickness of the microwafer layer are large, and the sodium storage mechanism is mainly expressed as carbon. Adsorption of Na^+ is on the edge of the layer, the surface of the carbon layer, and the microcrystalline gap. The experimental results show that the random layer microcrystals of the soft carbon anode material will expand with the change of the pyrolysis temperature, and the interlayer spacing increases from about 0.36 nm to about 0.42 nm, thereby improving the electrochemical performance. The soft carbon negative electrode material obtained by pyrolysis at 900 °C showed a reversible capacity of 114 mAh/g at a current density of 1000 mA/g with excellent rate performance and cycle performance.

Hard carbon is a carbon material that is difficult to graphitize at temperatures above 2800 °C. Its internal graphite crystallite arrangement is more disorderly than soft carbon, and it contains a part of micro-nanopore regions, which has a high sodium storage capacity. It is the ideal sodium storage carbon-based anode material.

6.3.2.2 Titanium-Based Anode Material

Titanium-based materials are based on Ti^{3+}/Ti^{4+} valences, but also have a class of deintercalation materials in LIBs. $Li_4Ti_5O_{12}$ and TiO_{12} materials are the most studied titanium-based materials in sodium-ion batteries. Due to the greater variety of sodium compounds, there are more options for active sodium-ion titanium-based materials.

Among many titanium-based materials, $Na_2Ti_3O_7$ has the lowest embedded sodium potential and has a specific capacity of 178 mAh/g at 0.3 V (vs. Na^+/Na) voltage platform. It also has good cycling and rate performance. $Na_2Ti_3O_7$ is a layered structure composed of TiO_6 octahedra, and the two Ti–O bonds shared by the three TiO_6 octahedra slowly move to form a parallel axis. This will form a zigzag $(Ti_3O_7)^{2-}$ transition metal layer. The sodium ions can be reversibly deintercalated in the middle Na layer, and since the transition metal layer is zigzag, sodium ions can occupy two positions in the Na layer. The transition metal layer can form stable covalent bonds, and sodium ions can form strong ionic bonds with TiO_6 octahedra.

Therefore, the structure of $Na_2Ti_3O_7$ is very stable. As another oxide of titanium, the large-diameter amorphous TiO_2 nanowire can also releasably resorb sodium and reversibly deintercalate 0.4Na at a voltage platform of 1.5 V (vs. Na^+/Na), in other words, it has a reversible specific capacity of 150 mAh/g.

$Li_4Ti_5O_{12}$ in LIBs is superior to its "zero strain" characteristics, and it also has a higher voltage, this can avoid the production of SEI film, so it has the best cycle stability, which is most widely used as anode material. Although it is a spinel type material, its lithium insertion mechanism is that lithium ions intercalate into excess oxygen octahedral voids, and thus can also be applied in SIBs.

P2 materials have received extensive concern as cathode materials in SIBs, but due to the lack of sodium in the initial composition of the material, the first Coulomb efficiency is affected (above 100%), which limits the application of the material in full cells. On the other hand, this initial sodium vacancy can also be used as a place for the sodium storage of the negative electrode. Based on the low voltage potential of Ti^{3+}/Ti^{4+}, the P2 type titanium-based material can also be used as anode electrodes in SIBs. For example, P2 – $Na_{0.66}[Li_{0.22}Ti_{0.78}]$ has a specific capacity of 116 mAh/g with a 75% capacity retention rate after 1200 cycles. The Coulomb efficiency is nearly 100% during the entire charge–discharge cycle. Further in situ XRD testing shows that the material is removed. Embedded sodium is only 0.77% of the volume change, and it can also be a "zero strain" anode material, which has a certain application prospect.

$NaTi_2(PO_4)_3$ is a titanium-based NASICON compound with a theoretical capacity of 133 mAh/g, which has attracted attention because of its low cost and environmental friendliness. At the same time, due to its high sodium-insertion potential (~2.1 V) and the stability of the water environment, this material is also used in the negative electrode materials of new generation aqueous sodium-ion batteries.

6.3.2.3 Metal and Alloy Anode Materials

Metals can store sodium ions by forming an alloy with sodium ions. The alloy as anode materials in SIBs can provide a very high specific capacity, but it can also bring about a corresponding problem of large volume expansion. This is in contrast to the anodes in LIBs. In fact, the effect of volume change is even more pronounced than that of alloy anode materials in LIBs because of the larger radius of sodium ions. Cheerier et al. found that the same volume expansion is found by calculating the relationship between volume energy density and volumetric expansion in the alloy. Under the conditions, the volumetric energy density of the sodium alloy compound is half that of the lithium alloy compound, which means that if it is desired to reach the specific capacity (~360 mAh/g) of the graphite anodes in LIBs, the sodium alloy compound needs to have 150% volume expansion rate.

In addition to traditional metallic materials, black phosphorus also has the alloying reaction characteristic of sodium inlay. NaP_7, NaP, Na_2P, and Na_3P are formed in sequence during charge and discharge. The insertion of up to three sodium ions can

Table 6.4 Comparison of maximum volume change after lithium intercalation and sodium insertion in various alloy reaction mechanism materials

	Sn	Sb	P
Embedded Li volume change	$Li_{22}Sn_5$-350%	Li_3Sb-260%	Li_3P-330%
Embedded Na volume change	$Na_{15}Sn_4$-520%	Na_3Sb-390%	Na_3P-490%

bring a theoretical ratio of 2600 mAh/g. Capacity is the highest theoretical sodium-ion electrode material among all materials. With the high capacity of the three sodium ions, it will also cause a huge volume change of 490%. The black phosphorus material prepared by high-energy ball milling, although it can reach a high specific capacity of 1750 mAh/g for the first time, and retains only about 1200 mAh of capacity after 30 cycles (Table 6.4).

6.3.2.4 Metal Oxide Anode Material

Tirado et al. of the University of Kent in the UK first discovered that transition metal oxides can also be applied to sodium-ion batteries. They first found that spinel-structured $NiCo_2O_4$ has a specific discharge capacity of 600 mAh/g, and subsequently showed a reversible ratio capacity of 200 mAh/g, and showed a reversible specific capacity of 250 mAh/g in a full battery ($NiCo_2O_4/Na_{0.7}CoO_2$).

The Fe_3O_4 material has a high specific capacity as anode material in LIBs. At the same time, the iron-containing element means a low raw material cost and excellent environmental protection. It can also be applied as an anode in SIBs. In the case of a sodium-ion battery half-cell test, it shows a reversible electrochemical performance of charge and discharge, and a first discharge specific capacity of 643 mAh/g at a magnification of 0.06 C, but the first Coulomb efficiency. Only 56% have no practical value, so further improvements are needed.

As an anode in SIBs, SbO_4 can react with a total of 14 sodium atoms according to the two-step reaction mechanism of the conversion reaction and alloy formation reaction:

(1) $Sb_2O_4 + 8Na \rightarrow 2Sb + 4Na_2O$;
(2) $2Sb + 6Na \rightarrow 2SbNa_3$.

It means that there is a theoretical specific capacity of 1227 mAh/g. It was found that Sb_2O_4 thin films prepared by magnetron sputtering have a reversible specific capacity of 896 mAh/g, which is the highest specific material among the anode material in SIBs. According to the foregoing reaction mechanism, the second-stage alloy formation reaction plays a major role in the reversible capacity.

As an emerging two-dimensional material, MoS_2 also has two kinds of charge and discharge mechanisms. When the cutoff voltage is higher than 1.5 V, the deintercalation mechanism is exhibited, and the crystal structure remains unchanged. When the cutoff voltage is set to 0 V, the conversion reaction mechanism is exhibited, and at the same time, high specific capacity is achieved.

6.3.2.5 Non-metallic Single Anode Materials

I. Red phosphorus

Compared with Sn and Sb, in addition to the larger volume expansion problem of red phosphorus, there is a problem that the red phosphorus conductivity is too low, so most researchers mainly combine red phosphorus with carbon (PC) or other. The metal composite forms an alloy (PM) to enhance its cyclic stability. The alloys that have been studied so far include Zn−Ge−P, Ge−P, Fe−P, CoP, and Sn−P. The P–C compound mainly combines P with graphene, carbon nanotubes or other carbon matrices to enhance the conductivity of the material.

II. Phosphorene

Phosphorene is a new type of two-dimensional material with only one or several layers of black phosphorus, which has a wrinkled layered structure. Common preparation methods include mechanical peeling, liquid phase peeling, and chemical vapor deposition. Sun et al. first used phosphorene as a sodium storage anode material, sandwich structure of phosphorene and graphene composites at 50 mA/g, and the first week charging capacity was as high as 2440 mAh/g, and the capacity retention rate was 83% after 100 cycles. The authors believe that after reducing the number of black phosphorus layers to form phosphorene, the layer spacing increases, which is conducive to the diffusion of sodium ions between layers, and graphene provides a fast channel for the transmission of electrons, which makes the sandwich structure of phosphorene-graphene composites have good electrochemical properties. Recently, Huang et al. used electrochemical cation embedding to prepare phosphorene as a sodium storage anode material. The preparation method is simple and rapid, and the number of phosphorene layers can be controlled by adjusting the voltage. The phosphene produced by this method had a specific charge capacity of 1968 mAh/g at 100 mA/g, remaining 60.5% after 50 cycles. At present, the application of phosphene in sodium-ion battery is less, and it needs further exploration. At the same time, the application of phosphorene also provides new ideas for developing the new high-performance materials.

The specific capacity of P is much higher than that of Sb- and Sn-based materials. It has a stable circulation for over 1000 cycles while maintaining a high specific capacity. However, P has certain toxicity and flammability. At the same time, its reduced product, Na_3P, will be easily hydrolyzed. Burning and toxic PH_3 gas, which limits the application of P in sodium-ion batteries.

6.3.2.6 Organic Anode Material

As shown in Fig. 6.8, only the following two types of free radical anode materials have been reported (Suga et al. 2009). The theoretical specific capacity of Polymer 1 is 59 mAh/g. Considering that a certain amount of cross-linking agent is added during the polymerization process, and its theoretical specific capacity is only 42 mAh/g.

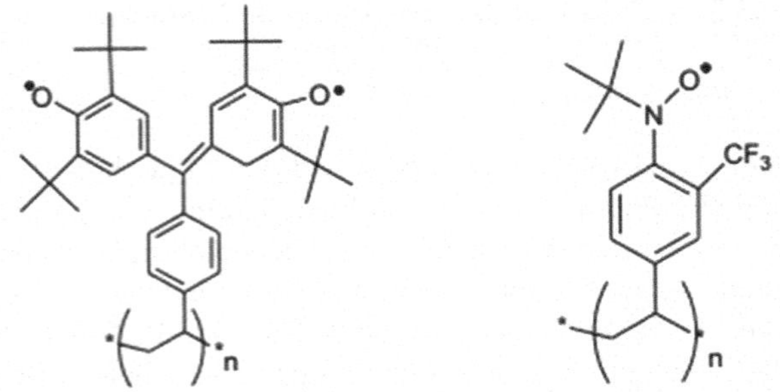

1

59 mAh/g

2

103 mAh/g

Fig. 6.8 Polymer anode material, reprinted from Ref. (Suga et al. 2009), copyright 2009, with permission from WILEY-VCH

At 10 C, the polymer can achieve a theoretical capacity of 100% with a discharge platform of 0.06 V (vs. Ag/AgCl) and a 500-cycle cycle capacity loss of less than 5%. Polymer 2 has a higher theoretical specific capacity (103 mAh/g) and a lower electrode potential (−0.76 V vs. Ag/AgCl) than Polymer 1, but it has not been reported as a battery anode.

6.3.3 Sodium-Ion Battery Electrolyte

The electrolytes of sodium-ion batteries are classified according to their state of existence and can be divided into liquid and solid electrolytes. The liquid electrolyte used in sodium-ion batteries is also dissolved in organic solvents by sodium salts. The sodium salts are generally $NaPF_6$, $NaClO_4$, $NaAsF_6$, $NaNO_3$, etc. The organic solvents can also be analogized to the electrolyte solvents in LIBs. With alkyl carbonate compounds. Alexandre Ponrouch across measurement of the viscosity and conductivity and single electrolyte and complex electrolyte solvents, the high temperature stability of the integrated solvent, the stability of the electrochemical window, and the electrolyte solution of different sodium salts can be assembled to measure the cycle stability of the whole cell. The rate performance concludes that the EC:PC (v:v, 1:1) electrolyte solution of $NaClO_4$ exhibits the best overall performance. The polymer electrolyte is made by adding a salt substance into a polymer matrix in a doped form to form an ion-conductive polymer. The common polymer matrixes are linear, chain, and cross-linked polymers, such as commonly used PEO, PTFE, and PAN. The alkali metal salts are all sodium salts with negatively charged bulky anions such as $NaBH_4$, NaI, $NaBF_4$, and sodium polyphosphate. There is also a sodium superconductor of NASICON structure, whose structural formula is $A_xMM(XO_4)_3(M = V, Fe, Mn)$, which is a kind of composite electrolyte. Its earliest research history is the application of sodium -sulfur battery solid electrolyte.

In addition to these liquid and solid electrolytes, there are also some ion exchange membranes used in batteries to replace the traditional addition of inorganic salts such as lithium salts or sodium salts.

6.4 Metal-Air Battery

The metal-air battery is a "half-fuel" battery that is between a primary battery and a fuel cell and is a type of the chemical battery. The metal-air batteries use metals as anode-active materials, and these electrode potentials are more negative (such as magnesium, aluminum, zinc, or lithium) and oxygen or pure oxygen as a cathode active material (Cheng and Chen 2012). Oxygen diffuses in form of gas to the gas–liquid-solid three-phase interface and reacts with the metal anode to release electrical energy. Metal-air battery generally uses alkaline electrolyte solution as electrolyte solution, because lithium, sodium, and other more negative metals can react with water, and it can only use non-aqueous organic electrolyte such as phenolic solid electrolyte or inorganic electrolyte such as $LiBF_4$ salt solution. Metal-air batteries not only have the advantages of batteries and fuel cells, but also have the characteristics of rich raw materials, high specific energy, long life, complete pollution-free, no need for charging equipment, can quickly replace metal fuel and electrolyte, and excellent shelf properties. It is considered as a new energy source with great development and application prospects in the future. Metal-air batteries have high mass-to-energy and volume-to-volume ratios. The most important thing is that cathode materials are inexhaustible. Therefore, researchers and workers have made a lot of effort to develop metal-air batteries. According to the different kinds of metal anodes, metal-air batteries can be roughly divided into four categories: zinc-air batteries, magnesium-air batteries, aluminum-air batteries, and lithium-air batteries. Scientists have also studied iron-air batteries, but because of the lower discharge voltage, mass-to-energy ratio, and the higher development and use costs, there are relatively few studies on such batteries. The following will describe the development of the four types of batteries (Table 6.5).

Table 6.5 Comparison of electrochemical performance of anode materials for metal-air battery

Anode materials	Capacity (Ah g^{-1})	Theoretical voltage/V	Theoretical specific energy (Wh Kg^{-1})	Actual voltage/V
Lithium	3.86	3.4	13.0	2.4
Aluminum	2.98	2.7	8.1	1.2–1.6
Magnesium	2.20	3.1	6.8	1.2–1.4
Zinc	0.82	1.6	1.3	1.0–1.1

6.4.1 Working Principle of Metal-Air Battery

The positive plate is pressed by a metal current collector, a waterproof layer, and a catalytic layer in this order. The waterproof layer is a breathable hydrophobic film made of carbon black and polytetrafluoroethylene emulsion (PTFE) that has the function of preventing electrolyte leakage. The catalyst layer consists of polytetrafluoroethylene emulsion (PTFE), activated carbon black, and catalyst. For rechargeable metal-air batteries, the catalyst not only has the ability to reduce oxygen but also to have oxygen oxide ions. The positive electrode active material is oxygen in the air. During the discharge process, oxygen is catalytically reduced to hydroxide ions at the gas–liquid–solid three-phase interface by a catalyst.

The reaction equation is $O_2 + 4e^- + 2H_2O = 4OH^- \quad E = 0.401\ V$.

The anode material of metal-air battery will be continuously consumed during the discharge process, that is to say, how much the theoretical energy density is, only depends upon how much the energy of the negative metal is released. This is the only active material transferred in the battery. The discharge reaction on the metal electrode depends on metals, electrolytes, and other factors, while oxygen is introduced from the air. Discharge reaction is generally as follows:

$$M \rightarrow M^{n+} + ne$$

$$\text{battery discharge total reaction: } 4\text{M} + n\text{O}_2 + 2m\text{H}_2\text{O} \rightarrow 4\text{M(OH)}.$$

In the above formula: M represents the metal used in the battery, and the value of n depends on the degree in the oxidation of the metal in the battery. Most metals are unstable in the electrolyte solution and can erode or oxidize to H_2. $M + nH_2O \rightarrow M(OH) + (n/2)H_2$. Figure 6.9 shows the structure diagram of metal-air battery. The Coulombic efficiency on cathode fades away for self-discharge or half-etching reaction, which must be controlled to reduce this capacity loss of the battery. The metal battery also has a reactive negative electrode and an air electrode, and is formed by electrochemical reaction. The cathode electrode reactant is inexhaustible. In some cases, metal-air batteries have high mass-to-energy and volumetric energy, whose ultimate energy depends on the negative current capacity of the negative electrode and the storage and processing technology of the reaction product. Metal-air batteries that have been researched and developed include primary batteries, storage batteries, rechargeable batteries, and mechanical recharging batteries.

6.4.2 Electrolyte

Electrolyte solutions in metal-air batteries are mostly neutral salt solutions or strong alkaline solutions. When a neutral electrolyte (such as sodium chloride and potassium carbonate) is used, the corrosion of the anode is small, but the passivation of the metal

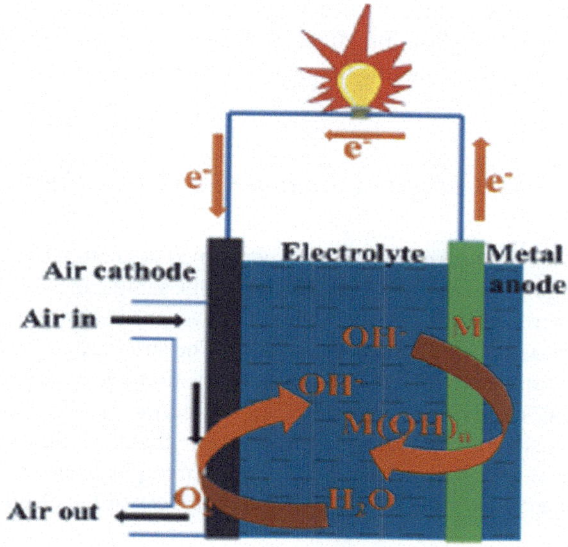

Fig. 6.9 Schematic diagram of metal-air battery, reprinted from Ref. (Yang et al. 2013), copyright 2013, with permission from The Royal Society of Chemistry

surface is severe, the operating voltage is reduced, the power and current of the battery are difficult to increase, and the voltage lag is also caused. The battery can only be used as a low-power power output device. When a strong alkaline electrolyte is used, the passivation of aluminum and magnesium is reduced, and the alkaline solution can absorb a certain amount of hydroxide of the reaction product. The performance of the battery is relatively good, but some metals will undergo strong hydrogen evolution corrosion in a strong alkali environment and release a lot of hydrogen, the output power and of the battery, the utilization of the anode will be reduced. The situation is worse at high current densities. The types of corrosion inhibitors include inorganic corrosion inhibitors, organic corrosion inhibitors, and composite corrosion inhibitors. Inorganic corrosion inhibitors mainly achieve the effect of corrosion inhibition by increasing the hydrogen evolution over potential of zinc electrodes. Organic corrosion inhibitors act by physical or chemical adsorption to create a physical barrier on the upper surface of the electrode to achieve corrosion inhibition. Each has its own advantages and disadvantages, so now organic–inorganic composite corrosion inhibitors are getting more and more attention. Inorganic corrosion inhibitors mainly include metal elements, metal oxides, metal hydroxides, and salts thereof, which gradually create a protective film on the metal's surface by alloying or substitution reaction to increase the hydrogen evolution overpotential of the metal electrode to achieve corrosion inhibition. Metallic indium has low resistivity, good flexibility, chemical stability, and high hydrogen evolution overpotential, and has good affinity with zinc. Adding metal particles can reduce the contact resistance between particles, so zinc is currently the most widely used mercury-inhibiting corrosion inhibitor. In addition, metals that have attracted wide attention include mercury, lead, cadmium, antimony, tin, aluminum, calcium, strontium, and barium. Among them, mercury, lead, and cadmium have a high hydrogen evolution overpotential to achieve a good corrosion inhibition effect. However, these metals with toxicity are harmful to the environment and the human body, it is limited to use.

6.4.3 Cathode Component and Its Influence

The air electrode reaction is performed at the three-phase interface of the gas–solid–liquid. Because the diffusion rate of the sample in the aqueous solution is small and the current density is small when the two-phase electrode is used, the three-phase electrode is used (Girishkumar 2010). The structure of the three-phase electrode consists of an active layer (hydrophilic activation layer), a hydrophobic layer (hydrophobic gas supply layer or waterproof gas-permeable layer), a conductive network (nickel mesh or nickel-copper mesh), and whether the interior of the electrode can be formed as much as possible. The effective three-phase interface will affect the utilization of the catalyst and the mass transfer process of the electrode.

6.4.3.1 Catalyst

The development of metal-air batteries is mainly the continuous renewal of oxygen electrode catalysts. In order to reduce the electrochemical polarization in the positive electrode reaction process, people have conducted extensive research on the electrocatalysts for oxidation–reduction reactions. The earliest carbon used in redox electrocatalysts was its catalytic activity, but its catalytic activity is currently limited to platinum and its alloys, silver, metal chelates, metal oxides (e.g., manganese oxides, calcium titanium), mineral oxides, etc. several series (Cheng and Chen 2012).

(1) Precious metal catalysts. Platinum and silver are the mostly used catalysts, although their catalytic performance is relatively high, few people use these precious metals because of expensive price.
(2) Metal macrocyclic compound catalysts. Organometallic macrocycles have good catalytic activity for oxygen reduction, especially when they are adsorbed on large surface area carbons, the stability and catalytic activity of organometallic macrocycles can be strikingly improved by heat treatment. Therefore, it is expected to replace the noble metal oxygen reduction catalyst. Common methods for synthesizing metal macrocyclic compounds include thermal decomposition and precursor preparation. However, the heat treatment process leads to the reaction of the metal macrocyclic compound with the carbon matrix, and the catalyst prepared by the precursor method has poor activity, so there are also some problems in the application.
(3) Perovskite-type oxide catalysts. Perovskite-type oxides not only have high catalytic activity for the reduction and precipitation of oxygen, but also are inexpensive. Therefore, Perovskite-type oxides are the promising materials

for aluminum-air batteries and fuel cells. Currently, the studies on perovskite oxygen electrode catalysts have focused on the improvement of the preparation method and research and development for new substituted elements to improve catalytic performance. People can produce fine-grained, large-area perovskite oxides by amorphous precursors, which greatly increase their catalytic activity. It is currently a good method to prepare perovskite oxides.

(4) Cheap catalysts. Manganese dioxide oxidizer is a major representative, which can be widely used in batteries with aqueous or non-aqueous electrolytes.

(5) AB_2O_4 spinel oxide catalyst. In the unit cell shall there be 32 close-packed O^{2-} ions, and 64 tetrahedral and 32 octahedral voids are occupied by metal ions. The dehydration activity of spinel is closely related to the fraction of B ions in the tetrahedral voids. The larger the fraction, the more acidic the surface of the catalyst and greater the dehydration activity. Generally, aluminum-air batteries do not use this catalyst.

(6) Other metal and alloy catalysts. Nickel is relatively inexpensive, and has high corrosion resistance in anodic polarization conditions in alkaline electrolytes. At the same time, the oxygen evolution efficiency of nickel is the highest in the metal element, so nickel is conventionally used as an alkaline aqueous electrolytic anode material. Nickel–iron, nickel–cobalt, and other alloy catalysts are also often used. They have good catalytic activity and corrosion resistance.

(7) Composite catalysts. The two or more catalysts are compounded together to better ameliorate the catalytic activity of the air electrode of the aluminum air.

6.4.3.2 Carbon Carrier

A good carrier can maximize the performance of the catalyst and reduce the amount of catalyst. Conventional catalyst preparation methods, active carrier materials (such as activated carbon, acetylene black, graphite, etc.) are prepared by mixing together. The carrier material should have excellent adsorption properties, large surface area as well as excellent electrical conductivity. The air electrode absorbs oxygen while also adsorbing other gaseous impurities. Acetylene black with excellent electrical conductivity, adsorption performance, and graphite with stable structure, and large specific surface area, corrosion resistance, electrical conductivity, and fine particle size can be used as carrier materials.

6.4.3.3 Adhesive

The electrode adhesive must have good viscosity, and the prepared electrode has a certain mechanical strength, and at the same time, it must be hydrophobic. It can construct a gas network channel, form a three-phase reaction interface, and has good conductivity to reduce the ohmic polarization. The commonly used adhesive PTFE (polytetrafluoroethylene), which is a non-polar adhesive, is very good in basicity, but its viscosity is very low. PTFE forms a three-dimensional network when hot pressed at about 300 °C, so that the active material does not fall off.

6.4.4 Advantages and Disadvantages of Metal-Air Batteries

(1) High specific energy. Since the cathode electrode active material comes from the outside of the battery, it does not need to occupy the space of the battery. In the same volume and weight, the metal-air battery can be loaded with more anode electrode active materials, so that the metal-air battery has a much higher capacity than the ordinary battery. The specific energy of zinc-air battery theory is over 1350 Wh/kg, and the actual specific energy is over 1000 Wh/kg. It belongs to high capacity and high-energy chemical power supply.
(2) Low prices. The resources of some metal (such as zinc, aluminum, and magnesium) are abundant and cheap. Oxygen is derived from the ambient air and is basically the same as the commonly used lead-acid battery.
(3) Stable performance and stable discharge. Since the cathode catalyst itself does not change during discharging, and the voltage of zinc electrode is stable, the change of the voltage is small during discharge.
(4) Excellent storage life. The metal-air battery currently produced is actually a reserve type battery. Because the sealing process is used in the storage process, the air hole of the battery is isolated from the outside air. Therefore, the battery capacity loss is minimal, less than 2% per year.
(5) Safe and reliable, no pollution. The whole process from production to use will not pollute the environment; it will not pollute the environment, nor will it burn and explode. It can be called green energy.

Although the zinc-air battery has many advantages mentioned above, there are still some deficiencies and restricts its large-scale commercial application. The first problem is self-corrosion of the zinc electrode. Since zinc is thermodynamically unstable in alkaline solutions, zinc will be accompanied by a small amount of hydrogen evolution corrosion reaction while discharging in the alkaline electro-hydraulic solution. The second problem is carbonation of the air electrode. Since the battery is discharged for a long time, the reaction product or alkaline electrolyte will produce a large amount of carbonate when it comes in contact with the outside CO_2. When the carbonate is dissolved and saturated, it will precipitate on the upper surface of the air electrode. This causes the air electrode to become clogged and degrades in performance and energy, and will no longer work by replacing the zinc negative electrode with energy. Third, because the air battery is a semi-open system, excessively high or low humidity environment will trigger the phenomenon of "drowning" or "drying up" of the air electrode, the phenomenon will cause damage to the battery structure.

6.4.5 *The Development of Metal-Air Battery*

6.4.5.1 Zinc-Air Battery

Zinc-air battery, a kind of metal-air battery, is known as zinc-oxygen battery (Fig. 6.10). Zinc-air battery uses a zinc alloy anode, cathode material is oxygen in the air, the battery reacts, oxygen is taken from the outside air at any time, and in the battery, the electrode material is mainly used in some high polymer (such as poly four Fluorine), activated carbon, catalysts (manganese dioxide, silver, platinum, etc.) and current collector materials, the finished product is a breathable, liquid impermeable, conductive, catalytically active film. Since the thin film constituting the cathode electrode is very thin, the space occupied by the entire battery is extremely small. The battery is generally kept sealed during storage, so there is essentially no self-discharge. The outer surface of the battery is a nickel-metal casing, which is a conductor with good corrosion resistance (Li and Dai 2014).

Zinc-air battery can be expressed as

$$(-)\ \mathrm{Zn}|\mathrm{KOH}|\mathrm{O_2(air)}\ (+)$$

The anode reaction when the battery is discharged is

$$\mathrm{Zn} - 2e^- \rightarrow \mathrm{Zn}^{2+} \tag{6.6}$$

$$\mathrm{Zn}^{2+} + 2\mathrm{OH}^- \rightarrow \mathrm{Zn(OH)_2} \rightarrow \mathrm{ZnO} + \mathrm{H_2O} \tag{6.7}$$

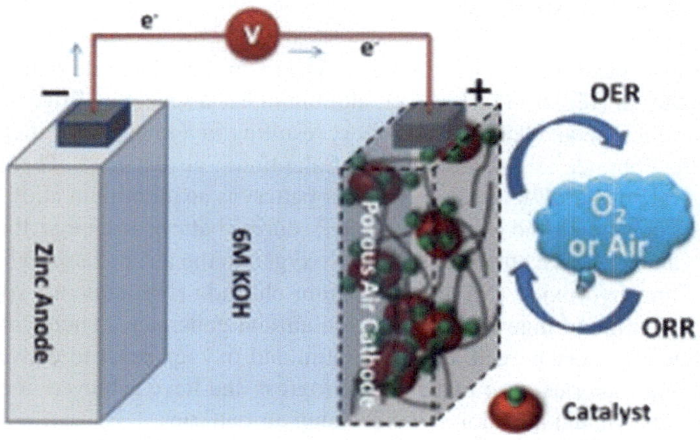

Fig. 6.10 Schematic diagram of metal-air battery, reprinted from Ref. (Li and Dai 2014), copyright 2014, with permission from The Royal Society of Chemistry

$$Zn - 2e^- + 2OH^- \rightarrow ZnO + H_2O \quad \varphi_a^0 = -1.245\,\mathrm{V} \tag{6.8}$$

The cathode reaction when the battery discharges:

$$O_2 + 4e^- + 2H_2O \rightarrow 4OH^- \quad \varphi_c^0 = 0.401\,\mathrm{V} \tag{6.9}$$

Battery reaction:

$$2Zn + O_2 \rightarrow 2ZnO \quad E_{cell}^0 = 1.646\,\mathrm{V} \tag{6.10}$$

Zinc is not easily eroded with the appropriate inhibitors in the electrolyte, and remains so stable that zinc-air batteries have become a concern of everyone in the world. The zinc-air battery has lots of advantages like the self-discharge rate, excellent energy density, stable capacity, terrific safety, low raw material cost, and environmental protection (Li et al. 2013). Zinc-air batteries have been commercialized successfully for many years. Initially, these products were large batteries using alkaline electrolytes. They were used in railway signaling, long-distance communication, and marine navigation devices requiring long-time and low-rate discharges. With the development of thin-layer electrode technology, this technology has been applied to small button batteries and high capacity primary batteries such as hearing aids, along with similar applications. The expanding of electric vehicles contributes to provide a vast space for development of zinc-air batteries during recent years. Zinc-air battery has significant advantages in terms of performance and cost. Based on various factors, we believe that despite the need to further improve the zinc-air battery, it is also very likely to become one of the best choices for the electric bicycle power supply, and has a far-reaching development prospect.

6.4.5.2 Aluminum-Air Battery

Aluminum is a cheap, lightweight metal, and has a higher electrode potential as an electrode material (Fig. 6.11). However, aluminum has a severe hydrogen evolution corrosion problem in an alkaline electrolyte, resulting in a low anode efficiency and hindering the commercial popularization of aluminum-air batteries. Therefore, the anode electrode material of the aluminum-air battery is an aluminum alloy, which is continuously consumed and generates $Al(OH)_3$ during battery discharge; the positive electrode is an oxygen electrode and ingests oxygen in the air; potassium hydroxide (KOH), sodium hydroxide (NaOH) or sodium chloride (Na) aqueous solution as electrolyte. Aluminum ingests oxygen in the air and generates a chemical reaction when the battery is discharged, and aluminum and oxygen react to convert it into alumina. Al-air batteries have made rapid progress and have achieved good results in EV applications, and are a kind of promising air batteries.

The aluminum alloy electrode continuously reacts with OH^- in the electrolyte to form $Al(OH)_4{}^-$ and emits electrons. The electrons flow into the air electrode (positive electrode) through the external line load, and the air electrode acquires electrons and

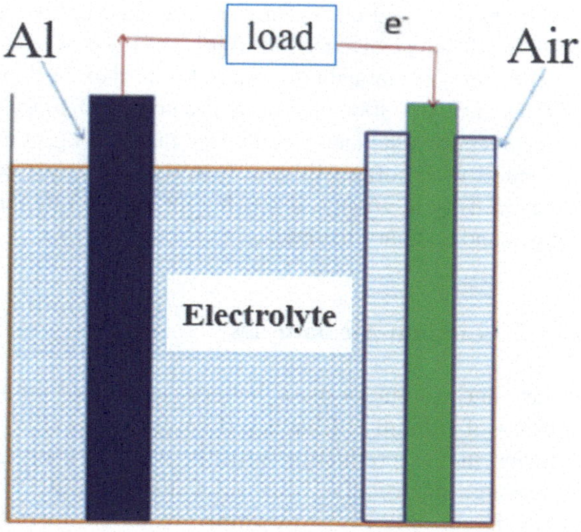

Fig. 6.11 Schematic diagram of aluminum-air battery, reprinted from Ref. (Feng et al. 2013), copyright 2013, with permission from Elsevier

undergoes a reduction reaction with water. OH^-. The chemical reaction continues and the aluminum electrode and oxygen are continuously consumed. The electrons continuously flow in the external circuit to form an electric current and generate electricity. Among them, when $Al(OH)_4{}^-$ reaches a certain concentration, $Al(OH)_3$ is naturally generated. The specific chemical reaction formula is

$$\text{Cathode reaction: } \frac{1}{2}O_2 + H_2O + 2e^- \rightarrow 2OH^-$$
$$\text{Anode reaction: } Al \rightarrow 3e^- + Al^{3+}$$
$$Al^{3+} + 3OH^- \rightarrow Al(OH)_3$$

Different electrolytes produce different electrochemical reactions. The total discharge of the+ battery in different electrolytes is

$$\text{Neutral solution: } 4Al + 3O_2 + 6H_2O \rightarrow 4Al(OH)_3$$
$$\text{Alkaline solution: } 4Al + 3O_2 + 6H_2O + OH^- \rightarrow Al(OH)_4^-$$

As a special type of fuel cell, aluminum-air batteries have great commercial application prospects such as military, civilian, and underwater power systems, telecom system backup power sources, and portable power supplies (Egan et al. 2013). Some aluminum-air products, such as the current research status and development trend of marine aluminum gas batteries, battery packs, reserve batteries for telecommunication systems, and aluminum-oxygen underwater power systems are being developed in the market. As a power system for underwater exploration, aluminum-air can provide ten times more power than nickel–cadmium batteries and greatly reduce the time spent on charging. As for medium-sized lightweight submarine power system, its energy density can reach 500 Wh/L and the power can up to 2.3 kW at 60 V. What's more, it can continuously discharge for 50 h at 2 kW, which ensures the underwater vehicles travel 50–300 km. Although a portable battery that is used directly in military communications devices weighs less than 4.5 kg, it can produce 300 W of power. Compared with current lithium batteries, the cycle consumption is lower and the energy density is greater; it has higher energy efficiency than internal combustion engines and ordinary batteries.

6.4.5.3 Lithium-Air Batteries

In the same volume or mass, battery capacity of lithium-air is 30% larger than the capacity of lithium-ion battery (Lee et al. 2011). The cost of lithium-air battery is lower in the production process without chemical processing and chemical processing. In particular, because of standardized 3 V voltage platform, lithium-air battery is not only easy to combine batteries suitable for various electrical appliances, but also is the best supporting battery for large-scale commercialization of 2 V semiconductor chips. Therefore, lithium-air batteries have become a research hotspot in recent years.

Lithium-air batteries currently have four configurations. Three configurations use liquid electrolytes, including inert organic electrolytes, aqueous electrolytes, and hybrids. The other configuration uses solid-state electrolytes. Currently, the studies are directed at polar aprotic electrolyte systems, lithium ions are transported between the anode electrode and the cathode electrode of the porous carbon structure and combined with oxygen, and the resulting Li_2O_2 stays on the positive electrode. Oxygen is provided by the external environment, so the theoretical energy density reaches 13 kWh/kg. For inert electrolyte systems (organic, ionic liquids) and all-solid electrolyte systems, the battery response is

$$2\mathrm{Li} + \mathrm{O_2} \leftrightarrow \mathrm{Li_2O_2} \tag{6.11}$$

$$2\mathrm{Li} + \frac{1}{2}\mathrm{O_2} \leftrightarrow \mathrm{Li_2O} \tag{6.12}$$

For aqueous (acidic and alkaline) electrolyte systems, the battery response is (Fig. 6.12)

$$2\mathrm{Li} + \frac{1}{2}\mathrm{O_2} + 2\mathrm{H^+} \leftrightarrow 2\mathrm{Li^+} + \mathrm{H_2O} \tag{6.13}$$

$$2\mathrm{Li} + \frac{1}{2}\mathrm{O_2} + \mathrm{H_2O} \leftrightarrow 2\mathrm{LiOH} \tag{6.14}$$

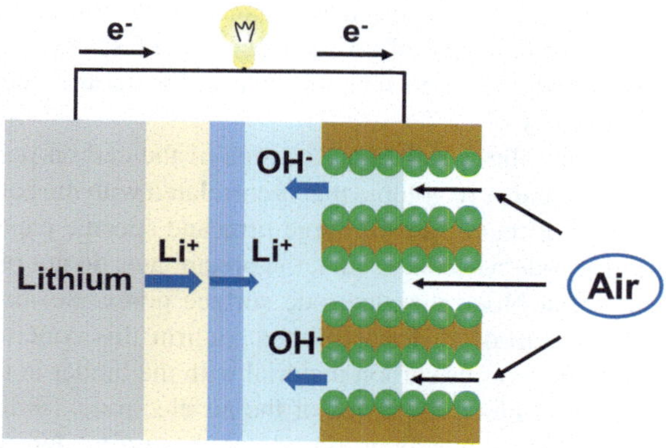

Fig. 6.12 Schematic diagram of lithium-air battery, reprinted from Ref. (Lee et al. 2011), copyright 2011, with permission from WILEY–VCH

The cathode reaction of the lithium-air battery not only transmits most of the battery energy, but also exists in the vast majority of the voltage drop. The cathode electrode of the air almost bears the voltage drop of the entire air battery. Therefore, the air, the positive electrode, has much effect on its performance. Key factors: A good air electrode must have (1) fast oxygen diffusion; (2) good electrical conductivity; (3) high specific surface; (4) stable electrode composition; (5) fast ion conductivity. The electrode surface pores are closely related to the battery capacity. When the electrode reaction occurs, the precipitates do not block the active charge transfer center, but block the electrode surface pores. The battery capacity is independent of the specific surface area of the carbon porous material, and the average pore size and volume. The discharge time and specific capacity raise along with the increase of the average pore size and volume. The results of impedance testing on the electrode material further confirm this conclusion. It has been found that after the porous carbon material is mixed with the binder in a certain proportion, too much binder will block the pores of the air electrode, leading to a sharp drop in battery capacity. Sandhu et al. proposed through mathematical simulations. The lower the flow density, the higher the utilization of the electrode; at a given current density, the thickness of the electrode gets reduced and the specific capacitance of the battery is greatly increased. In short, the air electrode material of the lithium-air battery does not only guarantee the normal transmission of oxygen and lithium ions, but also to ensure that the electrode surface pores are not blocked, and then to accommodate more lithium oxide. Mesoporous carbon materials and microporous carbon materials can well meet the above requirements. The study of air electrode materials mainly focuses on porous carbon materials, carbon nanotubes, and graphene.

Since both Li_2O_2 and Li_2O are insoluble in the organic electrolyte, the discharge product can only be deposited on the air electrode, the air electrode channel is blocked, and the discharge is terminated. Not only most of the energy of air cathode is transmitted, but also most of the voltage drop occurs. The cathode electrode of the air almost assumes the voltage drop of the entire air battery. Therefore, the air cathode plays a vital role in the performance of the lithium-air battery. A good air electrode must have the following aspects: (1) fast oxygen diffusion; (2) good electrical conductivity; (3) high specific surface; (4) stable electrode composition; (5) fast ion conductivity.

It is not the specific surface area of the carbon porous material but the average pore size and pore volume that is correlated with the battery capacity. In pace with the increase of them, the discharge time and specific capacity raise accordingly. When the electrode reaction occurs, the precipitates do not block the active charge transfer center, but block the electrode surface pores. Besides, the impedance test results of the electrode material further confirm this conclusion. It was found that after mixing the porous carbon material with the binder in a certain proportion, too much binder will block the pores of the air electrode, leading to a sharp drop in battery capacity. Sandhu et al. proposed that the current density is lower, the utilization rate of the electrode is higher through mathematical simulation; at a current density, the thickness of the electrode is reduced and the specific capacitance of the battery is greatly increased. In short, the air electrode material of the lithium-air battery is not only required to ensure the normal transmission of oxygen and lithium ions, but also does not block the electrode surface pores, and then accommodate more lithium oxide. Mesoporous carbon materials and macroporous carbon materials can meet the above requirements. Study about the air electrode material mainly concentrated in the porous carbon material, carbon nanotubes, graphene.

6.4.5.4 Magnesium-Air Battery

The theoretical voltage of the magnesium-air battery remains 3.1 V, and the theoretical specific energy gets 6.8 k Wh/Kg. Magnesium or magnesium alloy are used as the anode electrode active material for Magnesium/air battery, oxygen is used as the active material of positive electrode (Rahman et al. 2013). Under the different electrolytes used, reaction mechanisms differ as well. In an aqueous solution system (Fig. 6.13),

$$\text{Anode reaction: Mg} \rightarrow \text{Mg}^{2+} + 2e$$
$$\text{Cathode reaction: O}_2 + \text{H}_2\text{O} + 4e \rightarrow 4\text{OH}^-$$
$$\text{Battery reaction: Mg} + 1/2\text{O}_2 + \text{H}_2\text{O} \rightarrow \text{Mg(OH)}_2$$
$$\text{The battery reaction is in the non-aqueous system: Mg} + 1/2\text{O}_2 \rightarrow \text{MgO}$$

The advantage of this battery system is the low price and no pollution. The theoretical energy density of magnesium is second only to light metal lithium and aluminum, and therefore it is an ideal electrode material for the battery (Zhang et al. 2014). However, there are still many problems to overcome in the actual research process. The main problems currently faced by magnesium-air batteries for water systems are still the problem of hydrogen evolution from corrosion and the activation and passivation of magnesium alloys, because these problems determine the stability,

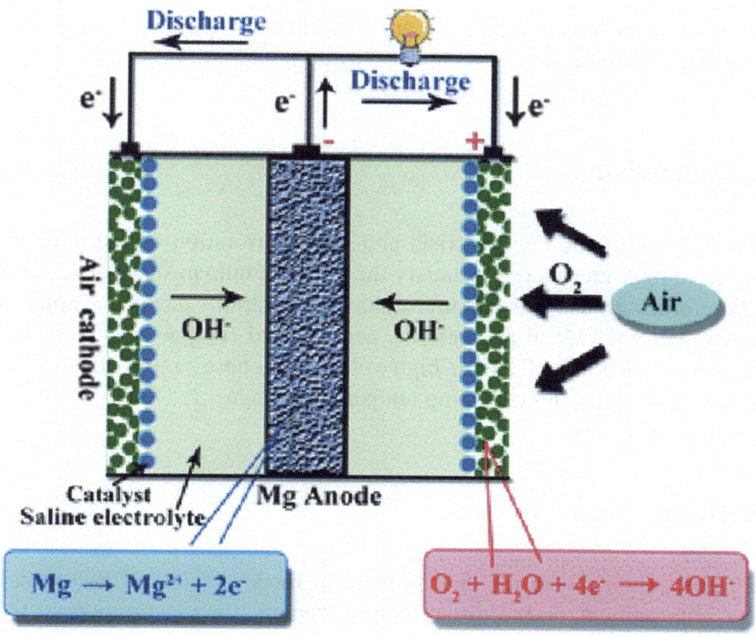

Fig. 6.13 Schematic diagram of magnesium-air battery, reprinted from Ref. (Zhang et al. 2014), copyright 2014, with permission from The Royal Society of Chemistry

reactivity, and energy efficiency of magnesium anodes in aqueous solutions, which determines the performance of the battery.

It is the hotspot to magnesium alloy anode materials with high anode utilization in the research of magnesium-air batteries in the world. For the sake of overcoming these defects of magnesium metal, magnesium and other alloying elements can be made into binary, ternary, or even multi-element alloys. On the other hand, the magnesium alloy grains can be refined and the over potential of the hydrogen evolution reaction can be increased to reduce the self-corrosion rate. On the other hand, the structure of the passivation film can be destroyed, and the relatively complete and dense passivation film becomes loose and porous which are easy to fall off. So the magnesium alloy passivation problem is reduced and electrochemical performance is improved.

Although there are some precedents about magnesium-air batteries for practical applications, they are limited to open-structured battery systems or battery systems that use an external electrolyte circulation method to solve gas production and exhaust problems during battery operation. In the same structural battery system, magnesium hydrogen peroxide fuel cells or magnesium seawater batteries have advantages over magnesium/air batteries. As a portable mobile power source, magnesium-air battery cannot be fully hermetically sealed because it needs to be vented. Judging from the current technical characteristics of the battery, the most likely applications of magnesium-air batteries in the future are fixed standby power sources and fixed power stations in remote areas.

6.4.6 Conclusion

With the increase of the energy crisis and the deterioration of the environment, people need to find new energy to satisfy the needs of human survival and development. Therefore, various types of batteries have been rapidly developed and received widespread attention. Metal-air batteries can be used for power cars and mobile power in no distant future for their high volumetric energy ratio, stable discharge voltage, low cost, and faster operating temperature range.

6.5 Lithium-Sulfur Battery

In the current century, with the development of modern science, the technologies of energy, environment, and information have become the most important fields in human society. The research progress of economic and eco-friendly energy is playing the vital role in improving our living standard sustainably. Nowadays, numerous energy conversion and storage devices have been designed and used successfully, such as the commercial solar cells, fuel cells, lithium-ion batteries, and supercapacitors, etc. Over the past 30 years, lithium-ion batteries have become the most considerable and useful storage devices in various energy storage systems, due to their lighter weight and higher capacity. Unfortunately, because of the poor low-temperature performance, lifetime, energy density, and safety, it cannot meet the requirements for large-scale storage in stationary power grids as well as the electric vehicles. Therefore, it arouses people's interests in investigating and designing new system of lithium-ion battery.

Lithium-sulfur batteries have a higher energy density than existing lithium-ion batteries, making them one of the most promising energy storage systems. The main differences between those two types of batteries are their energy storage mechanisms, for example, the Li-ion can insert into and take off from the layered electrode materials in lithium-ion batteries during charging and discharging process, which has the limited capacity of 387 Wh kg^{-1} according to the certain inserted sites. By contrast, the operation of lithium-sulfur batteries instead of intercalation, which is based on the metal plating and stripping on the lithium anode side and the conversion reaction on the sulfur cathode side, is thereby leading to the high theoretical capacities of lithium negative and sulfur positive electrodes as 3860 and 1673 mAh g^{-1}, respectively. Moreover, the average voltage of 2.2 V gives higher theoretical energy density of 2600 Wh kg^{-1} (Table 6.6) (Adelhelm et al. 2015; Seh et al. 2016b). For

Table 6.6 Electrochemical reaction and energy density of several energy storage devices

Battery system	Chemical reaction	Voltage (V)	Theoretical density (Wh kg^{-1})	Actual density (Wh kg^{-1})
Lithium-ion battery	$0.5LiC_6 + Li_{0.5}CoO_2 \leftrightarrow 3C + LiCoO_2$	3.8	387	200
Lithium-sulfur battery	$2Li + S \leftrightarrow Li_2S$	2.2	2600	350

the most important, the sulfur element on the earth is abundant, low cost, environmentally friendly, and other characteristics, which is very useful for the application of lithium batteries.

As early as 1962, Herbet and Ulam proposed the assumption of using sulfur element as electrode materials (Herbert and Ulam 1962). Subsequently, all kinds of alkali-sulfur battery systems were researched and reported, and most of according works are concentrated on the high temperature sodium-sulfur batteries and the room-temperature lithium-sulfur batteries (Abraham et al. 1978).

6.5.1 Working Principle

Lithium-sulfur batteries use sulfur element or sulfur composites as active materials in positive electrodes, and metallic lithium as negative electrodes. The working principle diagram is shown in Fig. 6.14. Lithium metal reacts with sulfur element and creates

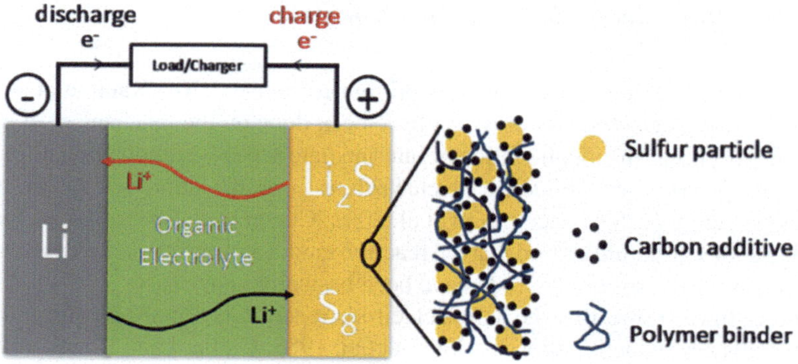

Fig. 6.14 Illustration of the charge (red)/discharge (black) process involved in a rechargeable Li_2S battery including lithium metal anode, organic electrolyte, and sulfur composite cathode, reprinted from Ref. (Manthiram et al. 2013), copyright 2013, with permission from The Royal Society of Chemistry

the Li_2S as the final discharge product. Due to the very poor conductivity of sulfur element, sulfur composites as the positive electrodes usually consist of sulfur element, conductive agents, porous carbon, and binders (Manthiram et al. 2013).

During the discharge process, the sulfur on the positive electrodes is gradually reduced to form an intermediate product of lithium polysulfide, which finally is completely reduced to lithium sulfide; on the negative electrode, lithium loses electrons and is oxidized into lithium ions to enter the electrolyte. During the charging process, lithium sulfide is re-oxidized to sulfur at the positive electrode, and liberated lithium ions migrate back to the negative electrode. In addition, lithium ions are reduced and deposited as lithium metal on the surface of the negative electrode. When the sulfur element is completely oxidized or reduced, the movement of two electrons is involved. Therefore, the sulfur electrode reaction can be expressed as

$$S_8 + 16Li^+ + 16e^- \underset{\text{charge}}{\overset{\text{discharge}}{\leftrightarrows}} 8Li_2S$$

According to the specific capacity formula:

$$C_0 = \frac{26.8N}{M}(\text{Ah g}^{-1})$$

(In which, $\boldsymbol{C_0}$ is the theoretical specific capacity of the battery, $\boldsymbol{N}$ is the number of electrons gained or lost during the reaction, $\boldsymbol{M}$ is the molar mass of the active materials). The theoretical specific capacity of sulfur element can be calculated as 1675 mAh g^{-1}. Based on the thermodynamic reaction potential of sulfur, the theoretical specific capacity of lithium metal is calculated as 3860 mAh g^{-1}.

6.5.2 Sulfur Electrode Reaction Process

The research of lithium-sulfur batteries originated in the 1970s (Rauh et al. 1979). In the following decades, lots of works have been done on the mechanism of sulfur electrode reactions, the identification of intermediate product composition, the study of the composition and influence of electrolytes, and the improvement of sulfur electrode cycling performance (Shim et al. 2002; Cheon et al. 2003; Mikhaylik and Akridge 2004). It should be said that the reaction model and performance degradation mechanism for the sulfur electrode have been basically understood.

Studies have shown that the sulfur electrode reaction is a multi-electron, multi-step progressive reduction process (Yamin et al. 1985; Gaillard and Levillain 1995; Leghié et al. 2002). The discharge process is extremely complicated, not only involves the formation of multifarious lithium polysulfide intermediates, but also involves a series of chemical transformation and coupling processes. Currently, the recognized reaction mechanism is shown in Fig. 6.15, where two voltage stages arise at 2.4 and 2.1 V during the discharge of sulfur electrode. Hence, the whole process can

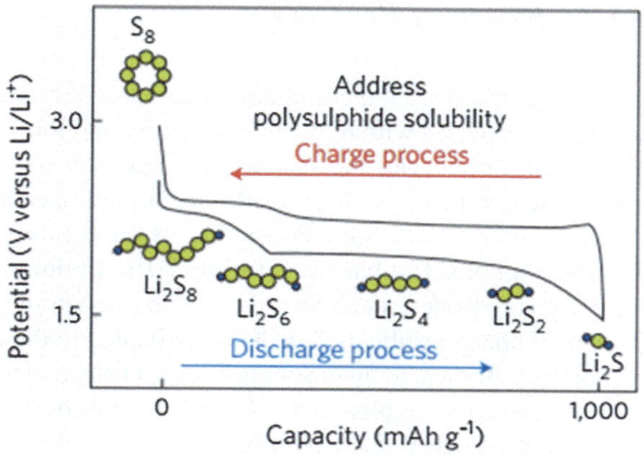

Fig. 6.15 Sulfur electrode depth of discharge and its corresponding electrochemical reaction process, reprinted from Ref. (Bruce et al. 2011), copyright 2011, with permission from Macmillan Publishers Limited

be divided into three stages, and the electrochemical reactions concerned with each stage are related to chemically coupled chemical reactions and the depth of discharge of the sulfur electrode (Jung and Kim 2007; Ji and Nazar 2010). The first stage is the conversion of solid sulfur to liquid polysulfide with a depth of discharge of 25%, corresponding to a high voltage platform area between 2.4 and 2.1 V. In this process, solid elemental sulfur deepens as the depth of discharge goes through successive conversion steps such as $S_8{}^0 \rightarrow S_8{}^{2-} \rightarrow S_6{}^{2-}$, and ultimately generates soluble polysulfides Li_2S_6. The nature of the first-stage reactants leads to a faster reaction kinetics, so the second-stage is the conversion of soluble polysulfides to solid Li_2S_2, which corresponding to a low voltage plateau of 2.1 V. At this stage, through successive conversion steps such as $S_6{}^{2-} \rightarrow S_4{}^{2-} \rightarrow S_2{}^{2-}$, Li_2S_6 is further reduced to insoluble Li_2S_2 for which the corresponding depth of discharge is 50% and the total depth of discharge of the sulfur electrode reaches 75%. The second-stage of solid phase nucleation leads to a slower reaction kinetics, so the third-stage is the conversion of solid phase Li_2S_2 to the final product solid phase is Li_2S which corresponds to a discharge area below 2.1 V and a depth of discharge of 25%. After completion of this stage, the depth of discharge of the sulfur electrode reaches 100%. Slow diffusion of the solid phase is in the bulk, where the slowest reaction occurs. Compared to the discharge process, the charging process of the sulfur electrode is relatively simple. In the cyclic scan of sulfur or polysulfide electrodes, only one oxidation peak is shown (Yamin et al. 1985). Numerous experimental phenomena indicated that the conversion of Li_2S to high valence polysulfides is a faster charge transfer process. However, the final oxidation product usually is Li_2S_8, and the sulfur discharge product cannot be completely restored to the elemental sulfur state by simple charging (Ryu et al. 2006).

6.5.3 Research Progress

In order to speed up the commercialization of lithium-sulfur batteries and develop secondary batteries with high energy density and long cycle life, the development of lithium-sulfur batteries has reached an important stage, which has made a series of major research progress. The results are summarized as follows:

The United States Sion Power Inc. is an earlier company in the research and development of sulfur batteries. In July 2010, lithium-sulfur batteries and solar cells were used on drones, and a record of 336 h of continuous flight was created. OXIS, a British company established in 2004, dedicated to the development of lithium metal secondary batteries, focusing on the high security and long cycle life of lithium-sulfur batteries since its establishment. At present, the energy density of its next-generation high-specific-energy lithium-sulfur battery has been higher than 300 Wh/kg. It is hoped that lithium-sulfur batteries with twice the current energy storage capacity will be developed and used in electric vehicles, aerospace, defense, and energy storage systems in 2018. Domestic research on lithium-sulfur batteries is later than foreign research, but with the support of national policies and funds, many achievements have also been achieved. In 2008, the energy density of lithium-sulfur flexible packaging batteries developed by Tsinghua University reached 246 Wh/kg. In 2012, the National University of Defense Technology reported a lithium-sulfur soft pack battery that attained 320–350 Wh/kg under 100% deep discharge for 100 cycles. In 2013, Dalian Institute of Materials reported a lithium-sulfur soft-bag battery with a current density of 350 Wh/kg after 40 cycles under 75% deep discharge. Then, it developed a 15 Ah lithium-sulfur battery with an energy density of more than 430 Wh/kg in August 2014.

The thermal research of lithium-sulfur batteries at home and abroad will certainly increase the level of research and development, and the energy density of batteries will continue to increase. The following describes the four aspects of the lithium-sulfur battery cathode material, metal lithium negative electrode stability, and modification, electrolyte solution, separator.

6.5.4 Cathode Material

Early sulfur cathode materials were simply a mixture of conductive agents, binders, and elemental sulfur powders which is a macroscopic composite material. The shuttle effect and the poor conductivity of the sulfur cathode have a greater influence on the sulfur utilization and cycle stability. Therefore, in the past 20 years, researchers have added some modified substances (carbon materials, polymers, or nano metal oxides) to form nanocomposites with sulfur to boost the electrochemical performance of lithium-sulfur batteries as well as the conductivity and stability of sulfur cathode materials.

6.5.4.1 Binary Metal Sulfides Materials

Binary metal sulfide, a cathode material for lithium-sulfur battery, has got some more researches in the early stage of development. The compound is simple in preparation, and the metal powder is mixed with the sulfur powder by a mechanical grinding method, a high temperature solid phase method, or an electrochemical deposition method to generate a simple binary metal sulfide. At present, there are many metal sulfides such as nickel, copper, and iron. As a cathode material for lithium-sulfur batteries, binary metal sulfides generate lithium compounds or oxides which are similar to oxides and Li_2S during discharge, and some can further generate lithium alloys. As shown in Fig. 6.16, NiS was obtained by mechanical grinding of spherical, flake, filamentous metallic nickel powders, and elemental sulfur powders, respectively. Figure 6.17 shows ex situ XRD profiles of NiS cathode active material obtained during electrochemical charge–discharge cycling. Figure 6.17a shows the XRD profiles of NiS before discharging. The XRD pattern in Fig. 6.17b shows that a new phase of Ni_3S_2 is formed after discharging at 1.8 V. During discharge to 1.3 V, Li_2S and nickel were formed as discharge products (Fig. 6.17c). During charging to 2.2 V, nickel and Ni_3S_2 are formed (Fig. 6.17d). Finally, at the end of charging, NiS is formed up to full charge, as seen in Fig. 6.17e. The total charging-discharging reaction of NiS can be explained in the following way (Han et al. 2003):

$$3NiS + 2Li \leftrightarrow Ni_3S_2 + Li_2S \tag{6.15}$$

$$Ni_3S_2 + 4Li \leftrightarrow 3Ni + 2Li_2S \tag{6.16}$$

$$3NiS + 6Li \leftrightarrow 3Ni + 3Li_2S \tag{6.17}$$

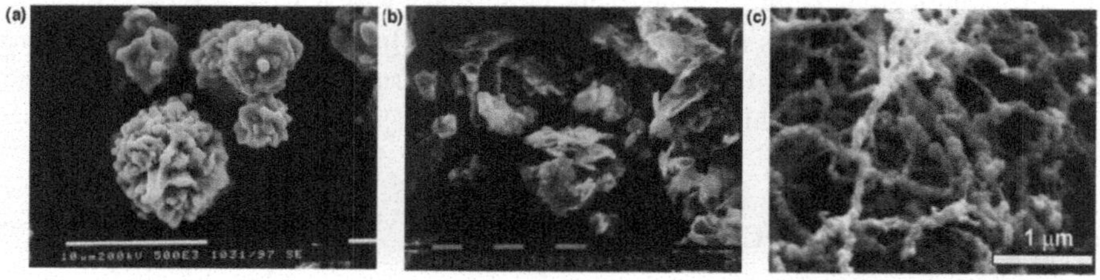

Fig. 6.16 Scanning electron microscopy photos of various types of nickel. **a** Spherical type, **b** flake type, and **c** filamentary type, reprinted from Ref. (Han et al. 2003), copyright 2003, with permission from Elsevier

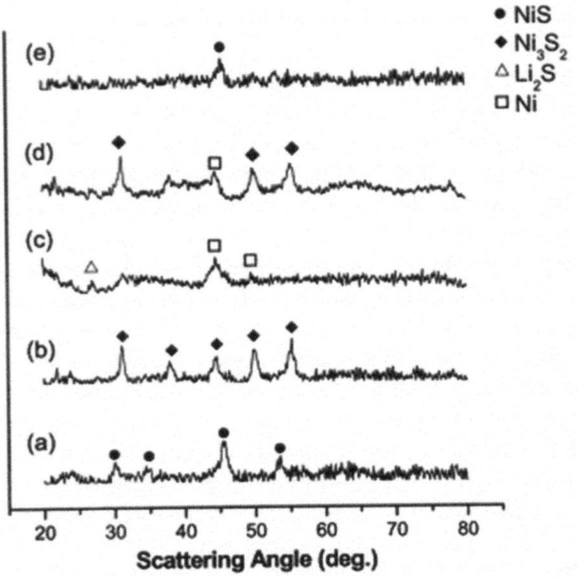

Fig. 6.17 Ex situ XRD profiles of NiS cathode during charge–discharge cycling, reprinted from Ref. (Han et al. 2003), copyright 2003, with permission from Elsevier

6.5.4.2 Organic Sulfide Materials

Organic sulfides are used as positive electrodes in electrochemical reactions. Energy storage and release of organic compounds or polymers by forming or breaking S–S bonds, including disulfides, organic polysulfides, and sulfide polymers, which are characterized by the absence of metallic elements that are lightweight and environmentally friendly. However, due to the instability of the S–C bond, sulfur easily detaches from the organic chain and loses its reactivity, resulting in poor cycle performance. In addition, the non-electrochemically active organic functional groups in the active molecules occupy a relatively large mass which is making the specific capacity of such materials relatively less noble than lithium-ion batteries, so there are few related researches after 2005. Table 6.7 summarizes several important organic sulfides as cathode materials (Sun et al. 2009).

6.5.4.3 Sulfur/Carbon Composites

As the cathode materials for lithium-sulfur batteries, sulfur/carbon composites are the hotspots. Because carbon materials have good conductivity, high porosity, strong adsorption capacity, low cost, etc. The use of carbon materials to form a conductive network can make up insulation defects of the sulfur element. The porosity of the carbon material allows the sulfur to be evenly distributed, thereby increasing the sulfur loading. At the same time, the abundant voids can also provide more active sites. Complex pore structures are distributed in the carbon material, which can

Table 6.7 The structure and performance of various organic sulfides

Category	Structure	Specific capacity (mAh g^{-1})
Disulfide	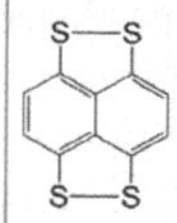	425
Poly-disulfide	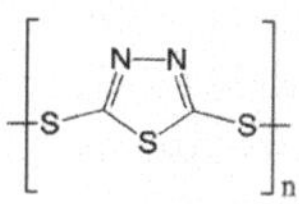	362
Dithio-polymer	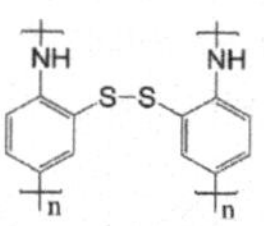	270
Organo-multi sulfide	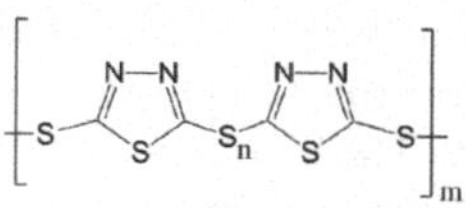	200–350
Sulfurized polymers	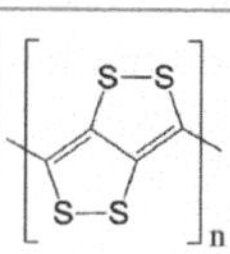	345–720
Sulfurized polymers	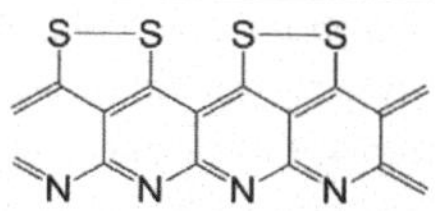	850
Sulfurized polymers	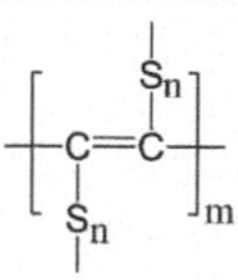	600

physically limit the polysulfide, prevent the dissolution and diffusion of polysulfides, and inhibit the "shuttle effect". In addition, utilizing the excellent mechanical strength and porous structure of the carbon material can reduce the volume expansion and contraction of the electrode caused by the charging and discharging of lithium-sulfur batteries At the moment, sulfur/carbon composites, sulfur/carbon nanotube composites, sulfur/graphene composites, and sulfur/carbon fiber composites are the most widely studied among sulfur/carbon composites.

Sulfur/Carbon Nanotubes Composites

Carbon nanotubes are the first carbon materials used in lithium-sulfur electrodes, they have many advantages, such as large specific surface area, high electrical conductivity, rich microporous structure, and high mechanical strength. Due to the low viscosity of sulfur at 155 °C, sulfur and carbon nanotubes can be prepared at 155 °C to obtain a good composite. Compared with the sulfur/carbon nanotube composite positive electrode is prepared by simply mixing at room temperature, the sulfur/carbon nanotube composite positive electrode mixed at 155 °C exhibits lower charge transfer resistance and better cycling ability. Therefore, sulfur-coated carbon nanotube composites (S@CNT) were obtained through heat treatment, which can greatly improve the utilization rate and cycle stability of sulfur (Yuan et al. 2009).

The structure of S-HMT@CNT is shown in Fig. 6.18. Hollow TiO_2 balls are composited with carbon nanotubes for sulfur loading, hollow TiO_2 is used for sulfur adsorption, and carbon nanotubes are used to increase electrical conductivity. The uniform composite structure ensures that the material has good strength, and the dual-function carbon paper sandwich improves the wettability of the material (Hwang et al. 2016).

In order to enhance the diffusion efficiency of Li^+ and the loading rate of sulfur, metal organic frameworks (MOFs) and the carbon nanotubes (CNT) are overlapped and used in the positive electrode of a lithium-sulfur battery without a binder as shown in Fig. 6.19. The highly porous metal organic framework not only improves the capacity of the positive electrode, but also improves the ion transmission efficiency. This layered with a porous three-dimensional conductive network has a very high volumetric energy density (Mao et al. 2017).

Sulfur/Graphene Composites

As a two-dimensional material made of monolayer carbon atoms, Graphene has the advantages of high specific surface area, high electrical conductivity, light weight, and high structural strength. Graphene can effectively coat the sulfur particles, thereby forming a conductive network to ensure the effective transmission of ions. At the same time, the interface impedance is reduced and the electrochemical activity of the cathode material is greatly improved. The introduction of graphene into the positive electrode material mainly includes the following ways: Graphene or graphene oxide is converted directly to a carbon matrix for supporting elemental sulfur. Introducing functional groups on graphene reduces the loss of positive active material. Change the interlayer structure of graphene to optimize the space structure and improve the anode sulfur loading rate.

Fig. 6.18 Schematic detailing the notion of the sulfur content in HMT and CNTs, reprinted from Ref. (Hwang et al. 2016), copyright 2016, with permission from WILEY–VCH

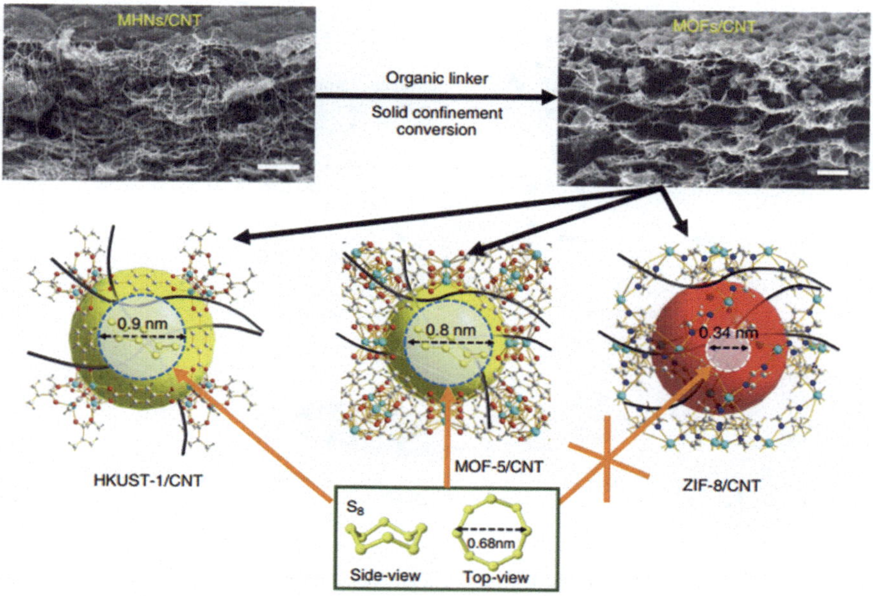

Fig. 6.19 Synthesis of S_8 loaded MOFs/CNT composite thin films. MHN stands for metal hydroxide nanochain. The scale bars of SEM images are 1 mm, reprinted from Ref. (Mao et al. 2017), copyright 2017, with permission from Nature

Graphene is used as a coating material for coating coat submicron-sized sulfur particles with polyethylene glycol (PEG) (Fig. 6.20a). Graphene coating improves the conductivity of the positive electrode. The PEG and graphene coating can suppress the dissolution of lithium polysulfide and relieve its volume expansion during charging and discharging (Wang et al. 2011). In addition, porous graphene (HPG) with a porosity of up to 3.51 cm^3/g achieves a sulfur loading of up to 80%. As shown in Fig. 6.20b, a highly conductive graphene (HCG) as a current collector, an oxygen-containing functional partial graphene oxide (POG) as a lithium polysulfide adsorption layer shows excellent cycle of the whole graphene cathode material stability (Fang et al. 2016).

Graphene materials with nanopore structure were prepared by modifying graphite nanosheets. The nanopore's bulk structure has much greater adsorption capacity for sulfur and intermediate products than surface adsorption. This pore structure is a good electronic conductor and ion conductor which can trap relatively large volume polysulfide, it not only effectively reduces the loss of sulfur, but also eases the damage caused by sulfur volume changes in charge and discharge cycles. In addition, nitrogen-doped graphene tends to exhibit better electrochemical performance, this is because there is a strong Li–N interaction between lithium polysulfide and the nitrogen atom according to calculations and experiments. The binding energy of pyrrole type nitrogen and lithium polysulfide can reach 1.33 to 2.09 eV, and the binding energy of pyridine type nitrogen and lithium polysulfide can reach 1.48 to

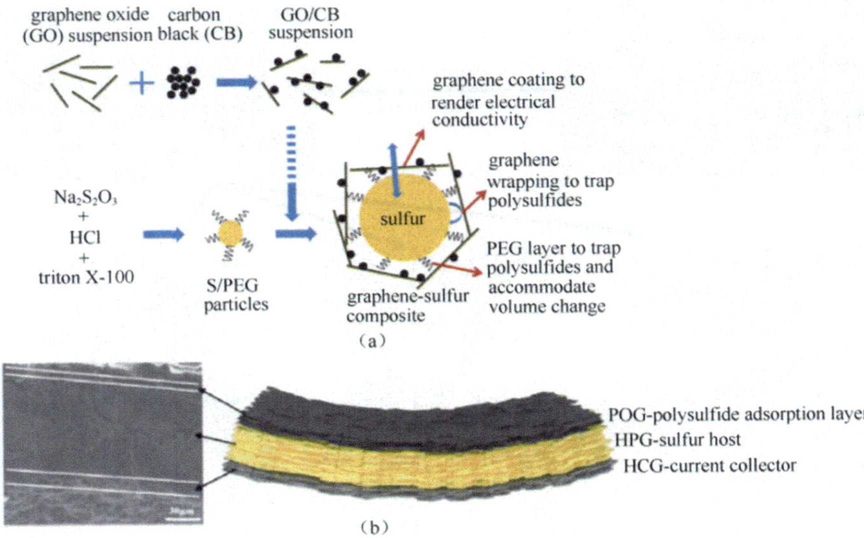

Fig. 6.20 **a** Schematic of the synthesis steps for a graphene/sulfur composite; **b** Schematic of the all-graphene structural design of the sulfur cathode, reprinted from Ref. (Fang et al. 2016), copyright 2016, with permission from American Chemical Society

2.10 eV, which is much higher than undoped graphene (The binding energy is usually between 0.25 and 0.60 eV) (Seh et al. 2016a).

Sulfur/Porous Carbon Composites

According to the pore size, porous carbon can be divided into micropores (<2 nm), mesopores (2–50 nm), and macropores (>50 nm). The porous structure of the porous carbon material can physically confine the active material, while the porous carbon has a higher specific surface area. It is used as the positive electrode material, in which the electrochemical performance of lithium-sulfur batteries can be improved obviously. The tiny sulfur particles are prepared by the liquid phase deposition method and serve as a positive electrode material after being compounded with active porous carbon. This method has a simple process when compared with the conventional sulfur-carrying method, convenient operation, and low cost. The initial discharge specific capacity of the obtained battery was 1093 mAh g^{-1} at C/10 (Wang et al. 2017a) (Table 6.8).

Mesoporous carbon has the advantages of high conductivity and high porosity. It can be used as a carbon matrix in positive electrode to significantly improve the battery's cycling performance. The results indicated that mesoporous carbon can inhibit the loss of sulfur and the diffusion of polysulfide. Further researches show that the pore size of mesoporous carbon is significant, and pore volume can effectively improve battery performance. The macropores are often present in the network structure of interwoven carbon fibers or carbon nanotubes. The macroporous

Table 6.8 Electrochemical performances of porous carbon materials with different textual parameters/sulfur composites

Material	Pore size (nm)	Pore volume (cm^3/g)	BET total surface area (m^2/g)	Sulfur content	Current density (mA/g)	1st discharge capacity (mAh/g)	Cycling capacity (mAh/g)
LMC/S	123	9.0	1215	70%	300	1380	30th, 800 100th, 615
CMK-3/S	3.3	2.1	1976	70%	168	1005	20th, 800
MPC/S	7.3	0.503	1566	11.7%	2500	1584	30th, 805
MPC/S	10–30	–	381	25%	50	1300	40th, 600
Active carbon/sulfur	1.76	–	1486	70%	100	1181	60th, 720
Carbon xerogel/sulfur	–	–	1774	<60%	220	1311	30th, 800
Elemental sulfur	–	–	–		300	1050	30th, 600

The capacity values are calculated according to the elemental sulfur

structure can well absorb the electrolyte, which inhibits the dissolution of lithium polysulfide to some extent. People use hollow-structured carbon fibers or carbon nanotubes to further improve their properties.

6.5.4.4 Sulfur/Conductive Polymer Composites

In the past 20 years, conductive polymers have developed greatly in terms of synthesis technology and performance enhancement, which is continuously applied to lithium-sulfur batteries.

Sulfur polymer composite mainly in two forms: one is to form a "polymer coated sulfur" core–shell structure, such as polypyrrole/sulfur (PPY/S), polyaniline/sulfur (PANi/S), poly (3,4-ethylene dioxythiophene) (PEDOT/S), polythiophene/sulfur (PTh/S) and other composite materials (Fig. 6.21) (Wu et al. 2011). The structure design of the hollow nano-sulfur covered by the conductive polymer effectively solved the problem of structural destruction attributed to the volume changing of sulfur in the charging and discharging process, and obtained excellent cycle performance. The other is a "sulfur-coated polymer" core–shell structure, such as the fusion of sulfur to the surface of a conductive polymer matrix by means of melting or chemical deposition (Fig. 6.22) (Fu and Manthiram 2012). Such conductive polymers mainly include PPY nanowires, nanotubes, and other network structures, PANi

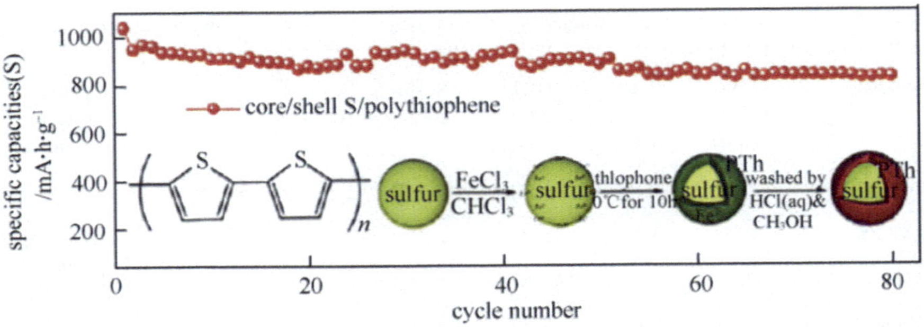

Fig. 6.21 Schematic of preparing PTh/S and its cycle performance, reprinted from Ref. (Wu et al. 2011), copyright 2011, with permission from American Chemical Society

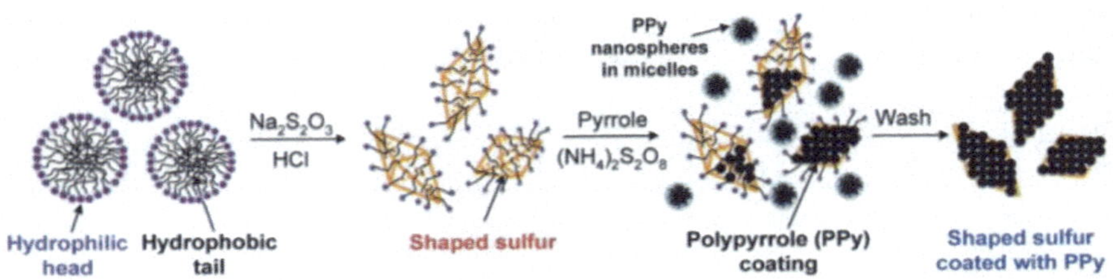

Fig. 6.22 Sulfur/polypyrrole nanosphere composites preparation process, reprinted from Ref. (Fu and Manthiram 2012), copyright 2012, with permission from American Chemical Society

nanotubes and hollow PANi balls. The polymer acts as a conducting agent, dispersant, adsorbent, and buffer volume change in the composite material and provides a better matrix for the deposition of sulfur.

However, conductive polymers have a limited ability to conduct electrons compared to carbon materials, and additional conductive agents are still required to be added in the production of pole pieces. Therefore, it is not suitable to be used in combination with sulfur alone in practical applications, it can be combined with carbon materials or metal compounds in some form as a carrier of sulfur.

6.5.4.5 Sulfur/Metal Oxide Composites

Metal Oxides

Nano metal oxides containing O^{2-} generally have a very polar surface and are not easily soluble in most organic solvents, by reason of the strong bonding between oxygen and metal. The metal oxides are compared to nanostructured carbon materials that provide abundant polar active sites for polysulfide absorption. Due to its inherent defects and unique band structure, metal oxides have better electrical conductivity. In addition, oxides can significantly increase the volumetric energy density of lithium-sulfur batteries. Metal oxides can not only be used as pure sulfur additives for lithium-sulfur batteries, but also be directly added to carbon composite materials to obtain composite positive electrode materials.

Metal Sulfide

With the gradual enrichment of metal supplied synthesis methods, researchers try to add it to the positive electrode of lithium-sulfur batteries. Metal sulfides have a strong affinity for sulfur-containing substances and can well limit the loss of positive electrode active materials. Lithium sulfide cathode materials have been the most widely studied of all metal sulfides. Since the sulfur composite positive electrode does not contain lithium in traditional lithium-sulfur batteries, the lithium metal negative electrode needs to provide a lithium source. In this way, the lithium negative electrode will undergo structural changes during the reaction process, causing potential safety hazards to the lithium-sulfur battery system. Researchers have developed sulfide materials to replace pure sulfur cathode materials, which has eliminated the impact of structural changes in lithium metal anodes (e.g., MnS, CuS, ZnS, SnS_2, and NiS_2). However, sulfides generally have poor conductivity and must be compounded with other conductive materials.

Other Metal Compounds

Most metal oxides are poorly conductive, so the use of metal oxides alone as cathode materials for lithium-sulfur batteries may impede electron transport, resulting in reducing sulfur utilization and rate capability. For example, TiN–S composite positive electrode has a high conductivity, a strong porous skeleton, and a good adsorption capacity, and the assembled lithium-sulfur battery shows excellent cycle stability (Cui et al. 2016). Recent studies have shown that nanostructured metal carbides have a strong adsorption effect on polysulfides, such as TiC and WC can adsorb polymers and promote polysulfide reduction reactions. Due to the microcrystalline size (about 25 nm) and a large number of active sites, TiC cathode materials can be used in lithium-sulfur batteries to improve the cycling performance. It has a steady specific capacity of 860 mAh g^{-1} in 100th cycle (Al Salem et al. 2016), which is instructive for the next step.

6.5.4.6 Lithium Sulfide Electrode Materials

Lithium sulfide (Li_2S) as the final discharge product of lithium-sulfur batteries has attracted attention due to its higher theoretical specific capacity (1166 mAh/g). Li_2S does not require metallic lithium as a negative electrode compared to the sulfur electrode material, which can reduce the safety problems caused by the lithium metal dendrite. However, its poor electronic and ionic conductivity, and its high solubility in electrolytes must be solved, the typical methods to use are carbon-coated or porous Li_2S-filled materials to improve its electrical conductivity and inhibit polysulfide from dissolving in the electrolyte.

The Li_2S–C nanocomposite synthesized by the in situ composite method has a uniform distribution in the mesoporous structure and the partially graphitized carbon matrix. The nanocomposite method effectively suppresses the shuttle influence and improves the cycling stability of Li_2S (Yang et al. 2013).

6.5.5 *Stability and Modification of Metal Lithium Negative Electrode*

Lithium metal is the most ideal anode material for lithium-sulfur batteries as a result of its extremely high theoretical specific capacity (3860 mAh/g), low potential (0.304 V), and low density (0.53 g/cm^3) (Cao et al. 2015). However, lithium metal anodes also have many problems, such as the growing lithium dendrites, instability of the solid electrolyte interface film, and almost unlimited volume expansion. Therefore, how to build a more stable lithium metal anode for lithium-sulfur batteries, promote the smooth deposition of lithium, inhibit the growth and volume expansion of lithium dendrites remain a challenge.

6.5.5.1 Lithium Dendrite Growth Mechanism

In lithium-sulfur secondary batteries, lithium metal needs to be continuously deposited and stripped, and the lithium dendrite will accumulate on the electrode surface after being reciprocated which caused many serious problems. Cations form a concentration gradient between the positive and negative electrodes during the electrodeposition process. When the current density between the two electrodes reaches the critical current density, the critical current will remain for a very short time called the transition time (τ). After that, the cations in the electrolyte will decrease, destroying the potential balance on the electrode surface. This process can lead to the generation of local space charge on the surface of the electrode and unstable metal deposition, thereby producing dendrites. However, SEI is generated on the metallic surface due to the deposition and exfoliation of lithium metal. Even if the current density is lower than the critical current density, lithium dendrite will still be generated, so the growth mechanism of lithium dendrite is more complicated and further research is needed (Lin et al. 2017; Xu et al. 2014). The current density has a greater influence on the growth of dendrites during lithium deposition and the growth of lithium dendrites was accelerated under high current densities.

6.5.5.2 Lithium Metal Anode Modification

Researchers have developed different methods to inhibit the growth of lithium dendrites. In addition to developing different electrolyte solvents, salts, and additives to form a stable SEI with lithium metal, many novel methods have recently been developed.

Manual SEI

SEI is generated by the reaction of lithium metal and electrolyte (including solvents, salts, additives, etc.), and its chemical composition is complex and unstable. Active material and the electrolyte are continuously consumed during cycling because of the SEI damage which caused low Coulombic efficiency and cycle life. Using in situ or ex situ methods can generate artificial SEI as a protective layer, thereby protecting the lithium metal anode and inhibiting the growth of lithium dendritic crystals. The ideal artificial SEI should have the following properties: (1) Chemically stable during the cycle; (2) It has excellent ion conductivity, which provides a channel for lithium-ion diffusion; (3) Electronic insulation, allows lithium ions to diffuse through the SEI and deposit on the lithium metal surface rather than the SEI surface; (4) Sufficient mechanical strength to inhibit lithium dendrite piercing (Fan et al. 2017).

Nano-Passivation Protection Layer

The nanometer-scale passivation layer can also effectively inhibit the growth of lithium dendrites other than artificial SEI, including hollow nanospheres, two-dimensional boron nitride, and graphene materials. These materials can facilitate the formation of stable SEI and the deposition of metallic lithium, thereby improving the Coulombic efficiency and cycling stability of the battery. The protection of lithium metal anodes is particularly important in Lithium-sulfur batteries. Due to the shuttling effect of lithium polysulfide, the sulfur that diffuses to the lithium metal anode reacts with lithium metal to produce solid lithium-sulfur compounds, which results in a decrease in battery capacity. However, using hexamethylditin and trimethylsilyl chloride as passivation protective films for the metal lithium negative electrode can effectively prevent polysulfide from reacting with the lithium metal and increase the cycling performance of lithium-sulfur batteries (Wu et al. 2016a,b).

Homogenization of Lithium-Ion Current

The uneven lithium-ion spatial distribution on the surface of the electrode also directly leads to the production of lithium dendrites, so homogenizing the lithium-ion flow has important significance for inhibiting the growth of lithium dendrites. Dispersing the current density by increasing the surface area of electrode is a valid method of homogenizing the flow of lithium ions (Lin et al. 2017). For example, a 3D porous copper mesh current collector/metal lithium composite negative electrode prepared by a mechanical pressing method is used. The 3D porous copper current collector has a large surface area that can homogenize lithium-ion flow, reduce local current density, and increase electrochemical reaction activity. In addition, 3D porous copper mesh can also meet the volumetric expansion of metallic lithium (Li et al. 2017). The use of 3D porous copper mesh current collector/metal lithium composite negative electrode can increase the cycle life, Coulombic efficiency and specific capacity of batteries, which provide new ideas for the research of lithium metal negative electrodes.

Stable Body Minimizes Volume Changes

The infinite volume expansion of lithium metal during deposition is also a serious problem that hinders the development of lithium metal negative electrodes. Compared to the volume expansion of 400% for silicon negative electrodes and 10% for graphite negative electrodes, the volume expansion of lithium metal is more severe and almost infinite, which causes the electrolyte/separator interface to float, generates internal stresses, and accelerates cell damage. The CUI team developed a series of composite lithium negative electrodes with a stable body to minimize the volume change of the lithium metal negative electrode, including silicon-coated carbon nanofibers, layered reduced graphene oxide, and zinc oxide coated carbon nanowires fiber (Lin et al. 2016; Liu et al. 2016; Liang et al. 2016). Besides, the surface area of the lithium negative electrode can also be increased, the effective current density can be reduced, and the lithium-ion current can be uniformized, thereby inhibiting the growth of lithium dendrite. The existence of a stable body can also decrease the internal stress of the battery and increase the safety performance and cycle stability of the battery.

6.5.6 Electrolyte Solution

6.5.6.1 Liquid Electrolyte

According to current literature reports, organic liquid electrolyte for the lithium-sulfur battery, the main electrolyte system includes two major categories of carbonate solvent system.

Carbonate Solvent System

This type of system is mostly used for polymer sulfur composites such as PAN and sulfur compounded by high temperature heat treatment (Wang et al. 2003, 2002), or use in microporous carbon–sulfur composites. Carbonate-based organic solvents have low solubility in polysulfide ions. Therefore, as the reaction progresses in a general carbon–sulfur composite system, the contact resistance between the discharge product and the conductive agent increases. Only when the sulfur and the composite materials are in a combined state or sulfur is in the form of small molecules can the efficient transport of ions and electrons in the electrode be satisfied, and carbonate organic solvents can work better. Under these conditions, the system has low solubility of polysulfide ions, resulting in excellent cycle performance.

Carbonate Solvent System

Due to the certain solubility of polysulfide in this system, the active material can have a high reaction activity in the reaction process, so as to ensure the smooth progress of the redox reaction. Currently used ether solvents mainly include Tetrahydrofuran (Cyclic ethers such as THF) and 1–3-dioxolane (DOL); ethylene glycol dimethyl ether (DME), triethylene glycol dimethyl ether (TGDME) and tetraethylene glycol dimethyl ether (TEGDME). The results show that chain ethers have good solubility of polysulfide ions, making the sulfur electrode reaction process proceed smoothly. In particular, DOL easily undergoes ring-opening polymerization reaction on the surface of the metal lithium negative electrode, so that it can form a layer of elastic and stable protective film, which is conducive to improving the stability of metal lithium during cycling. Therefore, chains and cyclic ethers are often used to improve the electrochemical performance of lithium-sulfur batteries (Choi et al. 2007). Among them, DOL and DME are the most commonly used mixed solvents in a volume ratio of 1:1.

6.5.6.2 Solid Electrolyte

Theoretically, the use of a solid electrolyte can completely avoid the deterioration of the electrochemical performance caused by the dissolution of polysulfide ions. However, the solid itself has intrinsic defects such as low room temperature conductivity and large interface contact resistance. It needs to work at relatively high temperatures, thus limiting its application in lithium-sulfur batteries. If the active material surface is evenly coated with a solid electrolyte membrane, the problem of high contact resistance between the active material and the electrolyte in the all-solid-state lithium-sulfur battery can be effectively overcome, and the electrochemical performance of the system is greatly improved. For example, the nanocomposite polymer electrolyte NCPE($PEO_{20}LiCF_3 - SO_3Li_2S + 10wt\%\ ZrO_2$) and the nanoscale $Li_2S - Li_3PS_4$ core–shell structure (Hassoun and Scrosati 2010; Suo et al. 2013).

6.5.6.3 Solvent-In-Salt Electrolyte System

A new type of dual-function electrolyte system "Solvent-in-Salt" (SIS) was proposed in 2013 and applied in lithium-sulfur batteries. At the same time, two key technical challenges of dissolving polysulfide ions and stabilizing metal lithium negative electrodes were solved (Fig. 6.23). By using a conventional ether system as a solvent, lithium salt concentration is greatly increased, and a large amount of free solvent molecules are complexed with a lithium salt. Also, the dissolution of polysulfide ions in the electrolyte is effectively suppressed, and the "multi-sulfide ion shuttle effect" of polysulfide ions dissolved in the electrolyte during charging is avoided. The serious overcharge of the battery was prevented, the circulating Coulombic efficiency was

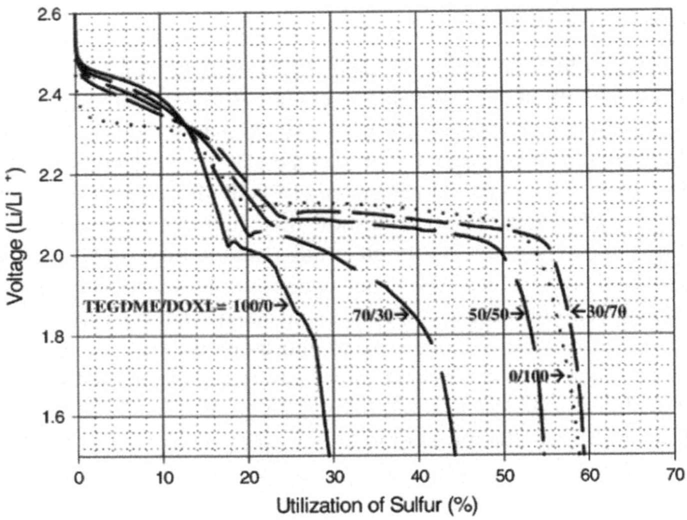

Fig. 6.23 Electrochemical Performance of Using $LiCF_3SO_3$/TEGDME-DOL Electrolyte System in Lithium-Sulfur Battery, reprinted from Ref. (Chang et al. 2002), copyright 2002, with permission from Elsevier

close to 100%, and the cycle stability was significantly improved. At the same time, compared with the conventional low-salt electrolyte system, the high-salt electrolyte system has a high concentration of anion and cation (7 mol LiTFSI/1 LDOL-DME), a high lithium-ion migration number (0.73), and comparison high viscosity (72 cP). The lithium dendrite growth due to the non-uniform deposition of metal lithium is effectively avoided, mainly because the high lithium-ion concentration is favorable for the uniform material exchange of the metal lithium negative electrode. The high anion concentration and viscosity help to reduce the space charge layer generated by anion depletion on the negative surface of metallic lithium, thereby reducing the electric field driving force of non-uniform deposition of metallic lithium. The high-viscosity system increases the resistance of lithium dendrite growth to a certain extent, thereby greatly improving the stability of the lithium metal negative electrode during cycling.

6.5.7 Separator

Separators of Lithium-sulfur battery are mainly based on polyolefin separators produced by Celgard and Ube of Japan, and their separator materials are mainly PP, PE and PP/PE/PP. PP/PE/PP three-layer separators are widely used in batteries due to their lithium-ion self-blocking function, good mechanical strength and puncture strength, and high safe operating temperature, which is compared to PP, PE

Table 6.9 Parameters of PP/PE/PP separator made by Celgard company

Sampling project	Unit	Skills requirement
Thickness	μm	25
Permeability	sec/100 mL	480–550
Porosity	%	38–41
Tensile strength, TD	MPa	102–106
Tensile strength, MD	MPa	91–96
Piercing strength	gf	461±2
Obturator temperature	°C	133±1
Rupture temperature	°C	152±1
Heat shrinkage (90 °C), TD	%	0.4±0.1
Heat shrinkage (90 °C), MD	%	2.8±0.1
Heat shrinkage (120 °C), TD	%	9±0.2
Heat shrinkage (120 °C), MD	%	15±0.2

single-layer separators. Table 6.9 shows the technical parameters of the three-layer diaphragm produced by Celgard, USA.

The lithium-sulfur battery separator mainly affects the cycle stability and specific capacity of the battery. Although separator has no significant improvement in the overall performance of lithium-sulfur batteries compared with electrode, it is still necessary to study. In recent years, researchers have been increasingly studying the diaphragm, because the modification of the diaphragm is also a better way to reduce the shuttle effect. Reducing the shuttle effect, in turn, can increase the Coulombic efficiency and lifetime of lithium-sulfur batteries. The main diaphragm material currently used is polypropylene, and the method used to reduce the shuttle effect is to add a coating to the diaphragm. Graphene layers and perfluorosulfonic acid membrane layers are two of the more common types (Sun et al. 2016). However, there are also many new materials, such as separators using lithium-sulfur batteries coated with conductive materials, which have significantly improved cycle stability and specific capacity. Finally, the initial specific capacity of the lithium-sulfur battery combined with sulfur nanometers was 1350 mAh/g at 0.5C, and the decay rate per cycle was only 0.09%.

6.6 Summary and Outlook

After many years of development, lithium-sulfur batteries are still in the research and development stage. Because of the differences in related hardware and software in countries' policies, technology bases, as well as research and development conditions, there is a large gap between their respective research and development stages. Among them, developed countries in Europe and the United States, South Korea, and Japan have established their own research and development teams, which formed effective lithium-sulfur batteries in cooperation with relevant government research institutes and universities.

At present, the research of lithium-sulfur batteries mainly focuses on the optimization and improvement of electrochemical performance. However, the basic scientific issues related to lithium-sulfur batteries are still relatively weak. At present, only a few research works are involved. In fact, lithium-sulfur batteries have a lot of fundamental scientific issues that need to be solved. First of all, the sulfur reaction process is extremely complicated and the reaction intermediates are diverse. Due to the limitations of research methods, it is still not well understood. The electrochemical performance of lithium-sulfur batteries is greatly influenced by the character of electrode materials, test conditions, and work environment. The internal reasons for the mutual effects still need to be further studied and summarized. Secondly, the mechanism of action of electrolyte on the positive and negative electrodes need to be further studied and explored, such as the dissolution of polysulfide ions in the electrolyte and the stability of lithium negative electrode in the electrolyte. Finally, the deposition-dissolution process of metal lithium, growth mechanism of lithium dendrite, and the formation mechanism and performance of solid electrolyte mesophase under long cycling are serious problems in terms of metal lithium negative electrodes. Although the study is more difficult, it is also an issue that cannot be avoided in the future. It is believed that the comprehensive performance of the lithium-sulfur battery system will continue to improve under the continuous attention and vigorous research in the scientific research and industrial colleagues around the world. The relevant mechanism research will be more in-depth, so as to achieve a full understanding of lithium-sulfur batteries.

References

Abraham KM, Rauh RD, Brummer SB (1978) ChemInform abstract: a low temperature sodium-sulfur battery incorporating a soluble sulfur cathode. Chemischer Informationsdienst 9(39):85027

Adelhelm P, Hartmann P, Bender CL, Busche M, Eufinger C, Janek J (2015) From lithium to sodium: cell chemistry of room temperature sodium-air and sodium-sulfur batteries. Beilstein Journal of Nanotechnology 6:1016–1055

Al Salem H, Chitturi VR, Babu G, Santana JA, Gopalakrishnan D, Arava LMR (2016) Stabilizing polysulfide-shuttle in a Li-S battery using transition metal carbide nanostructures. Rsc Adv 6(111):110301–110306

Arumugam D, Kalaignan GP (2008) Synthesis and electrochemical characterizations of Nano-SiO_2-coated LiMn2O4 cathode materials for rechargeable lithium batteries. 624(1):197–204

Aurbach D, Markovsky B, Rodkin A, Levi E, Cohen YS, Kim HJ, Schmidt MJEA (2003) On the capacity fading of $LiCoO_2$ intercalation electrodes: the effect of cycling, storage, temperature, and surface film forming additives 47(27):4291–4306

Bang HJ, Donepudi VS, Prakash JJEA (2003) Preparation and characterization of partially substituted $LiMyMn_2\text{-}yO_4$ (M=Ni, Co, Fe) spinel cathodes for Li-ion batteries. 48(4):443–451

Bruce et al. (2011) $Li\text{-}O_2$ and Li-S batteries with high energy storage. Nat Mater 2012, 11:19. https://doi.org/10.1038/NMAT3237

Cai J, Zhang YP, Shields LBE, Zhang ZZ, Liu N, Shields CBJJoPS (2001) Preparation and electrochemical/thermal properties of $LiNi_{0.74}Co_{0.26}O_2$ cathode material 92(1):35–39

Cao R, Xu W, Lv D, Xiao J, Zhang J-G (2015) Anodes for rechargeable lithium-sulfur batteries. Adv Energy Mater 5(16):513–537

Chang D-R, Lee S-H, Kim S-W, Kim H-T (2002) Binary electrolyte based on tetra(ethylene glycol) dimethyl ether and 1,3-dioxolane for lithium–sulfur battery. J Power Sources 112(2):452–460

Chang HH, Chang CC, Su CY, Wu HC, Yang MH, Wu NL (2008) Effects of TiO_2 coating on high temperature cycle performance of LiFePO4-based lithium-ion batteries. J Power Sources 185(1):466–472

Chang Y, Mao X, Zhao Y, Feng S, Chen H, Finlow D (2009) Lead-acid battery use in the development of renewable energy systems in China. J Power Sources 191(1):176–183

Chen CH, Liu J, Stoll ME, Henriksen G, Vissers DR, Amine K (2004) Aluminum-doped lithium nickel cobalt oxide electrodes for high-power lithium-ion batteries. 128 (2):278–285

Chen F, Yang B, Zhang W, Ma J, Lv J, Yang Y (2017a) Enhanced recycling network for spent e-bicycle batteries: a case study in Xuzhou, China. Waste Manage 60:660–665

Chen S, Wu C, Shen L, Zhu C, Huang Y, Xi K, Maier J, Yu Y (2017b) Challenges and perspectives for NASICON-type electrode materials for advanced sodium-ion batteries. Adv Mater 29(48):1700431

Cheng F, Chen J (2012) Metal-air batteries: from oxygen reduction electrochemistry to cathode catalysts. Chem Soc Rev 41(6):2172–2192

Cheng et al. (2012) Stable cycling of double-walled silicon nanotube battery anodes through solid-electrolyte interphase control. Nat Nanotechnol 7(5):309–314

Cheon S-E, Ko K-S, Cho J-H, Kim S-W, Chin E-Y, Kim H-T (2003) Rechargeable lithium sulfur battery. J Electrochem Soc 150(6):A796–A799

Choi J-W, Kim J-K, Cheruvally G, Ahn J-H, Ahn H-J, Kim K-W (2007) Rechargeable lithium/sulfur battery with suitable mixed liquid electrolytes. Electrochim Acta 52(5):2075–2082

Choi S, Jung G, Kim JE, Lim T, Suh KS (2018) Lithium intercalated graphite with preformed passivation layer as superior anode for Lithium ion batteries. Appl Surf Sci 455:367–372

Cui M, Bai P, Jiang Q, Sun S, Wang X (2014) A novel synthesis and characterization of $LiFePO_4$ and $LiFePO_4/C$ as a cathode material for lithium-ion battery 246(3):232–238

Cui Z, Zu C, Zhou W, Manthiram A, Goodenough JB (2016) Mesoporous titanium nitride-enabled highly stable lithium-sulfur batteries. Adv Mater 28(32):6926–6931

DiVincenzo DP, Mele EJ (1985) Cohesion and structure in stage-1 graphite intercalation compounds. Phys Rev B 32(4):2538–2553

Dunn B, Kamath H, Tarascon JM (2011) Electrical energy storage for the grid: a battery of choices. Science 334(6058):928–935

Egan DR, Ponce de León C, Wood RJK, Jones RL, Stokes KR, Walsh FC (2013) Developments in electrode materials and electrolytes for aluminium–air batteries. J Power Sources 236:293–310

Fan L, Zhuang HL, Gao L, Lu Y, Archer LA (2017) Regulating Li deposition at artificial solid electrolyte interphases. J Mater Chem A 5(7):3483–3492

Fan X, Wu Y, Ye S, Xu B (2008) Research and development of separator for lithium-ion batteries 22(12):11–15

Fang R, Zhao S, Pei S, Qian X, Hou P-X, Cheng H-M, Liu C, Li F (2016) Toward more reliable lithium-sulfur batteries: an all-graphene cathode structure. ACS Nano 10(9):8676–8682

Feng XY, Ding N, Wang L, Ma XH, Li YM, Chen CH (2013) Synthesis and reversible lithium storage of Cr_2O_5 as a new high energy density cathode material for rechargeable lithium batteries. J Power Sources 222:184–187

Fu Y, Manthiram A (2012) Orthorhombic bipyramidal sulfur coated with polypyrrole nanolayers as a cathode material for lithium-sulfur batteries. J Phys Chem C 116(16):8910–8915

Girishkumar G, McCloskey BD, Luntz AC, Swanson SA, Wilcke W (2010) Lithium-air battery: promise and challenges. J Phys Chem Lett 1(14):2193–2203

Gaillard F, Levillain E (1995) Visible time-resolved spectroelectrochemistry: application to study of the reduction of sulfur (S8) in dimethylformamide. J Electroanal Chem 398(1):77–87

Hamani D, Ati M, Tarascon J-M, Rozier P (2011) Na_xVO_2 as possible electrode for Na-ion batteries. Electrochem Commun 13(9):938–941

Han CP, He YB, Liu M, Li BH, Yang QH, Wong CP, Kang FY (2017) A review of gassing behavior in $Li_4Ti_5O_{12}$-based lithium ion batteries. J Mater Chem A 5(14):6368–6381
Han S-C, Kim K-W, Ahn H-J, Ahn J-H, Lee J-Y (2003) Charge–discharge mechanism of mechanically alloyed NiS used as a cathode in rechargeable lithium batteries. J Alloy Compd 361(1):247–251
Hassoun J, Scrosati B (2010) Moving to a solid-state configuration: a valid approach to making lithium-sulfur batteries viable for practical applications. Adv Mater 22(45):5198–5201
Herbert D, Ulam J (1962). U.S. Patent 3043896
Hong LYWJL (2014) Fundamental scientific aspects of lithium ion batteries (IX)—nonaqueous electrolyte materials. Energy Storage Sci Technol 3(3):262–282
Hwang J-Y, Kim HM, Lee S-K, Lee J-H, Abouimrane A, Khaleel MA, Belharouak I, Manthiram A, Sun Y-K (2016) High-energy, high-rate, lithium-sulfur batteries: synergetic effect of hollow TiO_2-webbed carbon nanotubes and a dual functional carbon-paper interlayer. Adv Energy Mater 6(1):1501480
Jeong ED, Won MS, Shim Y (1998) Cathodic properties of a lithium-ion secondary battery using $LiCoO_2$ prepared by a complex formation reaction 70 (1):70–77
Ji X, Nazar LF (2010) Advances in Li–S batteries. J Mater Chem 20(44):9821–9826
Jin K, Katayama Y, Miura T, Kishi T (1998) Lithium insertion behaviour of Li1+xV_3O_8 prepared by precipitation technique in CH_3OH. 110(3–4):199–207
Jung Y, Kim S (2007) New approaches to improve cycle life characteristics of lithium-sulfur cells. Electrochem Commun 9(2):249–254
Kakuda T, Uematsu K, Toda K, Sato M (2007) Electrochemical performance of Al-doped $LiMn_2O_4$ prepared by different methods in solid-state reaction. 167(2):499–503
Kanno R, Kubo H, Kawamoto Y, Kamiyama T, Izumi F, Takeda Y, Takano M (1994) Phase relationship and lithium deintercalation in lithium nickel oxides 110(110):216–225
Kim S-W, Seo D-H, Ma X, Ceder G, Kang K (2012) Electrode Materials for rechargeable sodium-ion batteries: potential alternatives to current lithium-ion batteries. Adv Energy Mater 2(7):710–721
Kumagai N, Ooto H, Kumagai N (1997) Preparation and electrochemical characteristics of quaternary Li-Mn–V–O spinel as the positive materials for rechargeable lithium batteries 68(2):600–603
Kumar V, Kameswara Rao PV, Rawal A (2017) Amplification of electrolyte uptake in the absorptive glass mat (AGM) separator for valve regulated lead acid (VRLA) batteries. J Power Sources 341:19–26
Lee J-S, Tai Kim S, Cao R, Choi N-S, Liu M, Lee KT, Cho J (2011) Metal-air batteries with high energy density: Li-Air versus Zn-Air. Adv Energy Mater 1(1):34–50
Lee J, Wu Y, Peng Z (2018) Hetero-nanostructured materials for high-power lithium ion batteries. J Colloid Interf Sci 529:505–519
Lee JI, Choi NS, Park S (2012) Highly stable Si-based multicomponent anodes for practical use in lithium-ion batteries. Energy Environ Sci 5(7):7878–7882
Lee SW, Kim KS, Moon HS, Kim HJ, Cho BW, Cho WI, Ju JB, Park J (2004) Electrochemical characteristics of Al_2O_3-coated lithium manganese spinel as a cathode material for a lithium secondary battery 126(1):150–155
Leghié P, Lelieur JP, Levillain E (2002) Comments on the mechanism of the electrochemical reduction of sulphur in dimethylformamide. Electrochem Commun 4(5):406–411
Li H, Wang Z, Chen L, Huang XJAM (2010) Research on Advanced Materials for Li-ion Batteries. 21 (45):4593-4607
Li Q, Zhu S, Lu Y (2017) 3D porous cu current collector/Li-metal composite anode for stable lithium-metal batteries. Adv Func Mater 27(18):1606422
Li Y, Dai H (2014) Recent advances in zinc-air batteries. Chem Soc Rev 43(15):5257–5275
Li Y, Gong M, Liang Y, Feng J, Kim JE, Wang H, Hong G, Zhang B, Dai H (2013) Advanced zinc-air batteries based on high-performance hybrid electrocatalysts. Nature Commun 4:1805
Liu et al. (2012) In situ atomic-scale imaging of electrochemical lithiation insilicon[J].Nat Nano 7(11):749–756

Liang Z, Lin D, Zhao J, Lu Z, Liu Y, Liu C, Lu Y, Wang H, Yan K, Tao X, Cui Y (2016) Composite lithium metal anode by melt infusion of lithium into a 3D conducting scaffold with lithiophilic coating. Proc Natl Acad Sci USA 113(11):2862–2867

Lin D, Liu Y, Cui Y (2017) Reviving the lithium metal anode for high-energy batteries. Nat Nanotechnol 12:194–206

Lin D, Liu Y, Liang Z, Lee H-W, Sun J, Wang H, Yan K, Xie J, Cui Y (2016) Layered reduced graphene oxide with nanoscale interlayer gaps as a stable host for lithium metal anodes. Nat Nanotechnol 11(7):626–632

Lingzhi Z (2009) Summaries of lithium-ion battery and progress of its anode materials 36(5):106–107

Liu H, Cao Q, Fu LJ, Li C, Wu YP, Wu H (2006) Doping effects of zinc on $LiFePO_4$ cathode material for lithium ion batteries 8(10):1553–1557

Liu H, Cui Y (2018) Microwave-assisted hydrothermal synthesis of hollow flower-like $Zn_2V_2O_7$ with enhanced cycling stability as electrode for lithium ion batteries. Mater Lett 228:369–371

Liu H, Zhang Z, Lin Z (2001) New progress in studies of lithium nickel oxide as positive electrode materials of lithium ion batteris 7(2):145–154

Liu P, Ru Q, Wang Z, Wang B, Guo Q, Zhang P, Hou X, Su S, Ling FC-C (2018) Harnessing the synergic lithium storage and morphology evolution of 1D bundle-like $NiCo_2O_4$@TiO_2 hybrid to prolong the cycling life for lithium ion batteries. Chem Eng J 350:902–910

Liu W, Sang J, Chen L, Tian J, Zhang H, Olvera Palma G (2015) Life cycle assessment of lead-acid batteries used in electric bicycles in China. J Cleaner Prod 108:1149–1156

Liu Y, Lin D, Liang Z, Zhao J, Yan K, Cui Y (2016) Lithium-coated polymeric matrix as a minimum volume-change and dendrite-free lithium metal anode. Nat Commun 7:10992

Ma X, Chen H, Ceder G (2011) Electrochemical properties of monoclinic $NaMnO_2$. J Electrochem Soc 158(12):A1307–A1312

Manthiram A, Fu Y, Su Y-S (2013) Challenges and prospects of lithium-sulfur batteries. Acc Chem Res 46(5):1125–1134

Mao Y, Li G, Guo Y, Li Z, Liang C, Peng X, Lin Z (2017) Foldable interpenetrated metal-organic frameworks/carbon nanotubes thin film for lithium-sulfur batteries. Nat Commun 8:14628

Miao F, Miao R, Wu W, Cong W, Zang Y, Tao B (2018) A stable hybrid anode of graphene/silicon nanowires array for high-performance lithium-ion battery. Mater Lett 228:262–265

Mikhaylik YV, Akridge JR (2004) Polysulfide shuttle study in the Li/S battery system. J Electrochem Soc 151(11):A1969–A1976

Muto S, Tatsumi K, Kojima Y, Oka H, Kondo H, Horibuchi K, Ukyo Y (2012) Effect of Mg-doping on the degradation of $LiNiO_2$-based cathode materials by combined spectroscopic methods 205(2):449–455

Ozawa K (1994) Lithium-ion rechargeable batteries with $LiCoO_2$ and carbon electrodes: the $LiCoO_2$/C system 69(3–4):212–221

Padhi AK, Nanjundaswamy KS, Goodenough JB (1997) Phospho-olivines as positive-electrode materials for rechargeable lithium batteries 144(4):1188–1194

Pan HL, Hu YS, Chen LQ (2013) Room-temperature stationary sodium-ion batteries for large-scale electric energy storage. Energy Environ Sci 6(8):2338–2360

Perkins JD, Bahn CS, Mcgraw JM, Parilla PA, Ginley DSJC (2010) ChemInform abstract: pulsed laser deposition and characterization of crystalline lithium cobalt dioxide ($LiCoO_2$) thin films 33(11):A1302–A1312

Picciotto LAD, Adendorff KT, Liles DC, Thackeray M (1993) Structural characterization of $Li_{1+x}V_3O_8$ insertion electrodes by single-crystal X-ray diffraction 62(3–4):297–307

Rahman MA, Wang X, Wen C (2013) High energy density metal-air batteries: a review. J Electrochem Soc 160(10):A1759–A1771

Rauh R, Abraham K, Pearson G, Surprenant J, Brummer S (1979) A lithium/dissolved sulfur battery with an organic electrolyte. J Electrochem Soc 126(4):523–527

Ryu H-S, Ahn H-J, Kim K-W, Ahn J-H, Lee J-Y (2006) Discharge process of Li/PVdF/S cells at room temperature. J Power Sources 153(2):360–364

Sathiya M, Thomas J, Batuk D, Pimenta V, Gopalan R, Tarascon J-M (2017) Dual stabilization and sacrificial effect of Na_2CO_3 for increasing capacities of Na-Ion cells based on P_2-Na_xMO_2 electrodes. Chem Mater 29(14):5948–5956
Sawai K, Ohmae T, Suwaki H, Shiomi M, Osumi S (2007) Idling-stop vehicle road tests of advanced valve-regulated lead-acid (VRLA) battery. J Power Sources 174(1):54–60
Seh ZW, Sun Y, Zhang Q, Cui Y (2016a) Designing high-energy lithium-sulfur batteries. Chem Soc Rev 45(20):5605–5634
Seh ZW, Sun Y, Zhang Q, Cui Y (2016b) Designing high-energy lithium–sulfur batteries. Chem Soc Rev 45(20):5605–5634
Shim J, Striebel KA (2007) Electrochemical characterization of thermally oxidized natural graphite anodes in lithium-ion batteries. J Power Sources 164(2):862–867
Shim J, Striebel KA, Cairns EJ (2002) The lithium/sulfur rechargeable cell. J Electrochem Soc 149(10):A1321–A1325
Song MY, Lee R (2002) Synthesis by sol–gel method and electrochemical properties of $LiNiO_2$ cathode material for lithium secondary battery 111(1):97–103
Soria ML, Valenciano J, Ojeda A (2004) Development of ultra high power, valve-regulated lead-acid batteries for industrial applications. J Power Sources 136(2):376–382
Sources W-J (2011) Structure and performance of $LiFePO_4$ cathode materials: a review 196(6):2962–2970
Suga et al. (2009) Emerging N-Type Redox-Active Radical Polymer for a Totally Organic Polymer-Based Rechargeable Battery. Adv Mater 21:1627–1630
Sun J, Sun Y, Pasta M, Zhou G, Li Y, Liu W, Xiong F, Cui Y (2016) Entrapment of polysulfides by a black-phosphorus-modified separator for lithium-sulfur batteries. Adv Mater 28(44):9797–9803
Sun WN, Ying JR, Huang Z, Jiang CY, Wan CR (2009) Organic sulfide electrode materials for lithium-ion batteries. Progress in Chemistry 21(9):1963–1968
Suo L, Hu Y-S, Li H, Armand M, Chen L (2013) A new class of solvent-in-salt electrolyte for high-energy rechargeable metallic lithium batteries. Nat Commun 4(2):1481
Taniguchi IJI, Research EC (2005) Physical and electrochemical properties of spherical nanostructured $LiCr_xMn_{2-x}O_4$ particles synthesized by ultrasonic spray pyrolysis 44 (17):6560–6565
Thackeray MM, David WIF, Bruce PG, Goodenough JB (1983) Lithium insertion into manganese spinels 18(4):461–472
Tu J, Zhao XB, Xie J, Cao GS, Zhuang DG, Zhu TJ, Tu JP (2007) Enhanced low voltage cycling stability of $LiMn_2O_4$ cathode by ZnO coating for lithium ion batteries 432(1):313–317
Wadsley AD (2010) Crystal chemistry of non-stoichiometric pentavalent vandadium oxides: crystal structure of $Li_{1+x}V_3O_8$ 10(4):261–267
Wang D, Li H, Shi S, Huang X, Chen L (2005) Improving the rate performance of $LiFePO_4$ by Fe-site doping 50(14):2955–2958
Wang DP, Fu M, Ha Y, Wang H, Wu R (2018) Metal-organic framework-derived mesoporous octahedral copper oxide/titania composites for high-performance lithium-ion batteries. J Colloid Interf Sci 529:265–272
Wang GX, Bewlay SL, Konstantinov K, Liu HK, Dou SX, Ahn J.-H (2004) Physical and electrochemical properties of doped lithium iron phosphate electrodes 50(2):443–447
Wang H, Yang Y, Liang Y, Robinson JT, Li Y, Jackson A, Cui Y, Dai H (2011) Graphene-wrapped sulfur particles as a rechargeable lithium-sulfur battery cathode material with high capacity and cycling stability. Nano Lett 11(7):2644–2647
Wang J, Yang J, Wan C, Du K, Xie J, Xu N (2003) Sulfur composite cathode materials for rechargeable lithium batteries. Adv Func Mater 13(6):487–492
Wang J, Yang J, Xie J, Xu N (2002) A novel conductive polymer-sulfur composite cathode material for rechargeable lithium batteries. Adv Mater 14(13–14):963–965
Wang JW, He Y, Fan F, Liu XH, Xia S, Liu Y, Harris CT, Li H, Huang JY, Mao SX, Zhu T (2013) Two-phase electrochemical lithiation in amorphous silicon. Nano Lett 13(2):709–715
Wang Y, Jin Y, Jia M (2018) Ultralong Fe_3O_4 nanowires embedded graphene aerogel composite anodes for lithium ion batteries. Mater Lett 228:395–398

Wang Y, Yan YL, Ren B, Yang R, Zhang W, Xu YH (2017) Activated porous carbon wrapped sulfur sub-microparticles as cathode materials for lithium sulfur batteries. IOP Conf Ser Mater Sci Eng 182:012013

Wang ZH, Cao XY, Ge P, Zhu LM, Xie LL, Hou HS, Qiu XQ, Ji XB (2017) Hollow-sphere ZnSe wrapped around carbon particles as a cycle-stable and high-rate anode material for reversible Li-ion batteries. New J Chem 41(14):6693–6699

Wen Y, He K, Zhu Y, Han F, Xu Y, Matsuda I, Ishii Y, Cumings J, Wang C (2014) Expanded graphite as superior anode for sodium-ion batteries. Nat Commun 5:5033

Wu F, Chen J, Chen R, Wu S, Li L, Chen S, Zhao T (2011) Sulfur/polythiophene with a core/shell structure: synthesis and electrochemical properties of the cathode for rechargeable lithium batteries. J Phys Chem C 115(13):6057–6063

Wu M, Jin J, Wen Z (2016) Influence of a surface modified Li anode on the electrochemical performance of Li–S batteries. RSC Adv 6(46):40270–40276

Wu M, Wen Z, Jin J, Chowdari BVR (2016) Trimethylsilyl chloride-modified Li anode for enhanced performance of Li-S cells. ACS Appl Mater Interfaces 8(25):16386–16395

Xu GJ, Han PX, Dong SM, Liu HS, Cui GL, Chen LQ (2017) $Li_4Ti_5O_{12}$-based energy conversion and storage systems: status and prospects. Coord Chem Rev 343:139–184

Xu W, Wang J, Ding F, Chen X, Nasybulin E, Zhang Y, Zhang J-G (2014) Lithium metal anodes for rechargeable batteries. Energy Environ Sci 7(2):513–537

Yamin H, Penciner J, Gorenshtain A, Elam M, Peled E (1985) The electrochemical behavior of polysulfides in tetrahydrofuran. J Power Sources 14(1):129–134

Yang H, Qi K, Gong L, Liu W, Zaman S, Guo X, Qiu Y, Xia BY (2018) Lead oxide enveloped in N-doped graphene oxide composites for enhanced high-rate partial-state-of-charge performance of lead-acid battery. ACS Sustain Chem Eng 6(9):11408–11413

Yang Y, Huang GY, Sun H, Ahmad M, Mou Q, Zhang H (2018) Preparation and electrochemical properties of mesoporous $NiCo_2O_4$ double-hemisphere used as anode for lithium-ion battery. J Colloid Interf Sci 529:357–365

Yang Z, Ding Y, Jiang Y, Zhang P, Jin H (2018) Hierarchical $C/SiO_x/TiO_2$ ultrathin nanobelts as anode materials for advanced lithium ion batteries. Nanotechnology 29(40):1361

Yang Z, Guo J, Das SK, Yu Y, Zhou Z, Abruna HD, Archer LA (2013) In situ synthesis of lithium sulfide-carbon composites as cathode materials for rechargeable lithium batteries. J Mater Chem A 1(4):1433–1440

Yanxia S, Lijuan Z (2017) Lithium rich ternary cathode materials for dynamical type lithium ion battery 80(1):34–40

Yu L, Qiu X, Xi J, Zhu W, Chen L (2006) Enhanced high-potential and elevated-temperature cycling stability of $LiMn_2O_4$ cathode by TiO_2 modification for Li-ion battery 51(28):6406–6411

Yuan L, Yuan H, Qiu X, Chen L, Zhu W (2009) Improvement of cycle property of sulfur-coated carbon nanotubes composite cathode for lithium/sulfur batteries. J Power Sources 189(2):1141–1146

Zhai J, Zhao M, Wang D, Qiao Y (2010) Effect of MgO nanolayer coated on $Li_3V_2(PO_4)_3/C$ cathode material for lithium-ion battery 502(2):401–406

Zhang LL, Liang G, Ignatov A, Croft MC, Xiong XQ, Hung IM, Huang YH, Hu XL, Zhang WX, Peng YL (2011) Effect of vanadium incorporation on electrochemical performance of $LiFePO_4$ for lithium-ion batteries. J Phys Chem C 115(27):13520–13527

Zhang SS, Jow TR (2002) Optimization of synthesis condition and electrode fabrication for spinel $LiMn_2O_4$ cathode 109(1):172–177

Zhang T, Tao Z, Chen J (2014) Magnesium–air batteries: from principle to application. Mater Horiz 1(2):196–206

Zhang Y, Wang X, Zeng L, Song S, Liu DJDT (2012) Green and controlled synthesis of Cu_2O-graphene hierarchical nanohybrids as high-performance anode materials for lithium-ion batteries via an ultrasound assisted approach 41(15):4316–4319

Chapter 7
Nanomaterials for Supercapacitors

Supercapacitors have become a research hotspot in the field of electrochemical energy storage due to their fast charging and discharging ability and long cycle stability. The electrodes are the core components of supercapacitors. Therefore, enhancing the electrochemical performance of electrode materials is the key to improve the overall performance of a supercapacitor device. Nanostructured materials have been widely studied as electrode materials for supercapacitors. Nanostructures can not only provide abundant electrochemical active surfaces, but also reduce the ion transport pathway. In this chapter, we summarized and evaluated the application of nanostructured materials in supercapacitors, including synthesis method, composition, structure, and electrochemical performance. Activated carbon, ordered mesoporous carbon, carbon nanotube, graphene, Co_3O_4, MnO_2, NiO, conductive polymer, and MXene are discussed in detail. Some relevant prospects are also proposed.

7.1 Introduction

The large-scale exploitation and continuous consumption of fossil energy result in more and more serious environment and climate problems. Therefore, the development and utilization of clean and renewable energy have become a common concern issue for all countries. The use of renewable and low-carbon clean energy, such as solar energy, wind energy, and tidal energy, is an important way to solve the current energy crisis. However, natural conditions lack stability and greatly influence the use of renewable energy. Therefore, it is particularly important to develop energy storage systems that can efficiently store the energy from renewable energy resources. Among various energy storage devices, electrochemical energy storage devices, mainly including lead-acid battery, nickel-metal hydride battery, lithium-ion battery, and supercapacitor, have attracted much attention due to their good reliability, high efficiency, ease of operation, and low or no pollution.

H. Pang et al., *Synthesis of Functional Nanomaterials for Electrochemical Energy Storage*,
https://doi.org/10.1007/978-981-13-7372-5_7

Supercapacitors can bridge the power and energy gap between batteries and traditional physical capacitors and combine the advantages of them, such as quick charge/discharge, long cycling life, high reliability, and wide range of working adaptability. On the basis of the charge storage mechanism, supercapacitors can be categorized into two types: electrical double-layer capacitors (EDLCs) and pseudocapacitors (Wang et al. 2016; Zheng et al. 2017).

The capacitive process of EDLCs is based on the charge separation caused by the directional arrangement of electrons and ions at the interface of electrode and electrolyte. During the charging process, electrons are transferred from the positive electrode to the negative electrode. Under the electric field force, positive ions and negative ions in the solution will move directionally to the two poles and form a tight electrical double layer on the electrode surface to generate a potential difference. During the discharge process, the charges on the electrode will release automatically through the external circuit to supply the power. Meanwhile, the ions on the electrode surface will migrate back to the electrolyte, and the tight electrical double layer will disappear.

For pseudocapacitors, the charge storage is due to the highly reversible redox reactions of the active material and its chemical adsorption/desorption. Compared to EDLCs, which occur only on the surface of the electrode, the charge storage of Faraday pseudocapacitor materials can occur not only on the surface of the electrode but also in the bulk phase of the materials. Therefore, Faraday pseudocapacitors can achieve a much higher specific capacitance than EDLCs, which can generally reach 10–100 times of the capacitance of EDLCs. However, the power density of the pseudocapacitors is lower than that of the EDLCs, because the Faraday process is generally slower than the non-Faraday process. In addition, the pseudocapacitor materials usually have a poor cycling stability, which is like the battery.

The mechanisms of EDLCs and pseudocapacitors indicate that the specific capacitance of supercapacitors is closely related to the interface of electrode materials and electrode. Therefore, the microstructure of electrode material is an important factor determining the performance of a supercapacitor. Nanostructured materials have been extensively studied as electrode materials of supercapacitor because they can provide more active surface area, facilitate the transport of electrolytes, and accommodate volume changes during charging/discharging.

In this chapter, we mainly discussed the application of nanostructured electrode materials in supercapacitors. The synthetic methods and electrochemical performance of nanostructured carbon materials, transition metal oxides, conductive polymers, and MXenes were systematically reviewed. The hybrid capacitor, which is composed of two different electrode materials, was also summarized. In addition, we put forward our own prospect for the further development of nanostructured materials for supercapacitors.

7.2 Double-Layer Capacitor

Nanostructured carbon materials are prospective electrode materials for supercapacitors due to their abundance, non-toxicity, high specific surface area, high electronic conductivity, and good chemical stability. Various carbon materials, including activated carbon, mesoporous carbon, carbon nanotubes (CNTs), and graphene, have been widely reported as electrode materials for EDLCs over the past years. Numerous studies have shown that specific surface area, pore size distribution, pore structure, pore volume, pore size, pore surface curvature, electrical conductivity, and surface functional groups of carbon materials have decisive influence on the performance of EDLCs. Herein, we briefly summarize the relevant advances in the development of carbon materials for EDLCs.

7.2.1 Activated Carbon

Activated carbons have been the earliest carbon materials for supercapacitor because of their good electrical conductivity, high specific surface area, abundant resources, and low cost. In recent years, the studies about the preparation of activated carbons mainly focused on raw materials and activation technologies. The precursors for the preparation of activated carbon mainly include natural renewable resources (such as wood, coconut shell), fossil fuel and its derivatives (such as asphalt, coal, coke), and synthetic polymers. The activation method includes chemical activation and physical activation. For the chemical activation, the carbon materials, which were obtained via the carbonization for precursors, were mixed with activating agents (such as KOH, K_2CO_3, H_3PO_4, $ZnCl_2$) and a subsequent heat treatment under inert atmosphere. For the physical activation, the carbon material was treated by oxidizing gas (such as CO_2 and H_2O) under high temperature. The activated conditions directly influenced the pore structure parameters of the products. The activated carbons obtained by the activation method have a dispersive pore distribution. Numerous studies have proved that there is no linear relationship between the specific capacitance and the specific surface area. The pore size and pore distribution also have an important influence on the specific capacitance.

Gogotsi et al. (Chmiola 2006) overturns the erroneous view that the pore smaller than solvent ionic diameter is invalid. They found an abnormal phenomenon by testing the specific capacitance of carbon materials with different pore sizes. When the pore size was more than 2 times of the solvent ionic diameter, the area specific capacitance was basically stable and did not change with the pore size. When the pore size was larger than the solvent ionic diameter and less than 2 times of the solvent ionic diameter, the area specific capacitance decreased with the decreasing pore size, owing to not forming effectively the dense layer and the dispersed double layer. When the pore size is less than the solvent ionic diameter, due to the existence of desolvation and the improvement of the storage charge of unit specific capacitance, the area specific capacitance increased sharply with the decrease of the pore size.

For typical activated carbon materials, the Brunauer-Emmett-Teller (BET) specific surface area is generally from 1000 to 2000 m^2 g^{-1}. The gravimetric capacitance of the materials is usually 100–120 F g^{-1} in organic electrolytes. For further improving the performance of activated carbon materials, Ruoff et al. (Zhu et al. 2011) synthesized a new active carbon (donated as activated graphene) via a chemical activation with KOH for microwave exfoliated graphite oxide (MEGO), which possesses a specific surface area of up to 3100 m^2 g^{-1}. When measured in an organic (BMIM BF_4) electrolyte, the capacitance of activated graphene was 200 F g^{-1} at 0.7 A g^{-1}. This material also shows stable cycling performance, 97% of its initial capacitance was retained at 2.5 A g^{-1} after 10,000 cycles.

7.2.2 Ordered Mesoporous Carbon Materials

Ordered mesoporous carbon materials not only possess high specific surface area, but also have large pore size and orderly pore arrangement, which is conducive to the diffusion of electrolyte. Therefore, it becomes a good electrode material for EDLCs. For instance, CMK-3 mesoporous carbon was prepared by the hard template method, exhibiting a fiber-like close stacking framework with a specific BET surface area of 900 m^2 g^{-1} and a pore diameter of 3.90 nm (Li et al. 2007). When utilized as the electrode material in organic electrolytes, CMK-3 showed a specific capacitance of 90 F g^{-1} at 50 mV s^{-1}. Dai et al. (Wang et al. 2010) effectively enhanced the specific surface area from 520 to 1940 m^2 g^{-1} via the KOH activation for the mesoporous carbon. The specific capacitance was increased from 71 to 188 F g^{-1}.

To further enhance the specific capacitance of mesoporous carbons, the introduction of surface functional groups is a promising way. The functional groups could proceed a reversible redox reaction, thus providing pseudocapacitance. Recently, Huang et al. (Lin et al. 2015) reported a breakthrough work about the mesoporous carbon with functional groups as the electrode materials of supercapacitor. N-doped ordered mesoporous few-layer-graphene carbon (OMFLC-N) was prepared via the combination of chemical vapor deposition (CVD) method and sacrificial template method (Fig. 7.1a). As shown in Fig. 7.1b, c, OMFLC-N showed ordered mesopores and few-layer-graphene like structure. When used as the electrode material, it exhibited nearly rectangular CV curves and symmetric features (Fig. 7.1d). The optimal OMFLC-N electrode achieved an ultrahigh specific capacitance of 790 F g^{-1} in 0.5 M H_2SO_4 electrolyte at 1 A g^{-1}, the corresponding value was 720 F g^{-1} in 2 M Li_2SO_4 electrolyte (Fig. 7.1e). The improvement mostly stemmed from the pseudocapacitance through robust redox reactions at the defects of nitrogen atoms. The inert graphene-like layered carbon was transformed into a material with electrochemical activity while maintaining good electric conductivity. This result provides an excellent low-cost carbon-based material for supercapacitors.

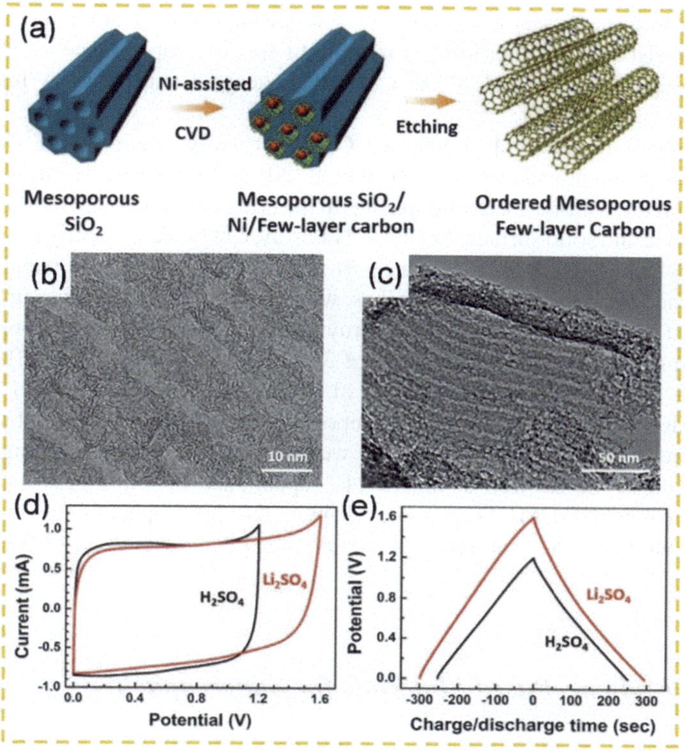

Fig. 7.1 **a** Schematic for the fabrication ordered mesoporous few-layer carbon (OMFLC). **b, c** TEM images of OMFLC-N. **d** Cyclic voltammetry of symmetric capacitor of OMFLC-N in 0.5 M H_2SO_4 (pH = 0) and 2 M Li_2SO_4 (pH = 1.8) at 2 mV s^{-1}. **e** Galvanostatic charge/discharge curves of OMFLC-N at 1 A g^{-1} for the first cycle. Reprinted from Ref. Lin et al. (2015), copyright 2015, with permission from Science

7.2.3 Carbon Nanotube for Supercapacitor

Owing to the one-dimensional tubular structure, high electrical conductivity, excellent mechanical performance, and good chemical and thermal stability, CNTs show good application potential in the field of supercapacitors. CNTs can be categorized as multi-walled CNTs (MWNTs) and single-walled CNTs (SWNTs). Comparing to activated carbons and mesoporous carbons, the specific surface area of CNTs is relatively low.

Niu et al. (1997) prepared a MWNT-based material with specific surface area of 430 m^2 g^{-1} as the electrode, which exhibited a specific capacitance of 102 F g^{-1} and a power density of 8 kW kg^{-1} in an acidic electrolyte. Lee et al. (An et al. 2001) investigated the key factors that determine the performance of supercapacitors based on single-walled CNT electrodes. The results show that the heat treatment at high temperature was necessary to increase the specific capacitance and reduce the CNT-electrode resistance. Furthermore, a maximum specific capacitance of 180 F g^{-1} with a large power density of 20 kW kg^{-1} at an energy density of 6.5 Wh kg^{-1} was realized.

Compared with randomly entangled CNT, vertically aligned CNT (VA-CNT) array possesses obvious advantages (Dai et al. 2012). The vertically aligned nanostructures and the well-defined tube spacing in the array are beneficial to the contact of the electrolyte and solid surface. For VA-CNT array, CNTs were directly grown on the substrate, which is beneficial to the rate performance. Moreover, the top end of CNTs can be opened via appropriate approaches, which results in more active surface area. Chen et al. (2004) fabricated VA-CNT array via CVD method using porous alumina as the template. A specific capacitance of 365 F g^{-1} was achieved at 210 mA g^{-1} in 1 mol L^{-1} H_2SO_4 electrolyte. Dai et al. (Lu et al. 2009) synthesized VA-CNT array on SiO_2/Si wafers that were pre-deposited with Fe nanocatalysts by vacuum CVD method. The end tips of nanotube were opened via Plasma etching method. The array electrode showed a remarkable capacitance of 440 F g^{-1} in the ionic liquid electrolyte ([EMIM][Tf2N]). Furthermore, the supercapacitor device exhibited a high cell voltage of 4 V and superior energy and power densities (148 Wh $kg^{-1}$and 315 kW kg^{-1}).

7.2.4 Graphene-Based Materials for Supercapacitor

Graphene possesses a honeycombed 2D crystal structure formed by the arrangement of sp2 hybrid carbon atoms. It has been widely studied as electrode materials for supercapacitors due to its good electric conductivity and large specific surface area. The theoretical capacity of graphene is up to 550 F g^{-1}, which is the largest value among all carbon-based materials. However, due to the characteristics of its two-dimensional plane structure and strong π-π interaction, graphene is easy to stack, which results in the decrease of specific surface area and the low utilization rate of electrolyte. Therefore, the actual specific capacitance value is far less than the theoretical value. Preventing the accumulation of graphene nanosheets is the focus of current research in this field.

Kaner et al. (El-Kady et al. 2012) presented a simple strategy to fabricate the graphene-based EDLCs by an all-solid-state laser-reducing approach (Fig. 7.2a). Graphene oxide (GO) was reduced directly to laserscribed graphene (LSG) on DVD, which avoided the restacking of graphene sheets. Figure 7.2b, c shows the SEM images of the well-exfoliated LSG sample and the initially stacked GO precursor. Moreover, the produced robust films possessed a high electrical conductivity of 1738 S m^{-1} and a large specific surface area of 1520 m^2 g^{-1}. In an aqueous electrolyte, the LSG electrode showed a nearly rectangular CV shape at 1000 mV s^{-1} (Fig. 7.2d). When LSG was designed and fabricated to a flexible and all-solid-state electrochemical capacitor, its performance was completely independent of the bending angle

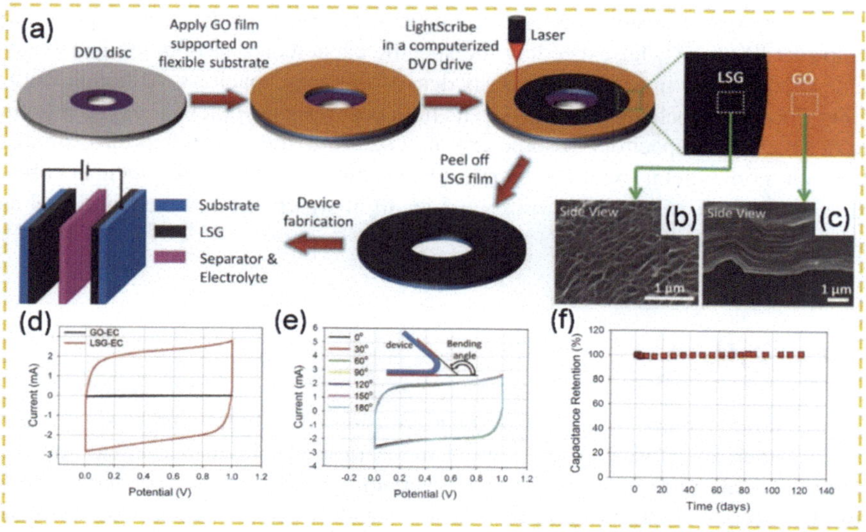

Fig. 7.2 **a** Schematic illustration of the fabrication of laserscribed graphene-based electrochemical capacitors. **b**, **c** SEM images of LSG and GO. **d** Cyclic voltammetry of LSG-ECs and GO-ECs at 1 V s^{-1} in 1.0 M H_3PO_4 electrolyte. **e** Cyclic voltammetry of the flexible all-solid-state LSG-EC at different bending angles at 1 V s^{-1}. **f** Cyclic stability test of the flexible all-solid-state LSG-EC using gelled electrolytes. Reprinted from Ref. El-kady et al.(2012), copyright 2012, with permission from Science

(Fig. 7.2e). And the LSG-EC capacitance retention was almost stable over 120 days of testing (Fig. 7.2f). Furthermore, the device exhibited excellent rate performance.

Duan et al. (Xu et al. 2014) prepared graphene foam with a large surface area of 1560 m^2 g^{-1} via H_2O_2 activation in solution. The specific capacitances of the sample were up to 310 and 298 F g^{-1} at 1 A g^{-1} in aqueous electrolyte and organic electrolyte, respectively. Furthermore, the energy density is 127 Wh kg^{-1} in organic electrolyte, which is close to the level of lithium-ion batteries. The high performance of the graphene foam was ascribed to the special preparation process, which didn't involve the drying treatment. The solvent between graphene interlayers could effectively prevent the stacking and aggregation of graphene.

7.3 Pseudocapacitors

Transition metal oxides are one kind of the most widely studied pseudocapacitor electrode materials. Among many metal oxides, ruthenium oxide (RuO_2) was first studied, which not only possesses excellent conductivity but also has three different oxidation states that can be used to store charge during 1.2 V. The cyclic voltammetry curve of RuO_2 is similar to that of the double electrode layer and presents an almost perfect rectangle. Currently, the specific capacitance of RuO_2 has been reported to be up to 1300 F g^{-1}, far exceeding the specific capacitance of double-layer materials, which greatly improves the energy density of supercapacitors. Although RuO_2 shows good electrochemical behavior, its high cost restricts practical application in the field of supercapacitors to some extent. Therefore, the researchers investigated other non-noble metal oxides as supercapacitor electrode materials, such as cobalt oxide, manganese dioxide, and nickel oxide, all of which showed their potential as pseudocapacitor materials.

7.3.1 Co_3O_4

Among various metal oxides, Co_3O_4 is considered a highly promising material for supercapacitors due to its good redox property, high theoretical specific capacitance (3560 F g^{-1}), and great reversibility. Co_3O_4-based materials with various structures/morphologies showed a clear difference in the electrochemical performance.

Two-dimensional (2D) nanostructure not only possessed high specific surface area but also facilitated the ion transfer during the charge-discharge process. Feng et al. (2015) fabricated sub-3 nm atomic layers Co_3O_4 nanofilms via a hydrothermal reaction without surfactants and substrates. Because of the ultrathin thickness and large size, the as-prepared atomic layers Co_3O_4 nanofilms achieved high specific capacitances of 1400 F g^{-1} at 1 A g^{-1} and 1276 F g^{-1} at 8 A g^{-1}. Furthermore, the Co_3O_4 nanofilms also showed a good cycling performance with a small decrease (less than 3%) after 1500 cycles. Meher and Rao (2011) synthesized 2D layered Co_3O_4 with high porosity under hydrothermal conditions. The pores of the obtained 2D flakes are quite narrow (~1.85 nm) and uniform. The specific capacitance of the sample was 548 F g^{-1} at 8 A g^{-1} and retained 66% at 32 A g^{-1}. Furthermore, the layered Co_3O_4 delivered outstanding cycling performance with 97.5% of initial capacitance after 2000 charge-discharge cycles at 16 A g^{-1}.

Inspired by oxygen vacancies, Wang et al. (2014) prepared reduced mesoporous Co_3O_4 nanowires via a facile solution reduction method. Due to the numerous oxygen vacancies on the nanowires' surface caused by the treatment of $NaBH_4$, the reduced Co_3O_4 nanowires achieved better electrochemical performance than the pristine Co_3O_4 nanowires. Supercapacitors assembled with the reduced Co_3O_4 nanowires showed a high capacitance of 978 F g^{-1} at 2 A g^{-1}. Furthermore, the capacitance of the reduced Co_3O_4 nanowires is about 883 F g^{-1} after 2000 cycles, which indicates that the oxygen vacancies are stable during the cycling test. Zhang et al. (2014a, b) fabricated hollow Co_3O_4 polyhedral particles with a porous shell by the heat treatment of ZIF-67 microcrystals. As the electrode, the material delivered a large specific capacitance of 1100 F g^{-1} at 1.25 A g^{-1} and excellent cycling performance with little capacitance decay even after 6000 cycles at 6.25 A g^{-1}.

Preparing composite materials can make use of synergistic effects of both components. Yang et al. (2018) prepared oxygen-vacancy abundant ultrafine

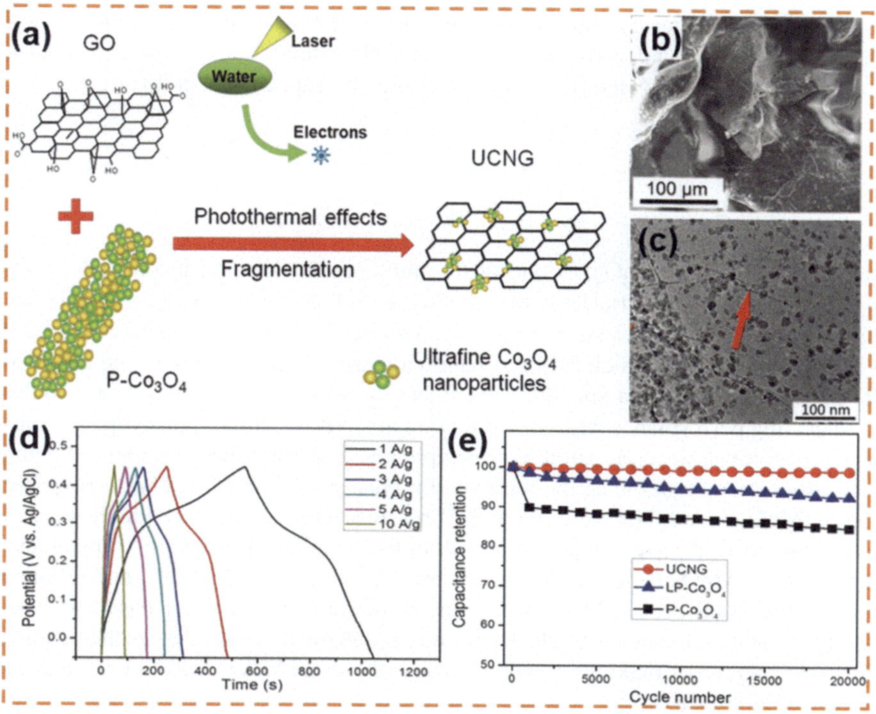

Fig. 7.3 **a** Schematic illustration of the fabrication of the UCNG composite. **b** SEM image of UCNG composite. **c** TEM image of UCNG composite. **d** Galvanostatic charge-discharge curves of the UCNG composite under different current densities. **e** Cycling performances of the LP-Co_3O_4, P-Co_3O_4, and UCNG composite electrodes under a current density of 10 A g^{-1}. Reprinted from Ref. Yang et al. (2018), copyright 2018, with permission from WILEY-VCH

Co_3O_4/graphene (UCNG) composites via a novel one-step laser irradiation route. As the illustration shows (Fig. 7.3a), the GO was reduced to laser-reduced graphene (LG) under laser irradiation, while the Co_3O_4 nanoparticles were fragmented from the porous Co_3O_4 nanorods and then were anchored on LG surface. In Fig. 7.3b, c, ultrafine Co_3O_4 nanoparticles ($\approx$10 nm) are uniformly distributed on graphene sheets, which is beneficial for short ionic diffusion distance and fast electron transport. Therefore, the as-obtained UCNG electrode exhibited a high specific capacitance of 977.1 F g^{-1} (135.8 mA h g^{-1}) at 1 A g^{-1} and excellent cycling stability with 99.3% retention even after 20,000 cycles (Fig. 7.3d, e). The hierarchical structure of Co_3O_4/nitrogen-doped carbon hollow spheres (Co_3O_4/NHCSs) was designed by Liu et al. (2018a, b) via a hydrothermal method and a subsequent calcination treatment. In the synthesis process, N-carbon/SiO_2 sphere was employed as the template to fabricate CoHC/NHCS and finally evolved to Co_3O_4/NHCS after the calcination. The as-prepared Co_3O_4/NHCS showed a higher specific capacitance of 581 F g^{-1}

at 1 A g^{-1} and better cycling performance with 91.6% of initial specific capacitance at 20 A g^{-1}, while pristine Co_3O_4 nanorods exhibited a specific capacitance of 318 F g^{-1} at 1 A g^{-1} and 67.1% of initial specific capacitance at 20 A g^{-1}.

7.3.2 MnO_2

Manganese dioxide (MnO_2) is one of the most potential supercapacitor electrode materials because of its rich polymorphisms, high theoretical specific capacitance, high natural abundance, and non-toxicity. Wei et al. (2013) designed a novel 2D β-MnO_2 networks, in which β-MnO_2 nanowires were highly ordered. The obtained 2D networks exhibited a specific capacitance of 453.0 F g^{-1} at 0.5 A g^{-1} and a good cycling performance with 97% retention at 2.5 A g^{-1} after 1800 cycles. Porous MnO_2 nanotubes were prepared via a simple one-step hydrothermal method using polycarbonate membrane as the template by Huang et al. (2015). The as-prepared porous MnO_2 nanotubes were about 200 nm in average diameter, which were in accordance with the average pore diameter of the template. The specific capacitance of the porous MnO_2 nanotubes electrode was 365 F g^{-1} at 0.25 A g^{-1} and retained 90.4% after 3000 cycles. Moreover, an asymmetric supercapacitor based on activated graphene as the negative electrode and porous MnO_2 nanotubes as the positive electrode achieved a maximum power density of 146.2 kW kg^{-1} and a large energy density of 22.5 W h kg^{-1}.

MnO_2-carbon composites were also reported as the electrode. Jia et al. (2018) synthesized graphene quantum dots (GQDs)/MnO_2 composites as electrodes for supercapacitors. In the two-step synthesis (Fig. 7.4a), MnO_2 nanosheet arrays were vertically grown on clean Ni foam substrate through a facile hydrothermal process, and then GQDs anchored in situ on the surface of MnO_2 nanosheets after the PECVD process. As shown in Fig. 7.4b, GQDs/MnO_2 heterostructure was vertical and highly porous. Among the samples with different PECVD deposition time, GQD/MnO_2-3 exhibited a much higher specific capacitance (1170 F g^{-1} at 5 mV s^{-1} in 0–1.3 V) than others. GQD/MnO_2-3 also delivered better cycle performance with 92.7% of initial capacitance even after 10,000 cycles (Fig. 7.4c, d). Furthermore, the 2.3 V aqueous asymmetric supercapacitor based on GQDs/MnO_2-3 as positive electrode and nitrogen-doped (NG) graphene as negative electrode achieved outstanding electrochemical performance with a high energy density of 118 W h kg^{-1} and a large power density of 12,351 W kg^{-1}. CNTs@MnO_2 core-shell structures grown on Ni foam were fabricated by Huang et al. (2014). The obtained binder-free capacitor electrodes of ultrathin layered MnO_2 nanosheets coated on the CNTs were prepared via a floating catalyst CVD process and followed by a facile hydrothermal approach. This unique CNTs@MnO_2 electrode showed a high specific capacitance of 325.5 F g^{-1} at 0.3 A g^{-1} and good cycling stability of 90.5% retention after 5000 cycles. Similarly, Huang et al. (2018) fabricated primary ε-MnO_2 nanosheets and secondary α-MnO_2 nanorod arrays on the surface of carbon fiber by a phase-controlled electrodeposition. The optimized hierarchical architecture (MnO_2-60, deposited at 60 °C) with

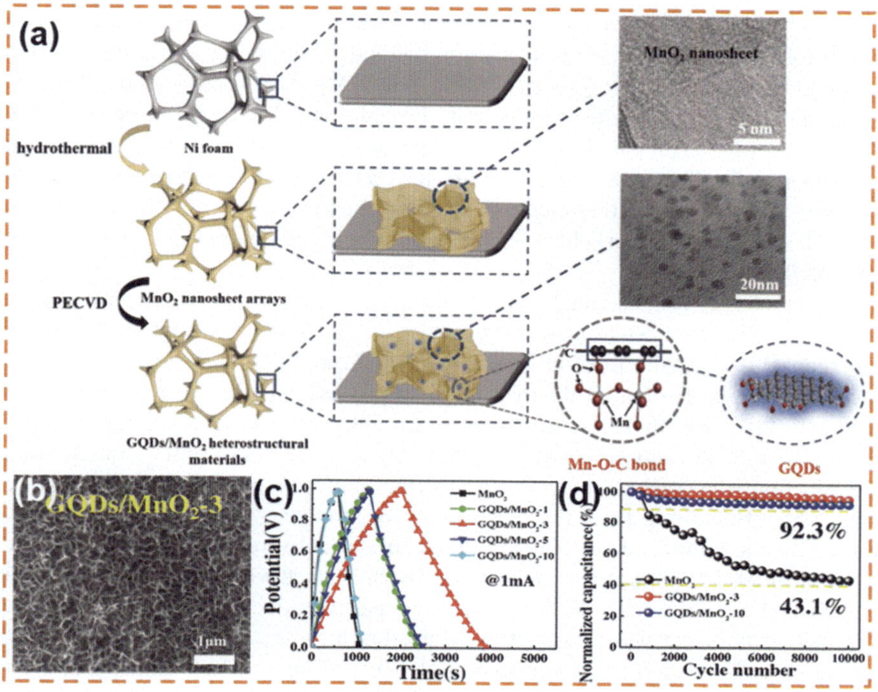

Fig. 7.4 **a** Schematic illustration of fabrication of GQD/MnO_2 heterostructural materials. **b** SEM image of GQDs/MnO_2-3. **c** Galvanostatic charge-discharge curves of all samples under a current of 1 mA. **d** Cycling performances of the MnO_2, GQDs/MnO_2-3, and GQDs/MnO_2-10 composites over 10,000 cycles. Reprinted from Ref. Jia et al. (2018), copyright 2018, with permission from WILEY-VCH

a large MnO_2 mass loading of 10 mg cm^{-2} achieved a high overall capacitance of 304 F g^{-1} at 3 mA cm^{-2}. The volumetric capacitance of asymmetric supercapacitors based on MnO_2-60 cathode and vanadium oxide anode was 14.85 F cm^{-3} (aqueous) or 14.41 F cm^{-3} at 1 mA cm^{-2} and retained 66% at a high current density of 20 mA cm^{-2}. At the same time, the devices showed excellent cycle performance with >90% capacitance retention after 8000 cycles.

7.3.3 NiO

Nickel oxide (NiO) has been extensively applied for use in supercapacitors because of its high theoretical capacity (2573 F g^{-1}). Various hierarchical nanostructures of NiO have been reported. Liang et al. (2012) synthesized a three-dimensional nanoporous NiO film by using an electrochemical method. The NiO film possessed a large specific surface area of 264 m^2 g^{-1} and its specific capacitance is up to 1776 F g^{-1} at

1 mV s^{-1}. Cao et al. (2014) fabricated hierarchical porous NiO nanotube arrays on the Ni foam substrate by electrodeposition method. The specific capacitance of the obtained NiO nanotube arrays was 675 F g^{-1} at 2 A g^{-1}. Similarly, inspired by the larger surface area and better conductivity brought by 3D network of nickel foam, Cheng et al. (2015) prepared ultrathin mesoporous NiO nanosheets in situ on the Ni foam. Due to the enhanced conductivity and ion transfer of the hybrid structure, the as-prepared additive/binder-free electrodes achieved an excellent specific capacitance of 2504.3 F g^{-1} and remarkable cycling stability with no obvious degradation even after 45,000 cycles.

Lin et al. (2018) designed and synthesized a novel core-shell structure of graphene@NiO (G@NiO) nanosheet arrays as electrodes for high-performance supercapacitors. As the illustration shows (Fig. 7.5a), NiO nanosheet arrays were anchored on g-Ni foam after the hydrothermal process, and then were changed into vertically nanosized NiO particles wrapped in graphene layers via a plasma enhanced chemical vapor deposition (PECVD) method. Among various samples with different PECVD time, the G@NiO-1 (1 min) nanosheets showed more flexible outer surface (Fig. 7.5b). Benefiting from the 3D conductive graphene frame-works and nano-sized NiO, the optimized G@NiO hybrid electrodes delivered a remarkable specific capacity of 1073 C g^{-1} and excellent cycling performance with ≈99% retention after 10,000 cycles (Fig. 7.5c, d). Another NiO nanoparticles encapsulated in carbon and SWCNTs composite was prepared by Majeed et al. via a floating catalyst chemical vapor deposition method (Majeed et al. 2017). The obtained binder-free electrode of carbon-encapsulated NiO nanoparticle decorated SWCNT films showed a high specific capacitance of 1422 F g^{-1} at 3 A g^{-1} and good cycling stability with 92% of initial specific capacitance after 5000 cycles. By using nickel acetylacetone and polyacrylonitrile as precursors, Li et al. fabricated NiO nanoparticles dispersed in N-doped porous carbon nanofibers (NiO/PCNF) via an electrospinning method followed by controlled heat treatment (carbonization at 800 °C) (Fig. 7.5e) (Li et al. 2018). The numerous multiscale nanopores can be obviously observed in Fig. 7.5f, g, which is beneficial for the diffusion distance of ion/charge and contact between electrolyte. Therefore, the NiO/PCNF-0.75 film electrode with superior surface area achieved the best specific capacitance of 850 F g^{-1} at 1 A g^{-1} and a high specific capacitance of 748 F g^{-1} when the current density was up to 10 A g^{-1}. Meanwhile, its specific capacitance was retained 96.7% of the initial capacitance even after 10,000 cycles (Fig. 7.5h–j). A hierarchical NiO/carbon hollow sphere composite (NiO/C-HS) was synthesized by Liu et al. through a semi-sacrificial template-assisted hydrothermal process and the following calcination treatment (Liu et al. 2018a, b). The prepared NiO/C-HS with carbon hollow spheres supporting well-dispersed NiO nanosheets delivered a high specific capacitance of 686 F g^{-1} at 1 A g^{-1} and good cycling stability of no obvious capacitance loss after 5000 cycles.

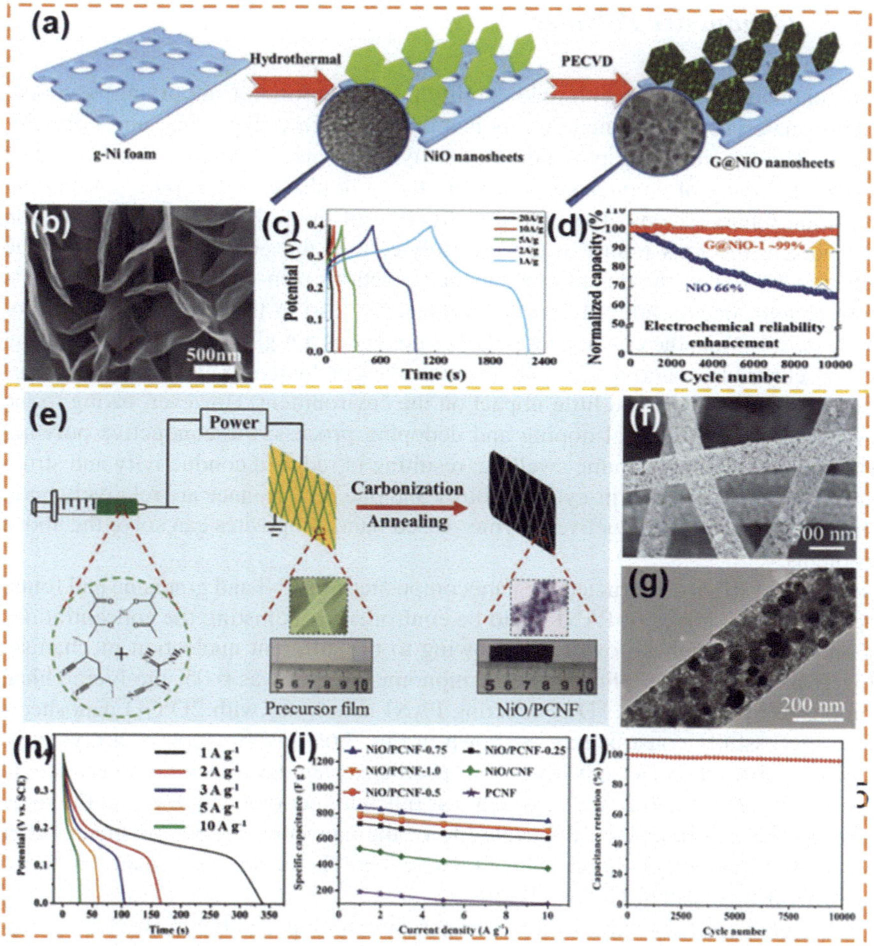

Fig. 7.5 **a** Schematic illustration of fabrication of G@NiO nanosheet arrays. **b** SEM image of G@NiO-1 nanosheet arrays. **c** Galvanostatic charge-discharge curves of G@NiO-1. **d** Cycling performances of NiO and G@NiO-1 electrodes at 10 A g^{-1} for 10,000 cycles. **e** Schematic illustration of fabrication of free-standing NiO/PCNF composite film. **f** FESEM image of NiO/PCNF-0.75. **g** TEM image of NiO/PCNF-0.75. **h** Galvanostatic discharge curves of NiO/PCNF-0.75 under various current densities. **i** Specific capacitances of all samples under various current densities. **j** Cycling performance of NiO/PCNF-0.75 under 10 A g^{-1}. **a–d** Reprinted from Ref. Lin et al. (2018), copyright 2017, with permission from WILEY-VCH. **e–j** Reprinted from Ref. Li et al. (2018), copyright 2018, with permission from WILEY-VCH

7.3.4 Conductive Polymer

The conductive polymer with redox ability is another kind of important pseudocapacitor material, mainly including polyaniline (PANI), polypyrrole, and polythiophene. These conducting polymers primarily store energy through the process of doping/dedoping of conjugated π bond in the bulk phase (redox process). During oxidation (charge), ions enter into the structure of the polymer. During reduction (discharge), the ions return to the electrolyte from the polymer's structure. The entire process (i.e., the doped and dedoped reactions of n-type or p-type elements store charge) occurs not only on the surface, but also in the whole bulk phase of the material. Thus the conductive polymer possesses a higher specific capacitance. As the electrode material of supercapacitor, the conductive polymer has the characteristics of low cost and little impact on the environment. However, owing to the reversible electrochemical doping and dedoping process, the conductive polymer will undergo obvious volume swelling, resulting in reduced conductivity and structural damage. Therefore, its cyclic stability and rate performance are relatively poor. The construction of conductive polymer-based nanocomposites can solve the above problems.

Xu et al. (2010) constructed the nanocomposites of PANI and graphene and found that the morphologies of PANI could be controlled by adjusting the concentrations of aniline and graphene oxide (GO) owing to the different nucleation mechanism (Fig. 7.6a). When the ratio of aniline monomer to GO was 0.05, the hierarchical composite by combining 1D conducting PANI nanowires with 2D GO nanosheets were successfully constructed. As shown in Fig. 7.6b, PANI nanowire arrays were vertically aligned on GO substrate. The specific capacitance of the composite could reach to 555 F g^{-1} at 0.2 A g^{-1}, which was much higher than 298 F g^{-1} at the same condition (Fig. 7.6c). Also, the cycle life of the nanocomposites was much better than that of randomly connected PANI nanowires. The better performance is due to the synergistic effect of GO and PANI.

Chen et al. (2016) synthesized porous perchlorate-doped polypyrrole electropolymerized on Ni nanotube arrays (NiNTAs@PPy). The NiNTAs were obtained by utilizing pre-synthesized ZnO-nanorod arrays as the sacrificial template via a simple electrodeposition method (Fig. 7.7a). As shown in Fig. 7.7b, c, NiNTAs@PPy presented the parallel arrangement of uniformly hollow nanotubes. Perchlorate-doped PPy was densely grown on both the outer and inner surfaces of the nanotubes. As the electrode of supercapacitor, NiNTAs@PPy showed good cyclic performance with 75.3% capacitance retention after 10,000 cycles (Fig. 7.7d).

7.3.5 MXene

MXene, as a new 2D layered material, has drawn researchers' attention due to its remarkable electrochemical properties. As electrode materials of SCs, MXene can

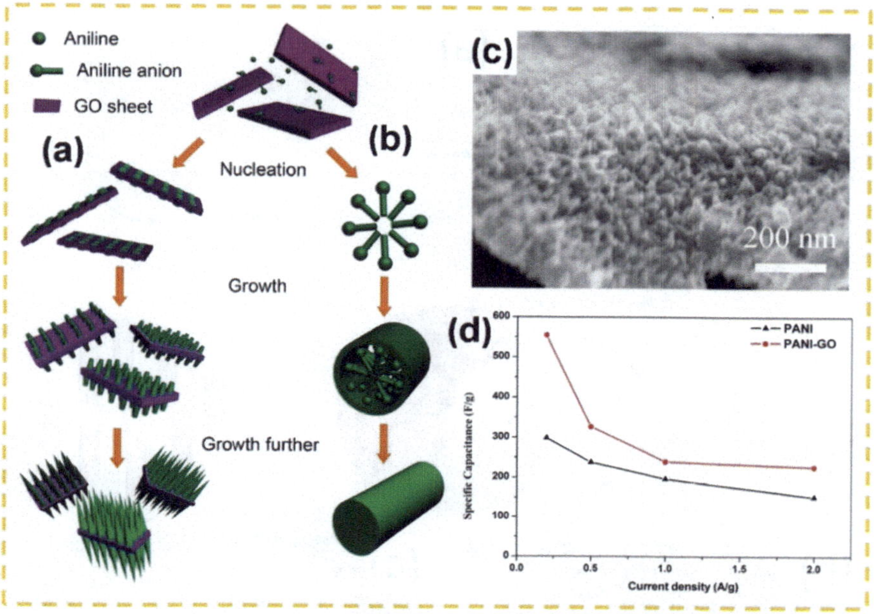

Fig. 7.6 Schematic diagram of the growth and nucleation processes of PANI nanowires: **a** heterogeneous nucleation on GO nanosheets; **b** homogeneous nucleation; **c** SEM image of PANI-GO nanocomposite obtained at 0.05 M of aniline; **d** Rate performance of PANI-GO and PANI. Reprinted from Ref. Xu et al. (2010), copyright 2010, with permission from American Chemical Society

achieve very high capacitance, which is comparable to the best-reported result of RuO_2. Gogotsi's group (Lukatskaya et al. 2013) firstly investigated MXene material as the electrode material in SCs and demonstrated that there existing the spontaneous intercalation of cations, including Li^+, Na^+, K^+, NH^{4+}, Mg^{2+}, and Al^{3+}, from aqueous salt solutions between $Ti_3C_2T_x$ MXene layers (Fig. 7.8a). They tested the MXene electrode in various electrolytes and found that the electrochemical properties varied with different electrolytes (Fig. 7.8b). When testing in KOH electrolyte, a volumetric capacitance of 350 F cm^{-3} at 20 mV s^{-1} (and 450 F cm^{-3} at 2 mV s^{-1}) was obtained. Furthermore, they also manufactured flexible $Ti_3C_2T_x$ paper electrodes (Fig. 7.8c). The results proved that a flexible electrode delivered a superior cycling performance with no measurable capacitance losses during 10,000 cycles at 1 A g^{-1} (Fig. 7.8d). Later on, Gogotsi et al. (Ghidiu et al. 2014) keep on studying the MXene electrode in SCs to pursue an improved electrochemical performance. The titanium carbide "clay" film was fabricated via a mild etching route (LiF in HCl) and acted as additive-free electrode material for SCs. It is worth noting that they replace the above electrolyte with H_2SO_4 electrolyte and get a remarkable enhancement compared to the previous work using KOH electrolyte. A high volumetric capacitance of up to 900 F cm^{-3} at 2 mV s^{-1} was obtained (Fig. 7.8e). Additionally, there is almost no capacitance decay during 10,000 cycles at 10 A g^{-1} (Fig. 7.8f).

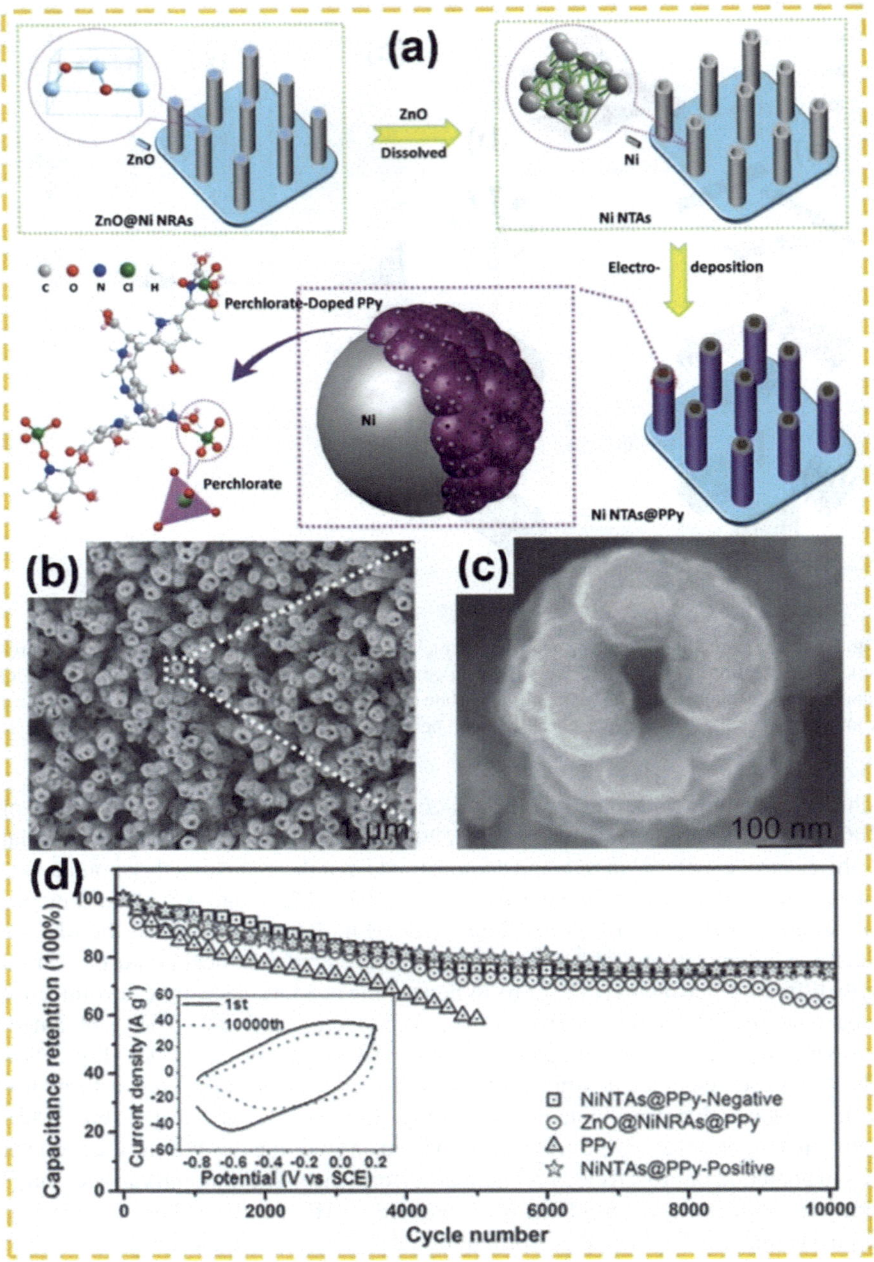

Fig. 7.7 **a** Schematic fabrication of NiNTAs@PPy. **b**, **c** SEM images of NiNTAs@PPy. **d** Cycling performance of four electrodes at 100 mV s^{-1}. The inset shows the 1st and 10,000th CV curves of NiNTAs@PPy-negative. Reprinted from Ref. Chen et al. (2016), copyright 2016, with permission from WILEY-VCH

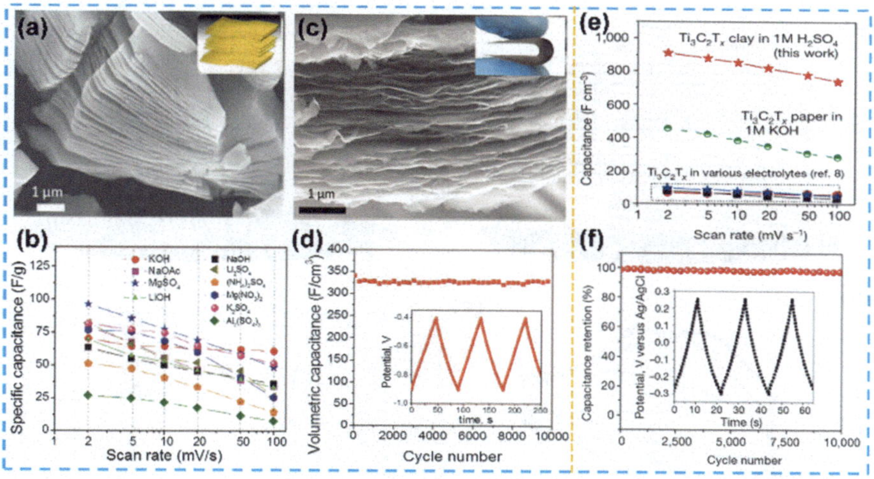

Fig. 7.8 **a** SEM image of the $Ti_3C_2T_x$ layered particle. **b** Rate performances in different aqueous electrolytes. **c** SEM image of paper electrode. **d** Capacitance retention of $Ti_3C_2T_x$ paper in KOH at 1 A g^{-1}. **e** Rate performance of different samples. **f** Capacitance retention of a 5-mm-thick rolled electrode in 1 M H_2SO_4. **a–d** Reprinted from Ref. Lukatskaya et al. (2013), copyright 2013, with permission from Science. **e, f** Reprinted from Ref. Ghidiu et al. (2014), copyright 2014, with permission from Nature

Owing to the large surface area-to-volume ratios, two-dimensional (2D) nanomaterials have become the favorable electrode materials for SCs. However, the traditional method for the fabrication of electrodes usually results in restacking of 2D nanomaterials, thus decreasing ion transport rate in thick films. Therefore, the thickness of the film can obviously affect the electrochemical performance of SCs. In this regard, Gogotsi's group (Xia et al. 2018) reported a suitable method to fabricate MXene ($Ti_3C_2T_x$) lamellar liquid crystal (MXLLC), which can lead to an excellent electrochemical performance regardless of the thickness of 2D nanomaterials. They prepared a vertical alignment of 2D flakes under an external mechanical shear force. As shown in Fig. 7.9a, the vertical alignment of MXene enabled fast ion transport. Thus, excellent electrochemical performance was achieved. MXLLC (with pseudocapacitive characteristics) got a distinct improvement of the electrochemical performance compared to the filtered MXene paper (lost most of the pseudocapacitive characteristics of $Ti_3C_2T_x$). Meanwhile, the electrochemical performance of MXLLC remained no significant changes when the film thickness changed from 40 to 200 µm, indicating a thickness-independent behavior (Fig. 7.9b). Additionally, the as-prepared MXLLC electrodes were extremely stable and retained almost 100% capacitance retention after 20,000 cycles at 20 A g^{-1} (Fig. 7.9c).

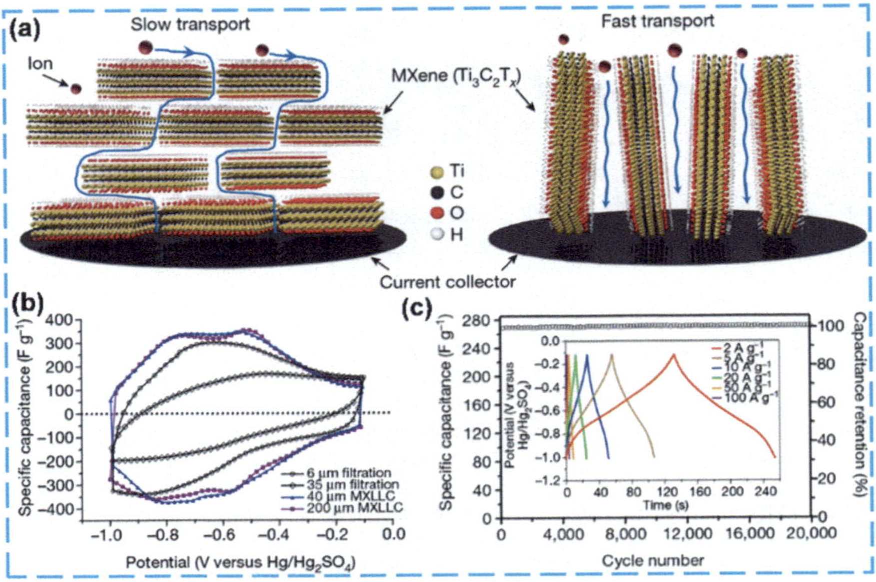

Fig. 7.9 **a** Schematic illustration of ion transport in horizontally stacked and vertically aligned $Ti_3C_2T_x$ MXene films. **b** Cyclic voltammograms of samples at 100 mV s^{-1}. **c** Capacitance retention of a 200-µm-thick MXLLC film at 20 A g^{-1}. The inset shows galvanostatic cycling profiles at various current densities. Reprinted from Ref. Xia et al. (2018), copyright 2018, with permission from Nature

7.4 Hybrid Capacitors

From the above analysis, we can see that the pseudocapacitors have higher specific capacitance and larger energy density than the EDLCs. However, compared with secondary batteries, such as lithium/sodium ion batteries and nickel-metal hydride batteries, which store energy through the body-phase electrochemical process, the energy density of pseudocapacitors is still low. Therefore, in recent years, to further increase the energy density of supercapacitors, researchers have developed a new kind of supercapacitors, hybrid supercapacitors, also known as asymmetric supercapacitors (ASCs). A hybrid capacitor is composed of two different electrode materials in a cell. In general, one pole is a battery-type electrode material, such as metal oxides and conductive polymers, which can store charge through redox reactions. Another one is carbon electrode material, which relies on an electrical double layer to store energy. At present, there are mainly two types of hybrid supercapacitors: (i) a metal oxide cathode and a carbon material anode; (ii) a lithium-ion or sodium-ion intercalation compound electrode and a carbon material electrode. Generally, the energy storage mechanism is to combine the pseudocapacitor materials or battery materials as the positive electrode with the carbon materials as the negative electrode. During the charging and discharging process, the hybrid supercapacitors can exert the advantages of both pseudocapacitances and EDLCs at the same time, thus increasing the operating voltage window (about 1.6–2.0 V) and improving the energy density of supercapacitors. In addition, it needs to be emphasized that the supercapacitor devices based on two different metal oxides, which serve as the cathode and the anode, respectively, also belong to the hybrid supercapacitor.

7.4.1 Metal Oxide Cathode and Carbon Material Anode/Metal Oxide Anode

Owing to the outstanding electrical properties and excellent electrochemical stability, graphene-based electrodes have attracted great interest. Pseudocapacitive materials are favorable materials to form nanocomposites with graphene. It was reported that a composite of graphene oxide and MnO_2 on nickel foam current collector achieved a high capacitance of 211.2 F g^{-1} (Chen et al. 2010). However, the actual capacitance is still far from the expected value. Sumboja et al. (2013) reported a flexible free-standing paper electrode of reduced graphene oxide (RGO)/MnO_2. The paper electrode combines the high conductivity of RGO and the large specific capacitance of MnO_2. A high areal capacitance of 897 mF cm^{-2} was achieved for the RGO/MnO_2 paper with a loading mass of 3.7 mg cm^{-2}. Additionally, the hybrid supercapacitor device based on RGO as anode and RGO/MnO_2 as cathode was also manufactured, which yield a high areal power density of 3.8 mW cm^{-2} at 1 A g^{-1} with an energy density of 11.5 μW h cm^{-2}. The device possessed a maximum areal energy of 35.1 μW h cm^{-2} at a power density of 37.5 μW cm^{-2}. Zhang et al. (2014a, b) reported a flexible, all-solid-state hybrid supercapacitor based on MnO_2/graphene as the cathode and CNT/graphene as the anode. The device could be operated under a high voltage region of 0–1.8 V and exhibited a maximum energy density of 31.8 W h kg^{-1}.

To realize high energy density, Xia et al. (2015) manufactured the hybrid supercapacitor device using a nanocomposite of functionalized graphene sheets (FGS) decorated with Fe_2O_3 quantum dots (≈2 nm) (Fe_2O_3/FGS) as anode and MnO_2/FGS as cathode (Fig. 7.10a). The Fe_2O_3/FGS composite not only possessed high surface area but also had excellent electronic conductivity, thus achieving good performance. Furthermore, MnO_2/FGS nanocomposite also exhibited high specific capacitance, which could match the anode materials, enhancing the specific energy of supercapacitors. As a result, a high energy density of 50.7 Wh kg^{-1} was obtained at a power density of 100 W kg^{-1} (Fig. 7.10b). In addition, the asymmetric device also exhibited a good cycling stability with 95% of initial capacitance retained after 5000 cycles (Fig. 7.10c).

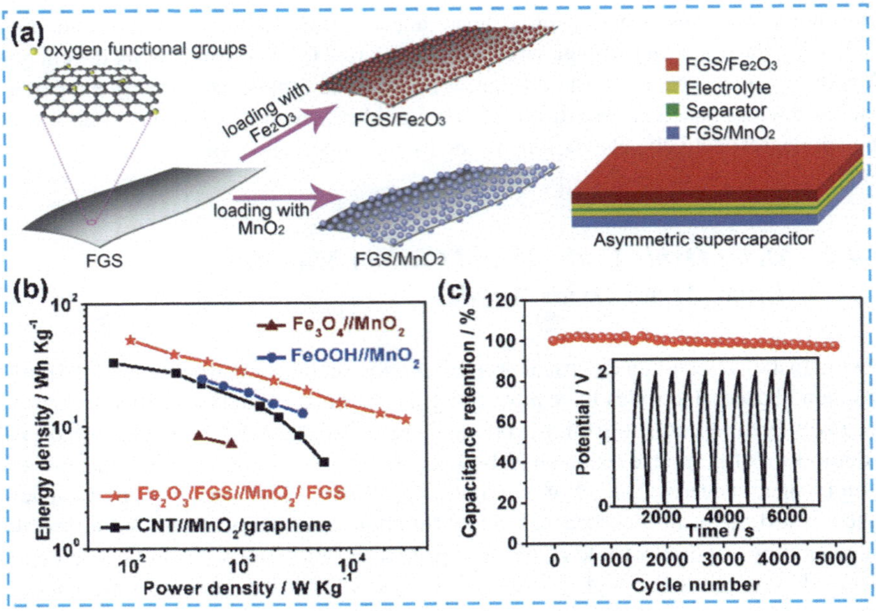

Fig. 7.10 **a** Schematic illustration of the preparation of the samples and the construction of the asymmetric supercapacitor. **b** The Ragone plots of the Fe_2O_3/FGS//MnO_2/FGS asymmetric supercapacitor compared with reported data in literature. **c** Cycle performance of the Fe_2O_3/FGS//MnO_2/FGS asymmetric supercapacitor. Reprinted from Ref. Xia et al. (2015), copyright 2015, with permission from WILEY-VCH

7.4.2 Li-Ion Hybrid Capacitor

Kong et al. (2015) used polyol-mediated solvothermal reaction to prepare the free-standing orthorhombic Nb_2O_5 (T-Nb_2O_5)/graphene composite paper (Fig. 7.11a) and applied it as the negative electrode for Li-ion asymmetric supercapacitor (activated carbon as the positive electrode). As shown in Fig. 7.11b, c, the resultant T-Nb_2O_5/graphene paper possessed a layer-stacked structure with some open sheet-like channels. Furthermore, it can be seen that Nb_2O_5 particles were highly dispersed on graphene (Fig. 7.11d). Based on the T-Nb_2O_5/graphene composite papers, the asymmetric supercapacitor achieved a high energy density of 47 W h kg^{-1} (Fig. 7.1e) and power density of 18 kW kg^{-1} (Fig. 7.11f).

V_2O_5 has also been studied as the anode material in Li-ion capacitors. Chen et al. (2011) utilized CNT/V_2O_5 nanocomposite as an anode and commercial activated carbon as the cathode to assemble Li-ion hybrid capacitor. The Li-ion hybrid capacitor delivered an energy density of 40 W h kg^{-1} and 6.9 W h kg^{-1} at a power density of 210 W kg^{-1} and 6.3 kW kg^{-1}, respectively. As an anode of Li-ion hybrid capacitors, the special crystalline structure of V_2O_5 allowed effective Li^+ diffusion, which can facilitate the energy storage property of Li-ion capacitors. Additionally,

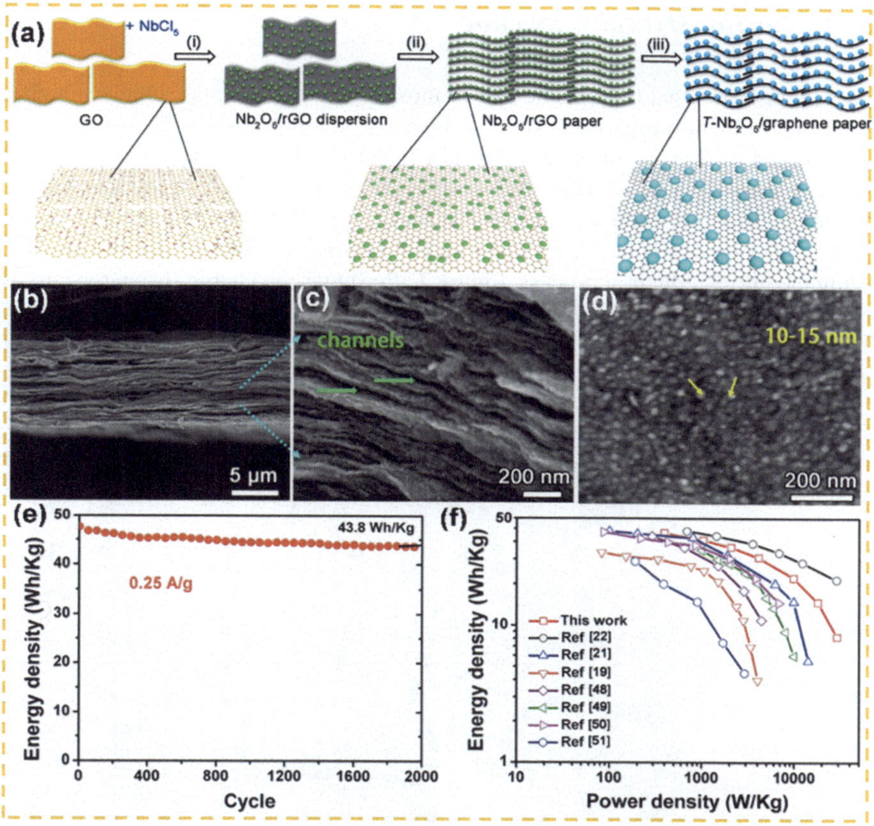

Fig. 7.11 **a** Schematic of the fabrication process for T-Nb_2O_5/graphene composite papers. **b**, **c**, **d** SEM images of T-Nb_2O_5/graphene composite papers. **e** Cycling performance **f** and Ragone plots of the T-Nb_2O_5/graphene//AC asymmetric supercapacitor. Reprinted from Ref. Kong et al. (2015), copyright 2015, with permission from American Chemical Society

the sintered V_2O_5 nanofibers were used as anode (SWNTs as cathode) to construct a hybrid Li-ion capacitor, which delivered maximum energy density and power density of 18 W h kg^{-1} and 315 W kg^{-1}, respectively (Aravindan et al. 2012).

Gao et al. (2015) prepared hierarchical $Li_4Ti_5O_{12}$ (LTO) nanosheet arrays decorated with rutile TiO_2 (denoted as RLTO) and applied it for Li-ion hybrid supercapacitor application. The unique hierarchical RLTO nanosheets possessed an interconnected 3D open microstructure, ensuring the robust penetration of electrolytes into the active materials. Moreover, the exposed (011) facets of LTO and (001) facets of rutile TiO_2 provided fast lithium ions diffusion pathways. The Li-ion hybrid supercapacitor based on LTO nanosheet arrays (negative electrode) and nitrogen-doped CNTs (positive electrode) exhibited an ultrahigh energy density of 74.85 W h kg^{-1} at a power density of 300 W kg^{-1}.

7.4.3 Na-Ion Hybrid Capacitor

It is demonstrated that the MXene Ti_2C nanosheet electrode could achieve a higher specific capacity in comparison to double-layer capacitor electrodes. To further verify the potential of Ti_2CT_x in Na-ion hybrid capacitors, the prototype Na-ion hybrid capacitor consisting of an alluaudite $Na_2Fe_2(SO_4)_3$ positive electrode and a Ti_2C MXene negative electrode was assembled (Wang et al. 2015). As shown in Fig. 7.12a, b, the expansion of the interlayer distance was consistent with the size of bare Na^+ during the initial reduction process. Figure 7.12c shows the charge-discharge curves of the full cell at different current densities. The average voltage is about 2.4 V. The Na-ion hybrid capacitor exhibited a high specific energy of 260 W h kg^{-1} at

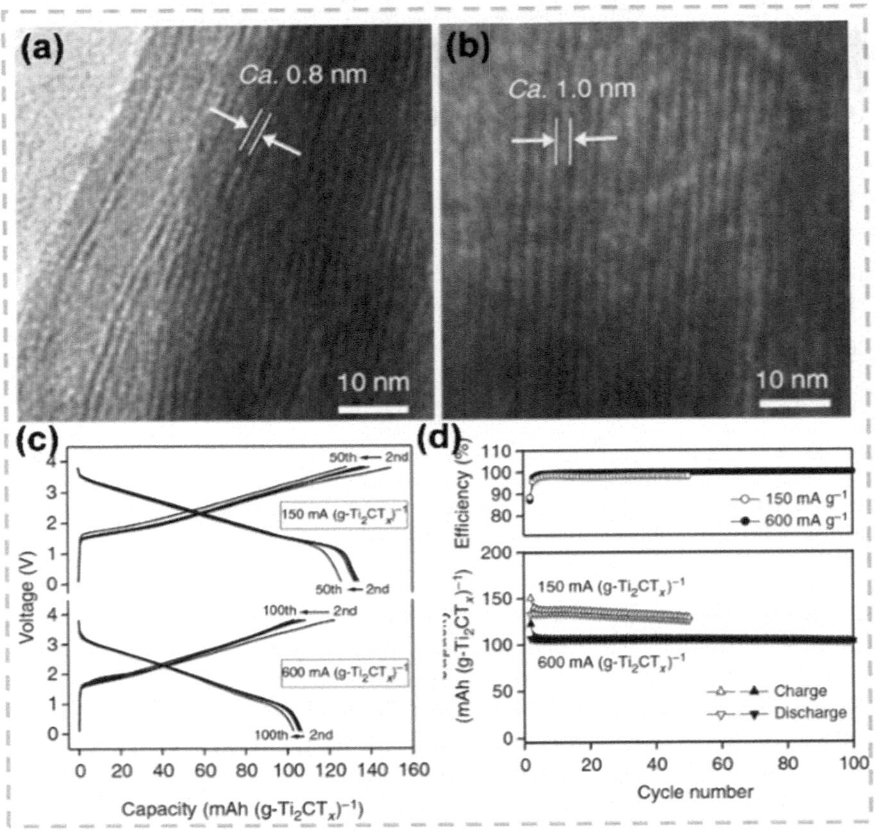

Fig. 7.12 **a** TEM image of the pristine Ti_2CT_X. **b** TEM image of the activated Ti_2CT_X after the first CV. **c** Charge-discharge curves of the Ti_2CT_X alluaudite $Na_2Fe_2(SO_4)_3$ full cell at 150 and 600 mA $(g\text{-}Ti_2CT_X)^{-1}$. **d** Cycle stability and Coulombic efficiency of the full cell with voltage range of 0.1–3.8 V at 150 and 600 mA $(g\text{-}Ti_2CT_X)^{-1}$. Reprinted from Ref. Wang et al. (2014), copyright 2014, with permission from Nature

1.4 kW kg^{-1} (based on the weight of Ti_2CT_x). Additionally, there was only 4% loss of the second discharge capacity after 100 cycles at 600 mA g^{-1} with a high Coulombic efficiency of ~99.7% after a few initial cycles (Fig. 7.12d). Dall'Agnese et al. (2015) studied the electrochemical behavior of V_2C MXene for the Na-ion hybrid capacitors. The XRD characterization results showed the continuous Na^+ intercalation between the V_2CT_x layers in a wide range of potentials. Furthermore, the Na-ion hybrid capacitor based on the V_2CT_x cathode and the hard carbon anode exhibited a specific capacity of 50 mA h g^{-1} after 100 cycles at 6 °C with a maximum cell voltage of 3.5 V.

7.5 Conclusions

Nanostructured materials have been extensively studied as electrode materials in supercapacitors. The nanostructure can not only provide a plentiful active surface for energy storage but also shorten the ion diffusion pathway, thus enhancing the performance effectively. For further improving the performance of supercapacitors, researchers could focus on the following several aspects.

The construction of nanocomposites of pseudocapacitive materials and nanostructured carbon materials can not only effectively improve the electrical conductivity of pseudocapacitive materials but also can inherit the nanostructures of carbon materials. The obtained nanocomposites can realize the synergistic effect of each component, thus effectively improves the performance of supercapacitors. This topic deserves further comprehensive study.

Abundant studies show that nanostructured materials exhibit high specific capacitance and good rate performance. However, the nanostructure usually causes the low tap density of the electrode materials. Hence, more reasonable nanostructures should be designed to ensure large volume energy densities of the supercapacitor device. MXenes have become very promising electrode materials for supercapacitors owing to their unique 2D structure and highly defined geometry. However, there are few reports about the electrochemical reaction mechanism of MXene materials. In order to further improve the performance of these materials, more research about the electrochemical reaction mechanism will be necessary. The free-standing flexible electrode is an important research direction in the field of supercapacitors. When the free-standing material is utilized as the electrode, the conductive carbon and adhesive are unnecessary, which can effectively improve the energy density of the device. Furthermore, the free-standing electrode can improve the cyclic stability of the device.

References

An KH, Kim WS, Park YS, Choi YC, Lee SM, Chung DC, Bae DJ, Lim SC, Lee YH (2001) Supercapacitors using single-walled carbon nanotube electrodes. Adv Mater 13:497–500

Aravindan V, Cheah YL, Mak WF, Wee G, Chowdari BVR, Madhavi S (2012) Fabrication of high energy-density hybrid supercapacitors using electrospun V_2O_5 nanofibers with a self-supported carbon nanotube network. ChemPlusChem 77:570–575

Cao F, Pan GX, Xia XH, Tang PS, Chen HF (2014) Synthesis of hierarchical porous NiO nanotube arrays for supercapacitor application. J Power Sources 264:161–167

Chen Q-L, Xue K-H, Shen W, Tao F-F, Yin S-Y, Xu W (2004) Fabrication and electrochemical properties of carbon nanotube array electrode for supercapacitors. Electrochim Acta 49:4157–4161

Chen S, Zhu J, Wu X, Han Q, Wang X (2010) Graphene oxide−MnO_2 nanocomposites for supercapacitors. ACS Nano 4:2822–2830

Chen Z, Augustyn V, Wen J, Zhang Y, Shen M, Dunn B, Lu Y (2011) High-performance supercapacitors based on intertwined CNT/V_2O_5 nanowire nanocomposites. Adv Mater 23:791–795

Chen G-F, Li X-X, Zhang L-Y, Li N, Ma TY, Liu Z-Q (2016) A porous perchlorate-doped polypyrrole nanocoating on nickel nanotube arrays for stable wide-potential-window supercapacitors. Adv Mater 28:7680–7687

Cheng G, Yang W, Dong C, Kou T, Bai Q, Wang H, Zhang Z (2015) Ultrathin mesoporous NiO nanosheet-anchored 3D nickel foam as an advanced electrode for supercapacitors. J Mater Chem A 3:17469–17478

Chmiola J (2006) Anomalous increase in carbon capacitance at pore sizes less than 1 nanometer. Science 313:1760–1763

Dai L, Chang DW, Baek J-B, Lu W (2012) Carbon nanomaterials for advanced energy conversion and storage. Small 8:1130–1166

Dall'Agnese Y, Taberna PL, Gogotsi Y, Simon P (2015) Two-dimensional vanadium carbide (MXene) as positive electrode for sodium-ion capacitors. J Phys Chem Lett 6:2305–2309

El-Kady MF, Strong V, Dubin S, Kaner RB (2012) Laser scribing of high-performance and flexible graphene-based electrochemical capacitors. Science 335:1326–1330

Feng C, Zhang J, He Y, Zhong C, Hu W, Liu L, Deng Y (2015) Sub-3 nm Co_3O_4 nanofilms with enhanced supercapacitor properties. ACS Nano 9:1730–1739

Gao L, Huang D, Shen Y, Wang M (2015) Rutile-TiO_2 decorated $Li_4Ti_5O_{12}$ nanosheet arrays with 3D interconnected architecture as anodes for high performance hybrid supercapacitors. J Mater Chem A 3:23570–23576

Ghidiu M, Lukatskaya MR, Zhao M-Q, Gogotsi Y, Barsoum MW (2014) Conductive two-dimensional titanium carbide 'clay' with high volumetric capacitance. Nature 516:78–81

Huang M, Mi R, Liu H, Li F, Zhao XL, Zhang W, He SX, Zhang YX (2014) Layered manganese oxides-decorated and nickel foam-supported carbon nanotubes as advanced binder-free supercapacitor electrodes. J Power Sources 269:760–767

Huang M, Zhang Y, Li F, Zhang L, Ruoff RS, Wen Z, Liu Q (2015) Self-assembly of mesoporous nanotubes assembled from interwoven ultrathin birnessite-type MnO_2 nanosheets for asymmetric supercapacitors. Sci Rep 4:3878

Huang Z-H, Song Y, Feng D-Y, Sun Z, Sun X, Liu X-X (2018) High mass loading MnO_2 with hierarchical nanostructures for supercapacitors. ACS Nano 12:3557–3567

Jia H, Cai Y, Lin J, Liang H, Qi J, Cao J, Feng J, Fei W (2018) Heterostructural graphene quantum dot/MnO_2 nanosheets toward high-potential window electrodes for high-performance supercapacitors. Adv Sci 5:1700887

Kong L, Zhang C, Wang J, Qiao W, Ling L, Long D (2015) Free-standing T-Nb_2O_5/Graphene composite papers with ultrahigh gravimetric/volumetric capacitance for Li-ion intercalation pseudocapacitor. ACS Nano 9:11200–11208

Li H-Q, Luo J-Y, Zhou X-F, Yu C-Z, Xia Y-Y (2007) An ordered mesoporous carbon with short pore length and its electrochemical performances in supercapacitor applications. J Electrochem Soc 154:A731

Li Q, Guo J, Xu D, Guo J, Ou X, Hu Y, Qi H, Yan F (2018) Electrospun N-doped porous carbon nanofibers incorporated with NiO nanoparticles as free-standing film electrodes for high-performance supercapacitors and CO_2 capture. Small 14:1–12

Liang K, Tang X, Hu W (2012) High-performance three-dimensional nanoporous NiO film as a supercapacitor electrode. J Mater Chem 22:11062

Lin T, Chen I-W, Liu F, Yang C, Bi H, Xu F, Huang F (2015) Nitrogen-doped mesoporous carbon of extraordinary capacitance for electrochemical energy storage. Science 350:1508–1513

Lin J, Jia H, Liang H, Chen S, Cai Y, Qi J et al (2018) In situ synthesis of vertical standing nanosized NiO encapsulated in graphene as electrodes for high-performance supercapacitors. Adv Sci 5:1700687

Lu W, Qu L, Henry K, Dai L (2009) High performance electrochemical capacitors from aligned carbon nanotube electrodes and ionic liquid electrolytes. J Power Sources 189:1270–1277

Liu T, Zhang L, Cheng B, You W, Yu J (2018a) Fabrication of a hierarchical NiO/C hollow sphere composite and its enhanced supercapacitor performance. Chem Commun 54:3731–3734

Liu T, Zhang L, You W, Yu J (2018b) Core-shell nitrogen-doped carbon hollow spheres/Co_3O_4 nanosheets as advanced electrode for high-performance supercapacitor. Small 14:1702407

Lukatskaya MR, Mashtalir O, Ren CE, Dall'Agnese Y, Rozier P, Taberna PL et al. (2013) Cation intercalation and high volumetric capacitance of two-dimensional titanium carbide. Science 341:1502–1505

Majeed A, Hou PX, Jiang S, Li JC, Ping LQ, Li GX, Zhang F, Liu C, Cheng HM (2017) Carbon-encapsulated NiO nanoparticle decorated single-walled carbon nanotube thin films for binderless flexible electrodes of supercapacitors. J Mater Chem A 5:24813–24819

Meher SK, Rao GR (2011) Ultralayered Co_3O_4 for high-performance supercapacitor applications. J Phys Chem C 115:15646–15654

Niu C, Sichel EK, Hoch R, Moy D, Tennent H (1997) High power electrochemical capacitors based on carbon nanotube electrodes. Appl Phys Lett 70:1480–1482

Sumboja A, Foo CY, Wang X, Lee PS (2013) Large areal mass, flexible and free-standing reduced graphene oxide/manganese dioxide paper for asymmetric supercapacitor device. Adv Mater 25:2809–2815

Wang X, Lee JS, Tsouris C, DePaoli DW, Dai S (2010) Preparation of activated mesoporous carbons for electrosorption of ions from aqueous solutions. J Mater Chem 20:4602

Wang Y, Zhou T, Jiang K, Da P, Peng Z, Tang J et al (2014) Reduced mesoporous Co_3O_4 nanowires as efficient water oxidation electrocatalysts and supercapacitor electrodes. Adv Energy Mater 4:1400696

Wang X, Kajiyama S, Iinuma H, Hosono E, Oro S, Moriguchi I, Okubo M, Yamada A (2015) Pseudocapacitance of MXene nanosheets for high-power sodium-ion hybrid capacitors. Nat Commun 6:1–6

Wang Y, Song Y, Xia Y (2016) Electrochemical capacitors: mechanism, materials, systems, characterization and applications. Chem Soc Rev 45:5925–5950

Wei C, Pang H, Zhang B, Lu Q, Liang S, Gao F (2013) Two-dimensional β-MnO_2 nanowire network with enhanced electrochemical capacitance. Sci Rep 3:2193

Xia H, Hong C, Li B, Zhao B, Lin Z, Zheng M, Savilov SV, Aldoshin SM (2015) Facile synthesis of hematite quantum-dot/functionalized graphene-sheet composites as advanced anode materials for asymmetric supercapacitors. Adv Func Mater 25:627–635

Xia Y, Mathis TS, Zhao M-Q, Anasori B, Dang A, Zhou Z, Cho H, Gogotsi Y, Yang S (2018) Thickness-independent capacitance of vertically aligned liquid-crystalline MXenes. Nature 557:409–412

Xu J, Wang K, Zu S-Z, Han B-H, Wei Z (2010) Hierarchical nanocomposites of polyaniline nanowire arrays on graphene oxide sheets with synergistic effect for energy storage. ACS Nano 4:5019–5026

Xu Y, Lin Z, Zhong X, Huang X, Weiss NO, Huang Y, Duan X (2014) Holey graphene frameworks for highly efficient capacitive energy storage. Nat Commun 5:4554

Yang S, Liu Y, Hao Y, Yang X, Goddard WA, Zhang XL, Cao B (2018) Oxygen-vacancy abundant ultrafine Co_3O_4/graphene composites for high-rate supercapacitor electrodes. Adv Sci 5:1700659

Zhang Y-Z, Wang Y, Xie Y-L, Cheng T, Lai W-Y, Pang H, Huang W (2014a) Porous hollow Co_3O_4 with rhombic dodecahedral structures for high-performance supercapacitors. Nanoscale 6:14354–14359

Zhang Z, Xiao F, Qian L, Xiao J, Wang S, Liu Y (2014b) Facile synthesis of 3D MnO_2-graphene and carbon nanotube-graphene composite networks for high-performance, flexible, all-solid-state asymmetric supercapacitors. Adv Energy Mater 4:1400064

Zheng M, Xiao X, Li L, Gu P, Dai X, Tang H, Hu Q, Xue H, Pang H (2017) Hierarchically nanostructured transition metal oxides for supercapacitors. Sci China Mater 61:185–209

Zhu Y, Murali S, Stoller MD, Ganesh KJ, Cai W, Ferreira PJ et al (2011) Carbon-based supercapacitors produced by activation of graphene. Science 332:1537–1541

Chapter 8
Conclusion

This book describes various functional nanomaterials for electrochemical applications. It consists of four chapters that show interesting examples of nanomaterials of different sizes. Two chapters show electrochemical energy storage applications that rely on modern electrochemistry. It also includes the synthesis of functional nanomaterials from quantum dots, one-dimensional nanostructures, two-dimensional nanostructures to three-dimensional nanostructures. A review of the collective work of functional nanomaterials in electrochemical applications illustrates the latest advances in many modern electrochemical energy storage hotspots: lithium batteries, lithium-ion batteries, sodium-ion batteries, other metal-ion batteries, halogen-ion batteries, and metal–gas batteries.

Quantum dots exhibit properties that are intermediate between bulk semiconductors and discrete atoms or molecules. Due to its unique tunable photoluminescence properties, excellent physicochemical properties, high light stability, biocompatibility, and small size, the emerging GQD and C-dots have received much attention due to their enormous potential for biomedical applications. The synthesis of C-dots and GQD can generally be divided into top-down and bottom-up methods. The 1D nanostructure can provide a DC path, shorten the ion diffusion distance, reduce the charge and discharge time, increase the electrolyte–electrode contact area, limit mechanical degradation, and accommodate volume expansion. This synergistic effect between each component imparts better conductivity to the heterostructure electrode, greater electrochemical cycle stability and reversibility, faster ion transport, improved mechanical stability, and the like. The synthesis of 1D nanostructures includes top-down and bottom-up methods as well. 2D nanomaterials have atomic thickness, large surface area, tunable electronic properties, excellent mechanical strength, and unique limiting effects. Methods for 3D nanomaterials include chemical precipitation, sol–gel, hydrothermal, solvothermal, thermal decomposition, microemulsion, and chemical vapor deposition.

For the batteries, the sulfur reaction process is extremely complicated and the reaction intermediates are various. The electrochemical performance of lithium–sulfur batteries is greatly influenced by the nature of the electrode materials, test conditions,

H. Pang et al., *Synthesis of Functional Nanomaterials for Electrochemical Energy Storage*,
https://doi.org/10.1007/978-981-13-7372-5_8

and the working environment. The underlying causes of mutual influence remain to be further studied and summarized. Further research and exploration of the mechanism of action of the electrolyte on the positive and negative electrodes, such as the dissolution of polysulfide ions in the electrolyte and the stability of the lithium negative electrode in the electrolyte, are required. In the case of a metallic lithium negative electrode, the metal lithium deposition-dissolution process, the lithium dendrite growth mechanism, and the formation mechanism and performance of the SEI film under long circulation are problems.

For supercapacitors, a large number of studies have shown that nanostructured materials have high specific capacitance and good rate performance. However, nanostructures generally result in a low tap density of the electrode material. Therefore, more rational nanostructures should be designed to ensure the bulk energy density of the ultracapacitor device. Due to its unique 2D structure and highly defined geometry, it has become a very promising electrode material for supercapacitors. However, there are few reports on the electrochemical reaction mechanism of MXene materials. In order to further improve the performance of these materials, more research is needed on the electrochemical reaction mechanism. Self-supporting flexible electrodes are an important research direction of supercapacitors. When a self-supporting flexible material is used as an electrode, conductive carbon and an adhesive are unnecessary for assembly of the ultracapacitor device; therefore, the energy density of the device can be effectively increased. In addition, self-supporting flexible electrodes can enhance the cycling stability of the device. In addition, the flexible all-solid-state asymmetric SCs device is critical because it has several advantageous properties such as a wide operating voltage range, high energy density, excellent flexibility, and excellent long-term cycle stability.

In summary, we provide a comprehensive description of the current synthesis of functional nanomaterials, describe the current applications of batteries and supercapacitors, highlight their practicality and practicality in everyday practice, and discuss the risks (current synthesis of functional nanomaterials and A safer alternative) electrochemical storage). We hope that readers will find this book a concise summary of the synthesis and application of functional nanomaterials.